CONCISE

EARTH FACTS

EARTHBOOKS, INC.

Concise EARTHFACTS

Author: **Bo Gramfors**
Maps International
Co-author: **Siv Eklund**
Production management: **Mats Halling**
Project manager: **Christina Bjorklund**
Esselte Map Service
Lay-out: **Kjeld Brandt**
Production:
Esselte Map Service, Stockholm
Production and photo research:
Lovell Johns, Oxford
Editor: **Fiona Sutcliffe**
Editor: **Alison Dickinson**
Production editor: **David Edwards**
Lovell Johns
Illustrations:
Hardlines, Oxford
Lidman Production, Stockholm
Cover illustration: **Randy Nelsen**

Concise EARTHFACTS ISBN# 1-877731-04-8

PUBLISHER: EARTHBOOKS, INCORPORATED
Denver, Colorado

This book is written to give you an understanding of many things, but not a great knowledge of any one thing.

Each subject that fills a double page could fill an entire book. Perhaps you will be inspired to find out more about subjects unknown to you.

This book is written to record our common heritage and history and to present possible scenarios for our future. "International English" is used throughout this book in recognition of the diverse nationalities of the readers.

The decisions we make and the actions we take today will affect not only ourselves but also our neighbors and the whole world.

People are more powerful today than ever recorded in our history. Our decisions and actions can damage and destroy or conserve and create. This book gives evidence of both.

Read the book like a novel or like a saga of the past, present and future. Remember, we hold the Earth's future in our hands!

EARTHBOOKS, INCORPORATED

Consultants

Alphabets	A. Gaur Dr. The British Library, London
Astrology	Örjan Björkhem Psychologist of Religion Simrishamn, Sweden
Astronomy	Gösta Gahm, Ph D University of Stockholm
Constructions	Eric G. Funegard Geotechnical Consultant Amoco Corporation Naperville, USA
	Space Media Network, Stockholm
Economy	Alberto Izquierdo Stockholm School of Economics
Environment	Björn Ganning, Ph D Lars-Erik Åse, Ph D University of Stockholm
	Greenpeace, Sweden FAO, Rome
Languages	Christina Edenås, M A Teacher of the deaf Manillaskolan, Stockholm
	Lars J Larsson, Ph D Alvesta, Sweden
Minerals	Swedish Goldsmith and Jewellery Association, Stockholm
Minorities and Refugees	Thomas Lundén, Ph D Swedish Institute, Stockholm
	Ragnhild Ek, B A University of Lund, Sweden
	UNHCR, Switzerland J. Crisp, United Nations High Commissionen for Refugees, Switzerland
Space	Forskning och Framsteg, Stockholm
	SSC, Satimage, Kiruna, Sweden
World heritage	UNESCO, Paris
Zoology	Bo Tallmark, Ph D University of Uppsala, Sweden
	The Swedish Society for the Conservation of Nature, Stockholm
	World Wild Life Fund
	J. Taylor World Conservation Monitoring Centre, Cambridge, UK

Within these covers you will find all this!

Inside front cover the World, political.

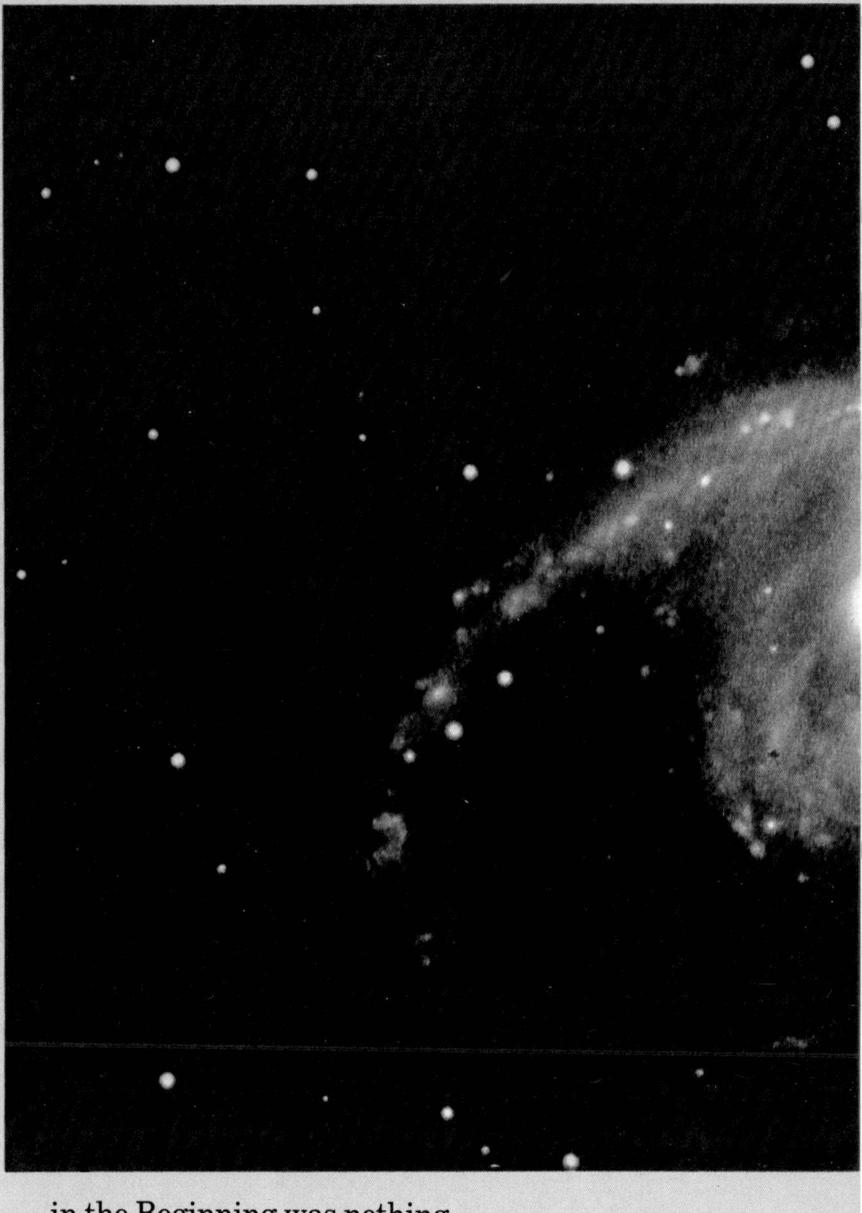

. . .in the Beginning was nothing

to look out into the Universe is to look back in time

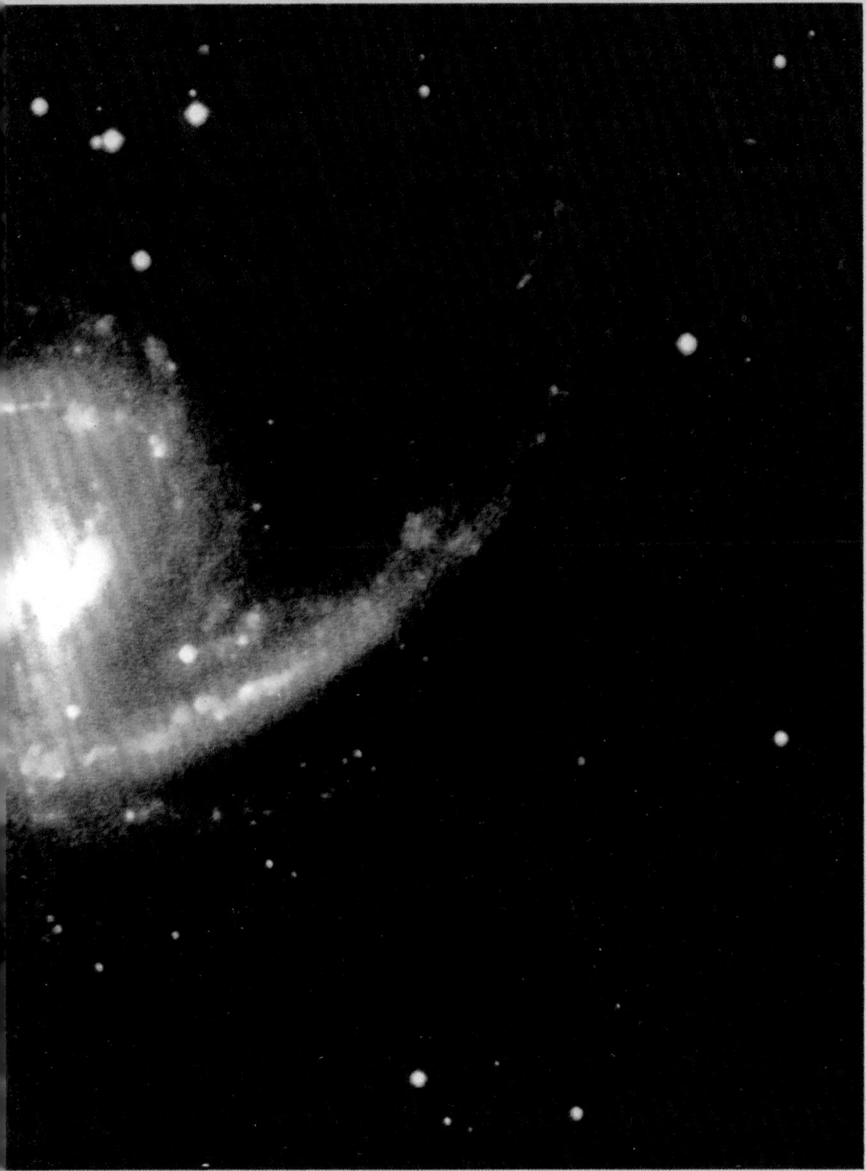

The galaxy NGC 1365, is at a distance of 60 million light years. We see the galaxy as it looked 60 million years ago.

Masses: • The Earth 6 × 10²⁴ kg • The Sun: 2 × 10³⁰ kg • The Galaxy: 3 × 10⁴¹ kg

Distances: 1 light year = the distance travelled by light in one year = 9.5 × 10¹³ kilometres (km)

Symbols: • 1000 = 10³ • 1000 000 000 (1 billion) = 10⁹ • 0.01 = 10⁻³ • 0.000001 = 10⁻⁶

- The Moon: 384 400 km (1.3 light seconds)
- The Sun: 150 million km (8 light minutes) = 1 astronomical unit (au)
- Sirius: 8 light years

Periods of light flashes of pulsars:

- The Crab pulsar: 0.033 sec.
- The pulsar related to the 1987 supernova: 0.0005 sec.

The primordial fire

No matter how often we gaze up at the stars we are unlikely to observe any changes from one night to the next. The idea of the universe being in a steady state with neither beginning nor end seems natural enough. Nevertheless every civilization has had its own legend about creation. The biblical account is found in the Book of Genesis; 'And God said: Let there be Light; and there was Light'.

The idea of the steady state was reinforced by natural science and this, despite biblical beliefs, was the predominant cosmological theory well into the century. In the 1920's the American atronomer Hubble discovered that light from distant galaxies is shifted further towards the red, longwave end of the spectrum, the farther away these galaxies are. Not until after prolonged debate did the scientists accept the most simple explanation for this; the shifting results from the Doppler effect and the galaxies are moving out and away from us at tremendous speed. Finally by the 1960's it was generally accepted that the universe is indeed expanding. It was concluded that the universe probably originated in a single point, a 'singularity', at about 12 to 15 billion years ago. This 'creation' was dubbed the 'big bang'. The universe has expanded ever since the big bang, and cooled from originally extremely high temperatures. Also, the density of the universe was extremely large at the beginning but is now very low, only about 10^{-30} grams per cubic centimetre on average.

The creation

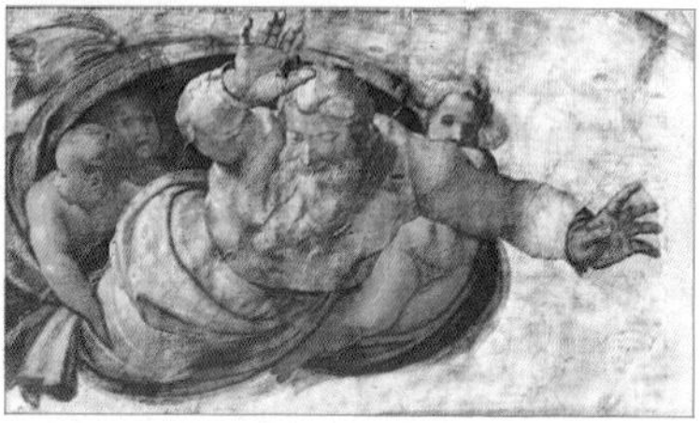

In the Bible's account of the creation (above, detail from Michelangelo's painting) God made the difference between light and darkness thereby establishing the visible universe.

The Andromeda Galaxy 2 million light years away is a large nearby galaxy of billions of stars similar to our own galaxy, the Milky Way. Two dwarf galaxies (seen as light balls) orbit this galaxy as satellites.

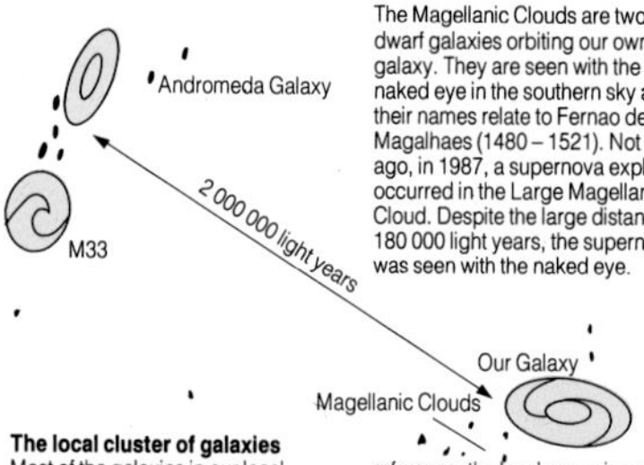

The local cluster of galaxies
Most of the galaxies in our local surrounding in the universe are small dwarf galaxies (the patches). Our galaxy is the largest one in the local group. Although the distances between the galaxies are enormous from an earth-bound view of reference, the local group is a very small entity in the extensive field of galaxies. The most distant galaxies known, so called quasars, are at more than 10 billion light years away, ten thousand times more distant than the Andromeda galaxy.

The Magellanic Clouds are two dwarf galaxies orbiting our own galaxy. They are seen with the naked eye in the southern sky and their names relate to Fernao de Magalhaes (1480 – 1521). Not long ago, in 1987, a supernova explosion occurred in the Large Magellanic Cloud. Despite the large distance, 180 000 light years, the supernova was seen with the naked eye.

Our galaxy

Our galaxy – The Milky Way. Viewed at a distance and at a certain angle, our galaxy may look like this spiral galaxy in the constellation Canes Venatici. The Milky Way contains about 250 billion stars, all moving around the galactic centre at a distance of 30 000 light years from our solar system. Most stars are located in a flat, circular disk around the central bulge. Young, massive stars are blue and outline the spiral arms. Older stars are yellow or red and spread over the galaxy. Interstellar clouds of dust are seen as black patches.

Stellar birth

New stars are continuously born out of fragments of gas and dust in interstellar clouds. The stars consist of 90% hydrogen, 8% helium (from the Big Bang) and only 2% of heavier elements like carbon, oxygen and iron (from the first generation of stars). When the massive stars start to shine, they excite the surrounding nebular material and magnificent bright nebulae appear. The Orion nebula (above) is an example of such a region in the sky.

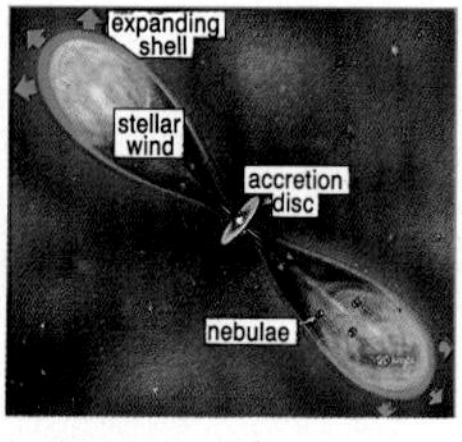

Gravitation governs the collapse of interstellar fragments. When a star is formed in the centre of a fragment it develops a stellar wind that blows away circumstellar material back in space. There are indications that these newborn stars are surrounded by a flat, thick disc in which matter accretes towards the star. The winds therefore often blow in two opposite directions, perpendicular to the discs, and they are called bipolar flows. When the wind hits small cloudlets it causes them to heat up and to accelerate.

Stellar clusters

When the interstellar material is dispersed in a region of star formation, a cluster of stars is left. Long ago, very large and dense clusters were formed, like this one in the southern sky. Clusters like this are called globular clusters and they have an almost spherical distribution relative to the galactic centre. It shows that at the beginning, our galaxy was not flat but spherical.

Stellar life

Solar mass star

A low-mass star, like the sun, has a lifetime of billions of years. During the last phase of evolution it grows in size and turns into a giant star with a radius of a hundred times that of the present sun. The surface then cools and the emitted light becomes red.

Massive star

A massive star appears blue in the sky. Compared to the sun its lifetime is very short. At the end the star grows to a thousand solar radii and becomes a red supergiant.

Stellar energy is drawn from nuclear reactions in the central regions of the stars. In these fusion reactions lighter elements build up heavier elements.

. . .and the Universe was created

birth of planets and stars, 'the Big Bang'

- Radius of the planetary system: 50 au
- Radius of our galaxy: 50 000 light years

Temperatures:
- The surface of the sun: 6000 K
- The solar interior: 15 × 10⁶ K

Energy flows:
- The Sun: 4 × 10²⁶ W
- Betelgeuze: 10³¹ W

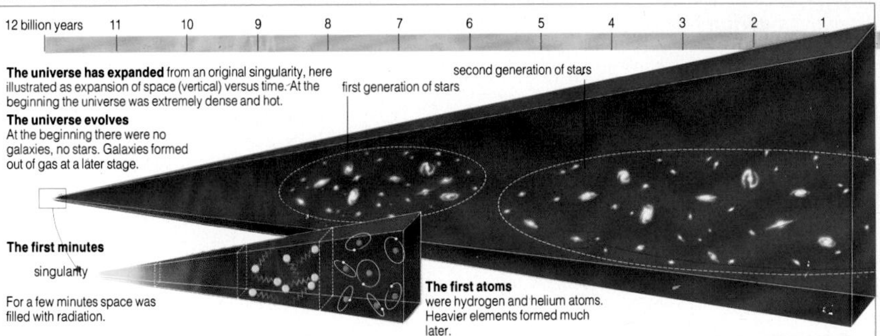

The universe has expanded from an original singularity, here illustrated as expansion of space (vertical) versus time. At the beginning the universe was extremely dense and hot.

The universe evolves
At the beginning there were no galaxies, no stars. Galaxies formed out of gas at a later stage.

The first minutes
For a few minutes space was filled with radiation.

The first atoms
were hydrogen and helium atoms. Heavier elements formed much later.

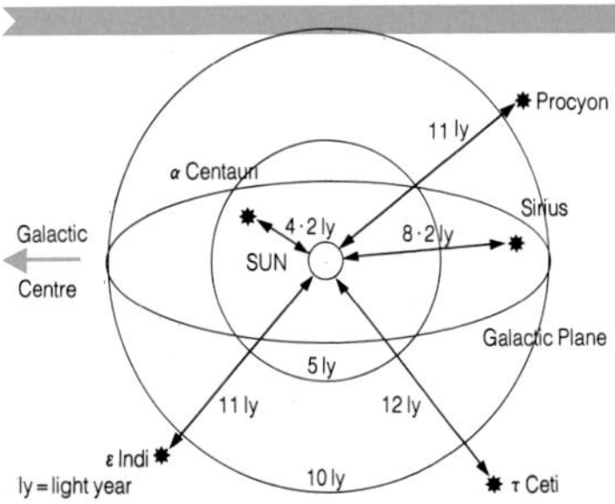

The brightest of the nearest stars
Within 15 light years from the sun there are about 25 stars known. The brightest of these are Sirius, Procyon and α Centauri. The latter consists of 3 stars, it is a triplet star, and it is one of the components, called Proxima Centauri, which in fact is closest to us, the distance being 4.2 light years. Many of the stars in the sky are double or multiple stars. The sun is a single star.

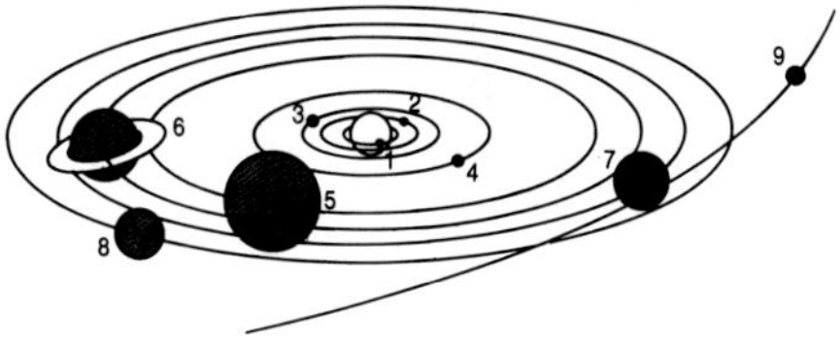

1 Mercury
2 Venus
3 Earth
4 Mars
5 Jupiter
6 Saturn
7 Uranus
8 Neptune
9 Pluto

Our own sun and its planetary system was formed between 4.5 and 5 billion years ago. The sun thus belongs to the second generation of stars in the universe. It is of medium size and medium heat and is middle aged. It has shone for some five thousand million years and should continue to do so for about as long again.
The planets in our solar system are of two different categories: the inner planets, Mercury, Venus, Earth and Mars are compact mineral spheres, whilst the outer planets, Jupiter, Saturn, Uranus and Neptune are gaseous. Unlike the other planets, Pluto's orbit is oblique and eccentric.

Stellar death

Red Giant
(Aldebaran)

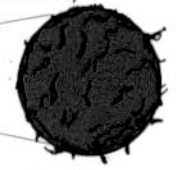

After 10 million years

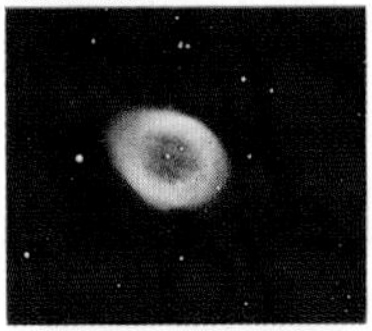

A planetary nebula in Lyra.

Finally, a solar mass star blows out its expanded envelope and forms, for a while a planetary nebula. It leaves its compressed centre, a white dwarf, to cool.

The white dwarf has a mass of 1 solar mass contained within the size of the earth. The nuclear energy sources are now exhausted.
The massive star leaves an extremely compact object, a neutron star, at its centre. Here, there is no space for individual atoms. Two light beams are radiated. An observer sees the flashes of light or radio emission.
If the central object is very massive it may collapse. Gravitation makes space and time close around such an object and not even light can escape from inside.

Red Supergiant
(Betelgeuze)

After 5 million years

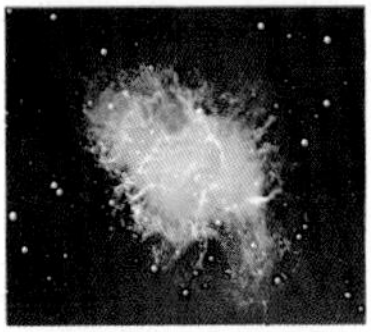

The Crab Nebula – the remnant of a supernova that exploded in 1054.

The massive star explodes dramically as a supernova, ejecting material at a speed of ten thousands km/sec into space. At a maximum the light of a supernova is comparable to the light of the whole galaxy.

The end products

White dwarf

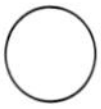

radius:- 7000 km
1 solar mass

Neutron star

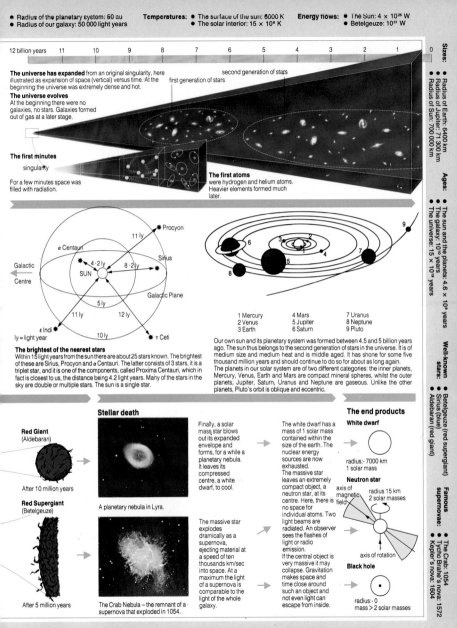

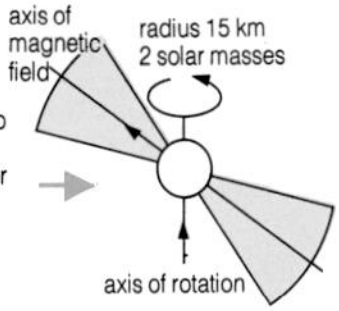

Black hole

radius:- 0
mass > 2 solar masses

Sizes:
- Radius of Earth: 6400 km
- Radius of Jupiter: 71 300 km
- Radius of Sun: 700 000 km

Ages:
- The sun and the planets: 4.6 × 10⁹ years
- The galaxy: 10¹⁰ years
- The universe: 15 × 10¹⁹ years

Well-known stars:
- Betelgeuze (red supergiant)
- Sirius (blue)
- Aldebaran (red giant)

Famous supernovae:
- The Crab: 1054
- Tycho Brahe's nova: 1572
- Kepler's nova: 1604

K (Kelvin). The SI unit of thermodynamic temperature; absolute zero is given the value of 0 k and the boiling point of water 273 k.

W (watt). The SI unit of power equal to one joule per second. It is the energy per second expended by a current of one ampere flowing between points on a conductor between which there is a potential difference of one volt.'

Major historical disasters and their victims:

- 1555 Shanxi Province, China — 83 000 dead
- 1755 Lisbon, Portugal — 50 000 dead
- 1855 Tokyo, Japan — 106 000 dead
- 1906 San Francisco, USA — 500 dead
- 1908 Messina, Italy — 83 000 dead
- 1934 Tokyo, Yokahama, Japan — 150 000 dead

Volcanic eruptions: date and casualties

- **Vesuvius, Italy:** The eruption in AD79 destroyed Pompeii under ashes and Herculaneum under a mudflow.
- **Etna, Italy:** In 1669 an eruption left 20 000 people dead.
- **Krakatoa (Pulau Rakata):** The 1883 eruption was one of the most catastropic in history. In Java 36 000 people died, many drowned by a tidal wave.
- **Cotopaxi, Ecuador:** 5897 m World's highest continuously active volcano.

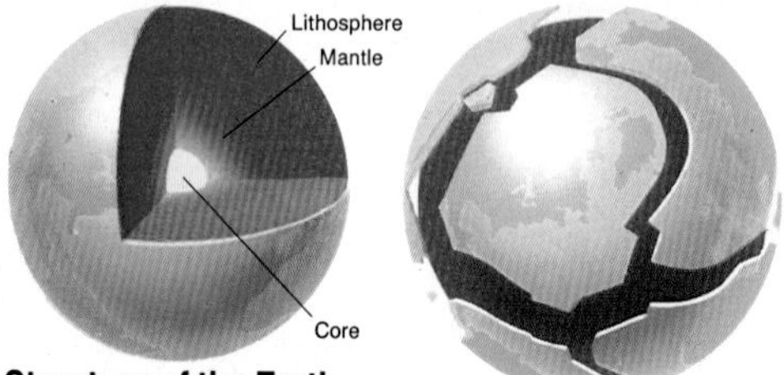

Structure of the Earth

Built up in layers. The main part of the Earth is a fluid mantle, consisting of silicates. In volcanic eruptions, this viscous material is recognised as magma. The core is mainly made up of iron, and despite the high temperature (about 5000°C, 9000°F) it is solid due to the high pressure of about 3 -5 million atmospheres. The surface layer is called the crust and is only some 5-15 km, 3-9 miles thick. It consists of rigid plates that move because of the currents in the mantle.

Drifting continents

The movements of the plates in the crust make the continents move. In past times continents have been located in other parts of the Earth, sometimes far apart, sometimes close together and sometimes even colliding, all as a result of currents in the viscous mantle. This theory was formed by a German climatologist, Alfred Wegener in the beginning of this century.

Permian/Triassic 230 million years ago

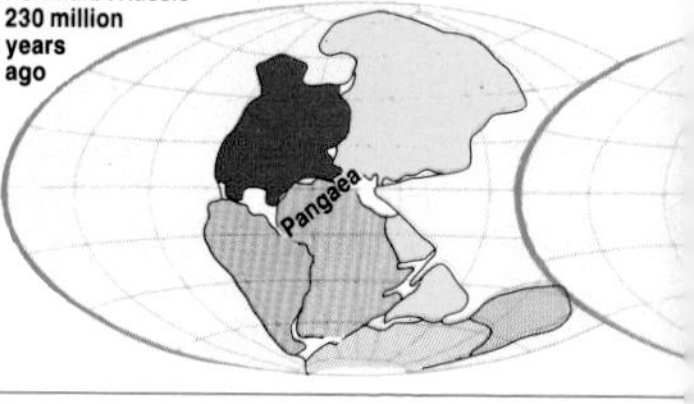

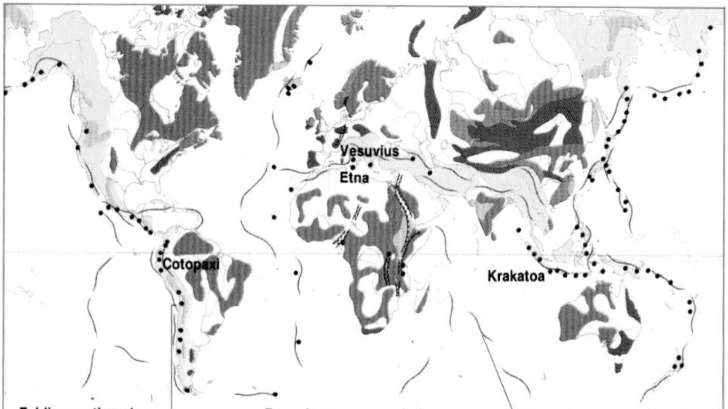

Folding, earthquakes and volcanoes

Zone of strong seismic activity

Active volcano

Other colours: see geological diagram below

Subduction

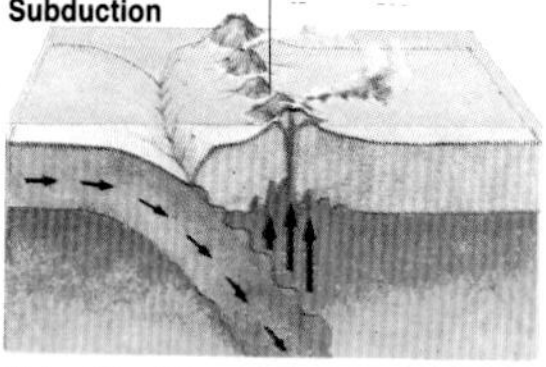

Diving plates. When oceanic and continental crusts clash, the softer and thinner oceanic crust bends downwards in what is called subduction, while the continental crust forms a range like the Andes and the Rocky Mountains. The subduction movements may cause deep earthquakes. The material of the oceanic crust melts and breaks down in the viscous mantle.

Continental drift

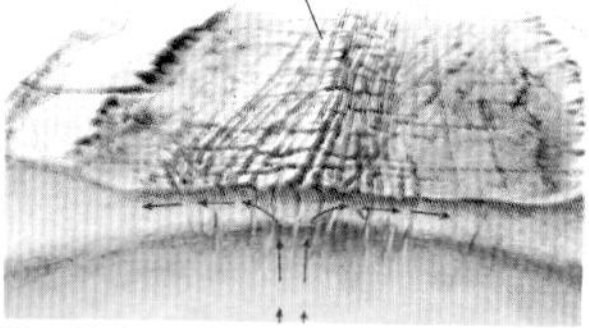

Currents in the mantle move magma towards the surface, that is the sea floor. Where the sea floor, or oceanic crust expands new crust is continuously formed by magma and marine sediments. Thus, oceans like the Atlantic have a ridge in the middle. The Mid-Atlantic Ridge is in fact the longest mountain range in the world and moves America away from Europe and Africa by about 2 cm, 1 inch a year.

Folded by collision

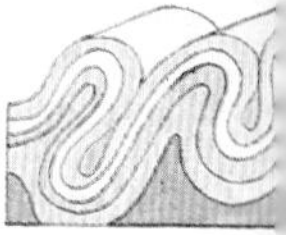

Sedimentary rocks form at the bottom o gravel, sand, clay and dead plants anc deposited. High temperature anc these particles together to form rocks. sedimentary rocks remain in horizonta of the crust collide, stresses, strains anc sedimentary rocks. Destructive forces like taneously with the folding. As the layers different parts erode at varying rates ridges.

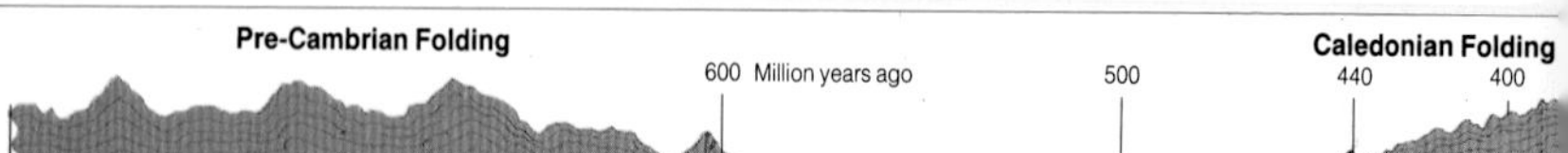

- 1060 Agadir, Morocco 12 000 dead
- 1970 Northern Peru 67 000 dead
- 1976 Tientsin, China 650 000 dead
- 1970 Northeast Iran 20 000 dead
- 1985 Mexico 8 000 dead
- 1988 Armenia 50 000 dead

Major volcanoes, elevation, latest eruption

- **Africa:** Kilimanjaro, Tanzania 5895 m; Mt. Cameroon 4070 m 1982
- **North America:** Citlaltepetl, Mexico 5700 m 1667; Popocatepetl, Mexico 5452 m 1943; Mt. Rainier, USA 4392 m 1882
- **South America:** Guallatiri, Chile 6060 m 1960; Colopaxi, Ecuador 5987 m 1943
- **Europe:** Etna, Italy 3340 m 1981
- **Asia:** Kiyuchevskaya Sopka, USSR 4750 m 1980; Fujiyama, Japan 3776 m 1707
- **Pacific:** Mauna Kea, Hawaii 4205 m

His strongest evidence was that the coastlines of Africa and South America nearly fit together and that the same kind of fossils were found in each. However, this theory was not generally accepted until the 1960's.

Jurassic/Cretaceous
140 million years ago

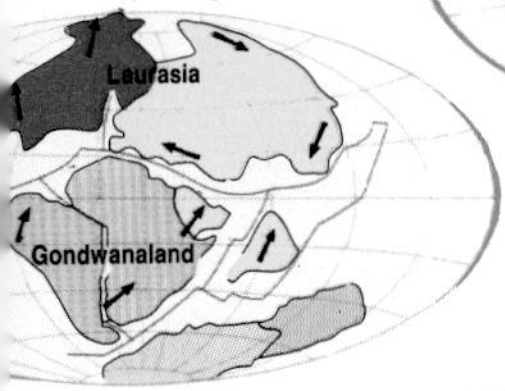

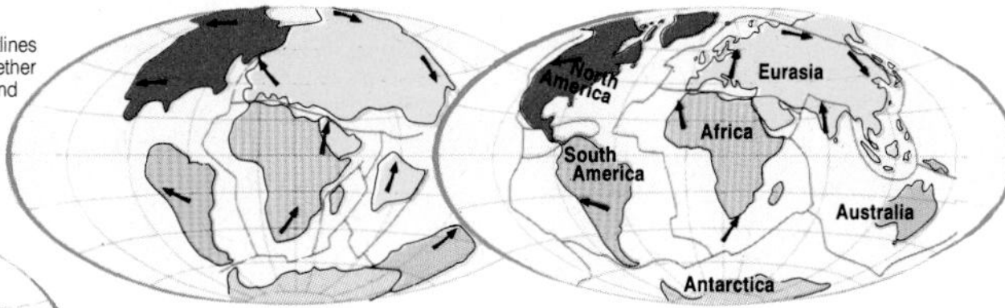

Tertiary
65 million years ago

Quaternary
Present day

In the Permian and Triassic eras, about 230 million years ago (far left), all the continents moved towards each other and formed a super continent which Wegener named Pangaea. Later, by the time of the dinosaurs, about 140 million years ago, the super continent split up into what was to become today's continents (left). The South Atlantic opened up, while India, which in fact is a continent of its own, moved northwards and collided with Asia, forming the Himalayas. Australia and the Antarctic were attached for a longer time.

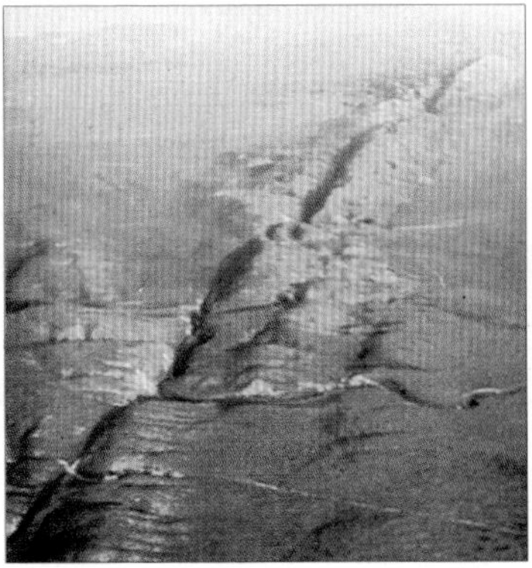

Earthquakes

Volcanic blowout

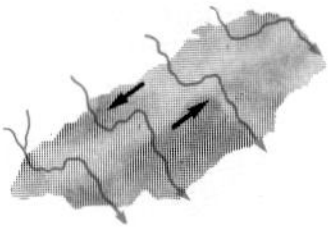

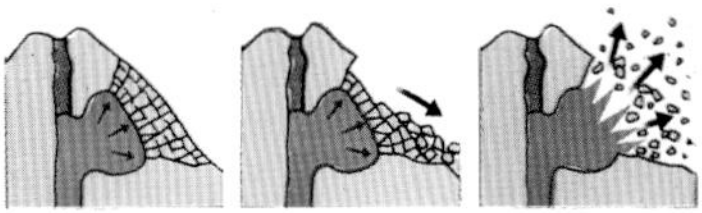

he oceans where animals are pressure fuse all undisturbed these ayers, but as plates tensions may fold the erosion work simul- differ in hardness, forming mountain

Movements between the plates cause tensions and strains. Earthquakes are a sort of discharge of such tensions and are most common in areas where different plates of crust meet or separate.
The San Andreas fault (photo above) in California is situated in such an area. The fault runs through cities like San Francisco, which was severely damaged by an earthquake in 1906. Great care must therefore be taken in town planning and house building in order to prevent catastrophes. Unfortunately some structures, such as the elevated highway, could not withstand the shocks during the most recent earthquake in October 1989.

Volcanoes are visible evidence that the major part of the Earth consists of hot viscous magma. In areas where plates meet or separate, and where new oceanic crusts are formed, volcanoes are most common. The magma is under high pressure and forces its way up through funnels in the crust and if the surface layers are not strong enough, the magma will well up into the air. When a volcano erupts, vast amounts of lava, boulders and ash are spread. The picture above shows how all the vegetation was completely destroyed up to 15 km, 10 miles from the violent eruption of Mount St. Helens in northwest USA in May 1980.

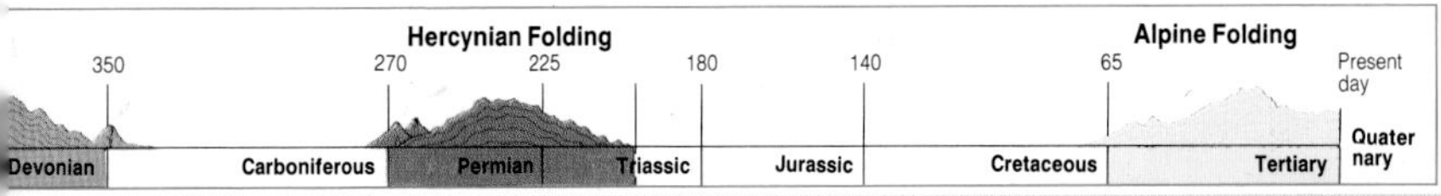

Volcanic plug

Stratovolcano

Shield volcano

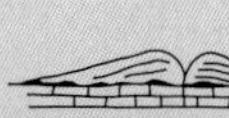

Explosion crater

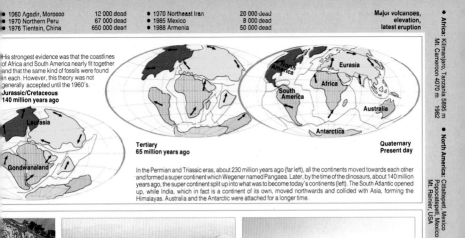

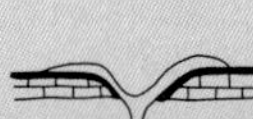

Hot and cold records
Temperatures are extremes

The Beaufort scale was originally used to measure winds at sea. The behaviour of smoke and trees provide an estimate on land.

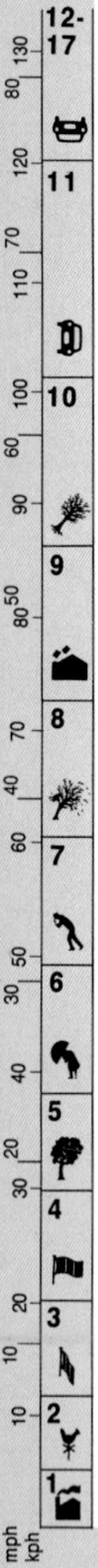

World climate

Temperature, January

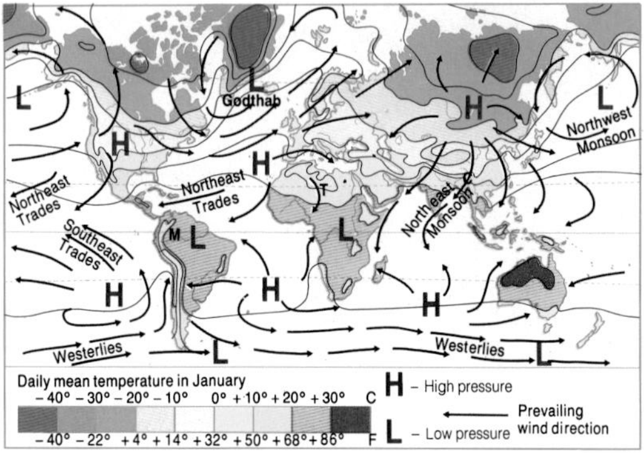

Temperature, July

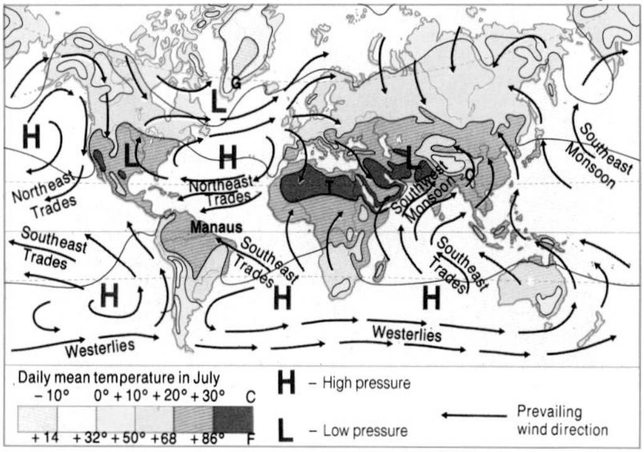

Annual precipitation

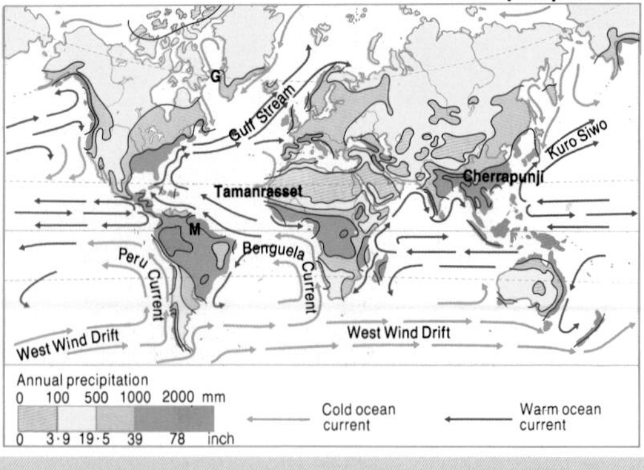

Cloud formations

Cirrus clouds form at 10 000 m, 30 000 ft. They look wispy and are made up of ice crystals which melt as they fall.

Cumulus clouds appear soft and downy, they may be several kilometres thick and are formed by rising thermals.

Stratus clouds are low, flat and extensive. Light continuous rain may be associated with them.

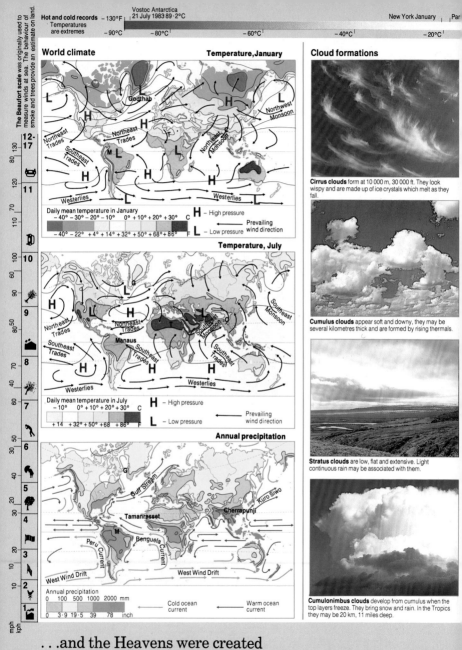

Cumulonimbus clouds develop from cumulus when the top layers freeze. They bring snow and rain. In the Tropics they may be 20 km, 11 miles deep.

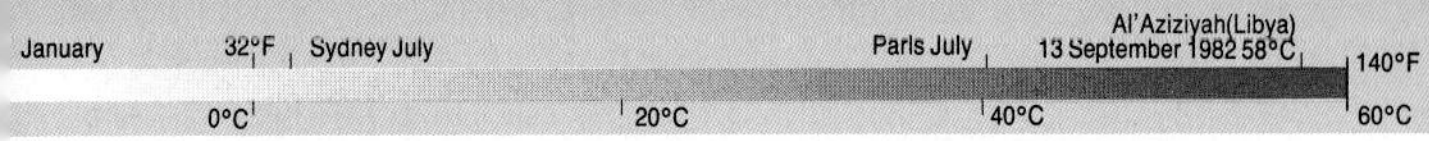

In Fahrenheit freezing temperature is 32° and in Celsius it is 0°. The Kelvin scale starts at −273°C.

Rain

Cold and warm fronts

Sometimes an air mass meets another air mass moving in the same direction; where the two masses at different temperatures meet is called a front. The movement of fronts determines changes in the weather in the temperate latitudes. When cold air (left) pushes its way beneath warm air a storm can be expected. A warm front (right) occurs when moist air moves in over the cold air resulting in continuous rain. The different fronts are heralded by typical cloud forms.

Mountains and rain

As moist air is forced upwards the reduced pressure at higher altitudes causes a fall in temperature and rain-bearing clouds to form.

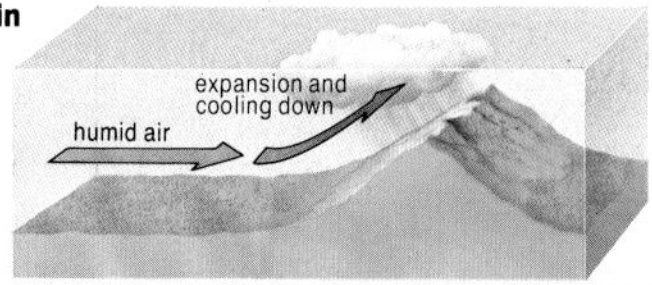

Cold and rain

A cooling process may occur as a result of heat loss to cold ground or cold water, or through radiation. This process may produce mist, rain and coastal fog.

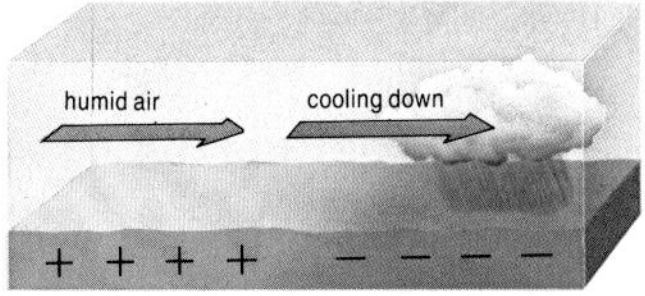

Local circulation

Winds blow clockwise round an anticyclone and anticlockwise round a depression. Directions are reversed in the southern hemisphere.

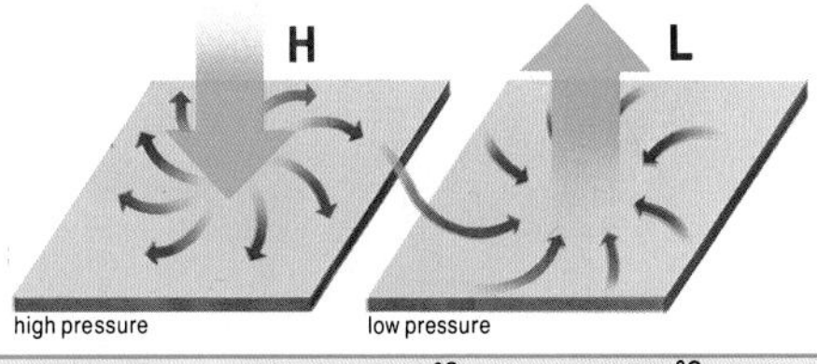

Picturing climate

The height of the bars shows the mean monthly rainfall at the weather stations on the left hand maps. The range of temperature is the gap between maximum (red line) and minimum (blue line).

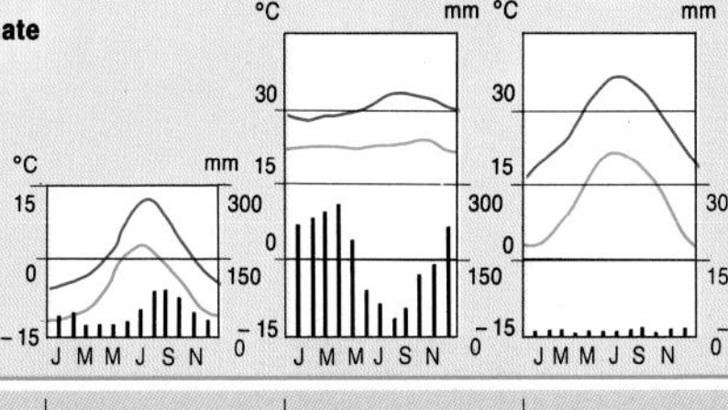

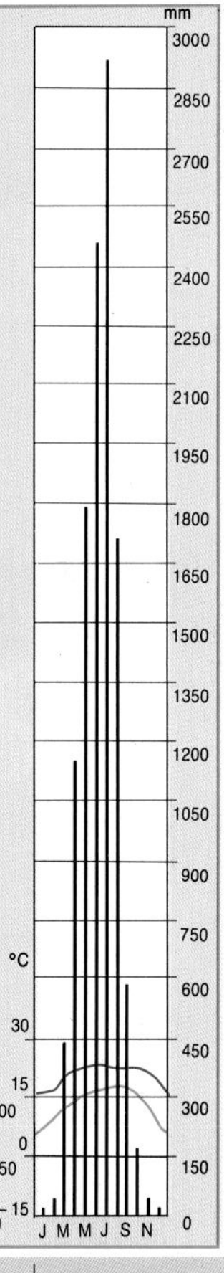

Godthab G
This station on the west coast of Greenland usually has very cold winters.

Manaus M
In the Amazon Basin temperatures are consistently high, but there is a wet and very wet season.

Tamanrasset T
A very dry station in the Sahara desert. On average rain falls on only 21 days in the year.

Cherrapunji C
The Southwest Monsoon brings very heavy rain to this station at a height of 1313 m, 4308 ft.

Worldwide weather

Hailstones weighing more than 1 kg, 2·25 lb fell in Bangladesh in 1985, while in Nebraska, USA in 1928 hailstones with a circumference of 43 cm, 17 inches were reported.

The 'garua' in coastal Peru. This cold mist condenses from the sea breezes blowing off the cold Peru (Humboldt) current, and supports patches of woodland called 'lomas' in the desert.

The Pole of Cold. At this Antarctic weather station near the South Pole the annual mean temperature is −57·8°C, −72°F.

In Western Australia there were more than 330 days in 1946 when temperatures topped 32°C, 90°F

Khartoum. After a number of very dry years the city had 200 mm, 7·5 inches of rain on the night of 4/5 August 1988. This was more than an average year's total rainfall.

Cherrapunji has heavy rain on 89% of the days between May and the end of August. 180 very wet days can be expected in the year.

Tutunendo in Colombia has an average rainfall of 11 770 mm, 463 inches making it the wettest place in the world.

The Atacama Desert in Chile has no rainfall at all. It is drier than the Sahara desert or the centre of Australia.

Kansas, USA. America's breadbasket had a rainfall 90% below average in 1988/89. On a few days winds gusting up to 72 mph lifted the topsoil, while dust clouds darkened Kansas City 530 km, 300 miles to the east.

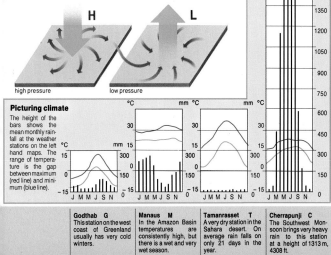

Celsius/Fahrenheit
[(°Cx9) ÷ 5] + 32 = °F
[(°F − 32)x5] ÷ 9 = °C

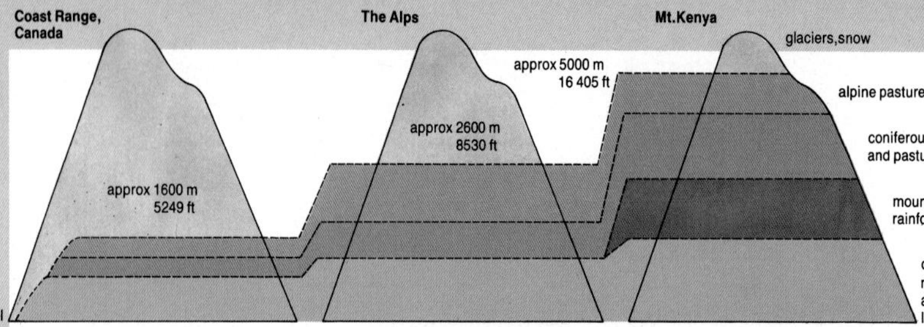

The new island of Surtsey appeared off the coast of Iceland in 1963 as a result of volcanic activity along the Mid Atlantic Ridge.

Everest reaches its peak? Mountains become unstable when they get high, but it is thought that a wedge is being squashed out of the lower Himalayas. So although Everest is building up from the bottom the top is collapsing.

The birth of mountains

The Earth's crust is often composed of horizontal sediments laid down under water. These layers are forced upward by compression to form folded mountains. Erosion by water and glaciers wears away the mountain to leave a typically rugged landscape.

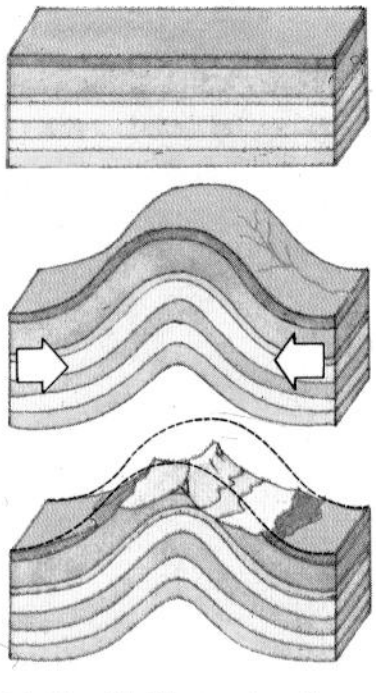

The birth of hills and valleys

A simple fault occurs through vertical movements on both sides of a fault line. Where a block is forced upwards between two faults, or the earth's crust sinks, a horst is formed. A graben occurs when the bedrock block slips downward.

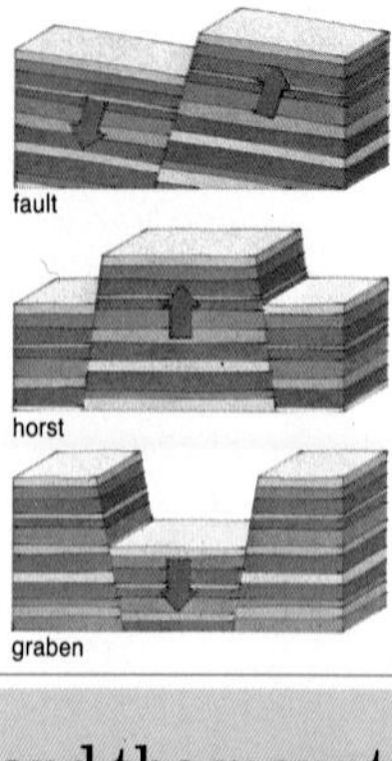

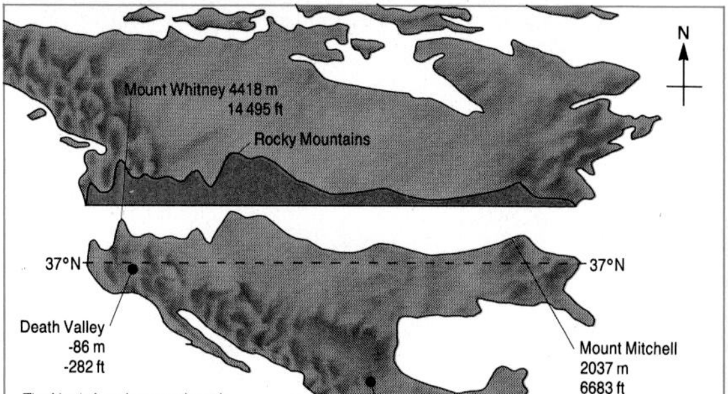

The North American continent is, broadly speaking, a central lowland bordered by the Cordillera and the Appalachians. The Canadian Shield contains some of the oldest rocks on Earth. They have been folded into mountains, perhaps not unlike the Rockies today, and then eroded to form a low lake-fringed plain.

North America

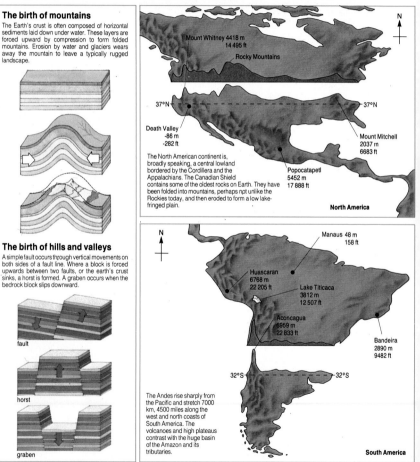

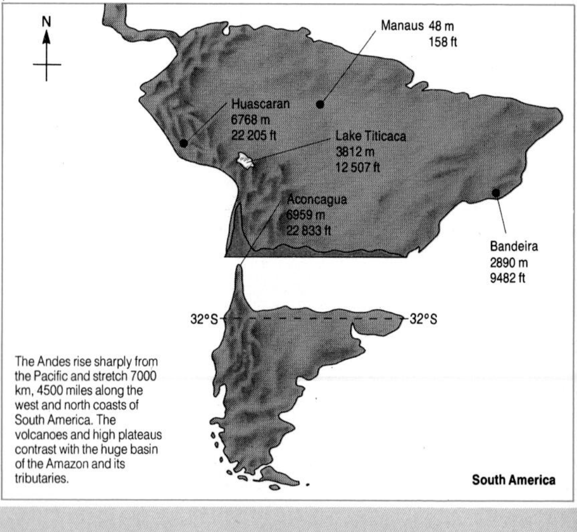

The Andes rise sharply from the Pacific and stretch 7000 km, 4500 miles along the west and north coasts of South America. The volcanoes and high plateaus contrast with the huge basin of the Amazon and its tributaries.

South America

. . .and the mountains were raised

physical world relief

Half Dome mountains, Yosemite, USA has an almost sheer rock face which is 670 m, 2200 ft high. In 1957 it took 5 days to climb.

The Dead Sea in the Jordan valley between Israel and Jordan is a highly saline lake with a surface level 400 m, 1300 ft below sea level.

The Challenger Deep drops 11 034 m (36 213 ft or more than 6 miles) and is part of the Mariana Trench, the deepest trough in the Pacific Ocean.

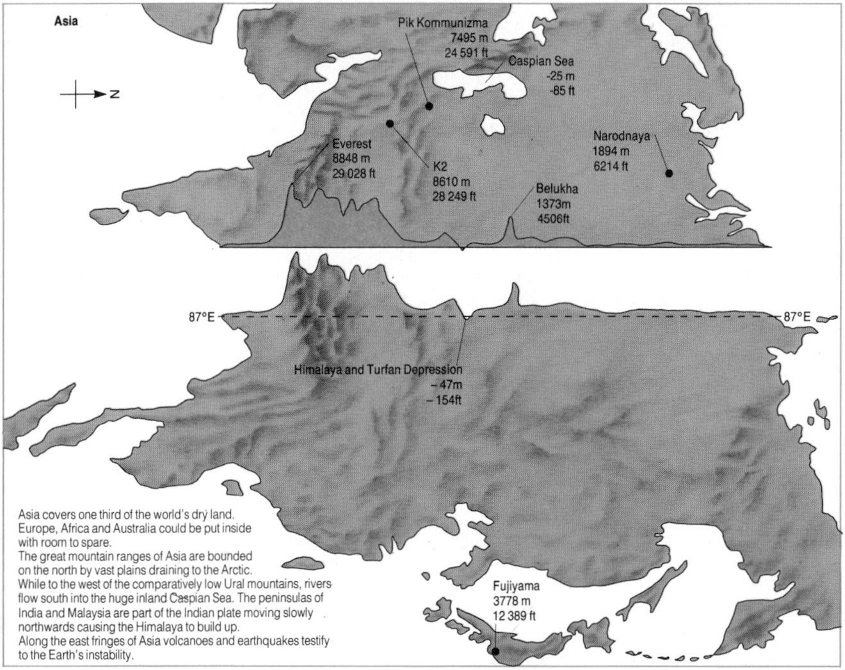

Asia covers one third of the world's dry land. Europe, Africa and Australia could be put inside with room to spare.
The great mountain ranges of Asia are bounded on the north by vast plains draining to the Arctic.
While to the west of the comparatively low Ural mountains, rivers flow south into the huge inland Caspian Sea. The peninsulas of India and Malaysia are part of the Indian plate moving slowly northwards causing the Himalaya to build up.
Along the east fringes of Asia volcanoes and earthquakes testify to the Earth's instability.

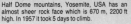

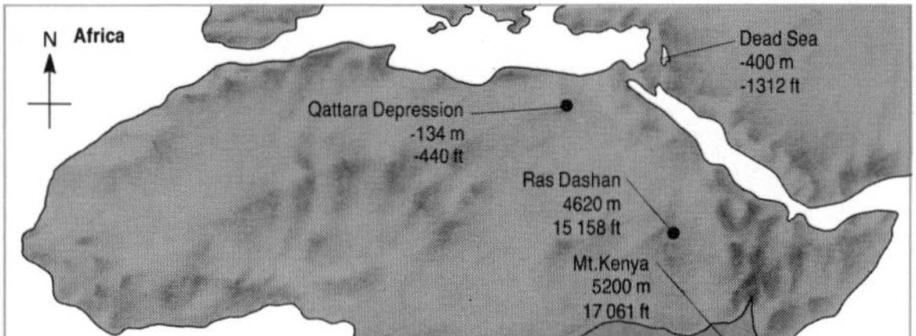

A line from Angola to Ethiopia divides the continent into Low and High Africa. Sedimentary plains rarely rising above 1000 m, 3300 ft characterise the northeast, while nearly all the southeast is above this height. Even Lake Victoria, lying between two arms of the Rift Valley has a surface level of 1130 m, 3700 ft.

The lifecycle of mountains

Mountains are born in the oceans, in deep subsiding basins called geosynclines where layers of sand and mud build up. The mountain chains surrounding the Pacific are associated with deep ocean trenches and volcanoes called the 'circum-Pacific ring of fire' where the ocean crust dives beneath continental crust.
The crashing together of these plates results in volcanic activity while the sediments are pressed into fold mountains. From the moment of birth these mountains are eroded and the sediments carried back to the sea though some geological formations are made of more resistant material and ridges and peaks will be a feature – for a while. Rivers are the main agents in wearing away the rock and wind too can be a powerful force of erosion, particularly when it carries abrasive particles. Sheets of sediment will be formed and eventually the mountains will be brought low, unless another cataclysm of mountain building restarts the cycle. It is thought that Ben Nevis in Scotland may once have been higher than Everest but erosion has had 400 million years longer to denude the Scottish mountains.

The cross-sections show how the continents would look if they were cut through along a selected line of latitude or longitude. The vertical scale is exaggerated in relation to the horizontal scale, but nevertheless a good impression of relative heights and slopes can be obtained.

The average depth of the sea is 3·729 km, 2·317 miles whilst the average height of the land is only 840 m, 2756 ft.

The continental margins comprise only 21% of the ocean floors but are economically most important.

From the highest elevation to the deepest ocean floor, the total vertical distance is a mere 0·31% of the Earth's equatorial radius.

Nearly 30% of the Earth's surface lies within 1000m, 1093yds above or below sea level.

The world's four oceans: 0% Atlantic Ocean 25% Arctic Ocean 4% Indian 22%

Percentage share of total sea area

Underwater landscapes

Topography of the ocean floor can be divided into two distinct features: the continental margins and the deep sea basins.

The character of the ocean basin depends on the extent to which sediments mask the crust and also the degree of volcanic activity. The sediments may be either pelagic or terrigenous. The latter are brought down by turbidity currents which are avalanches of silt and sand from the continental shelf. These powerful currents can cut channels in the continental shelf such as the Hatteras Canyon off North America and transport material thousands of kilometres.

On the continental shelf, sediments are affected by waves, tidal currents and changes in sea level.

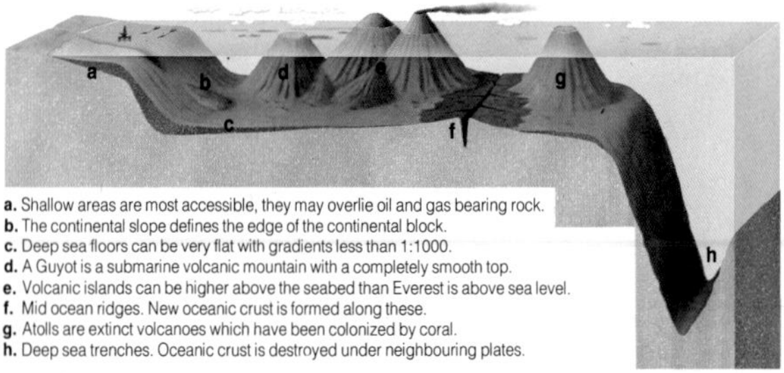

a. Shallow areas are most accessible, they may overlie oil and gas bearing rock.
b. The continental slope defines the edge of the continental block.
c. Deep sea floors can be very flat with gradients less than 1:1000.
d. A Guyot is a submarine volcanic mountain with a completely smooth top.
e. Volcanic islands can be higher above the seabed than Everest is above sea level.
f. Mid ocean ridges. New oceanic crust is formed along these.
g. Atolls are extinct volcanoes which have been colonized by coral.
h. Deep sea trenches. Oceanic crust is destroyed under neighbouring plates.

. . .and water filled the depths

world oceans

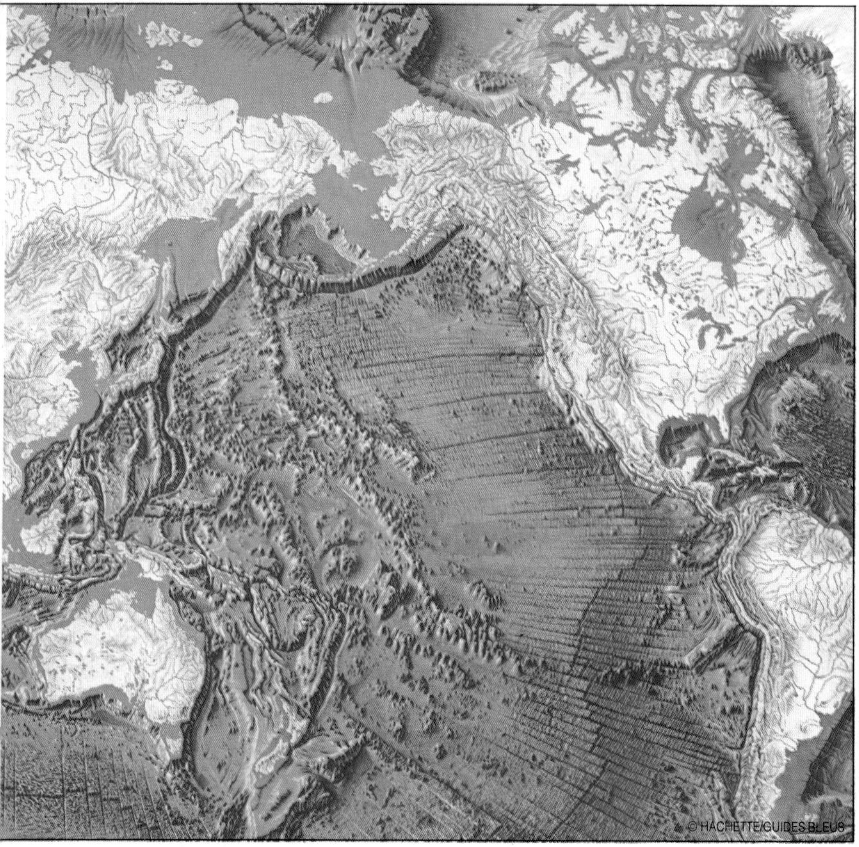

Seabed treasures

In the deeper sea regions mineral exploitation has concentrated on manganese nodules. These lumps grow at rates of between 3-8 mm, .25 in each million years, and they are valuable for the copper, nickel and cobalt they contain. Granules vary in size and may be up to 150 mm, 6 ins in diameter.

On the continental shelves and near coastal regions placer deposits are often commercially viable. They consist of heavy mineral particles which have been weathered from locally occurring ore bodies and deposited on beaches and in estuaries. Gold is extracted from placer deposits off Alaska.

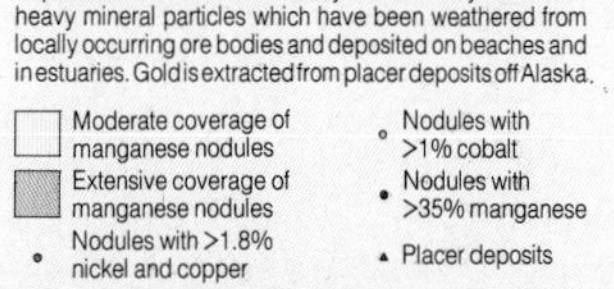

Terrigenous sediments. These consist of material which has been eroded from the land usually by rivers.

Pelagic sediments are deposited in the deep oceans and comprise the microscopic remains of plankton that have sunk to the bottom.

Metalliferous muds are sediments with high concentrations of metals which are often found at constructive plate margins.

Glacier, Ice cap

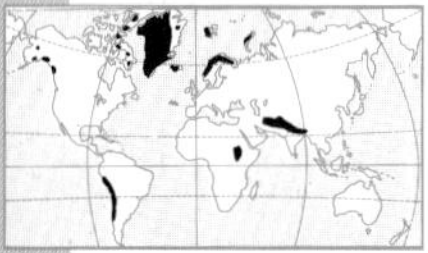

Tundra

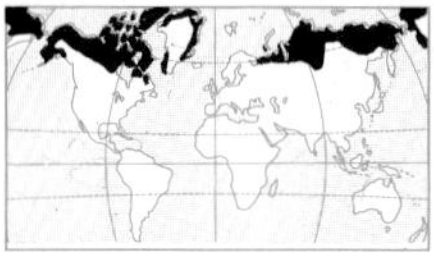

Tundra vegetation consists of low-growing perennial plants which bloom during the long days of the brief summer. **Coniferous forest** covers about 11% of the land surface. Often the same tree species grows over large areas. **Tropical rain forest** (19% cover) supports the greatest number of different plant and animal species. **Savanna grasses** (11% cover) may grow up to 2 m, 6 ft tall during the short wet season.

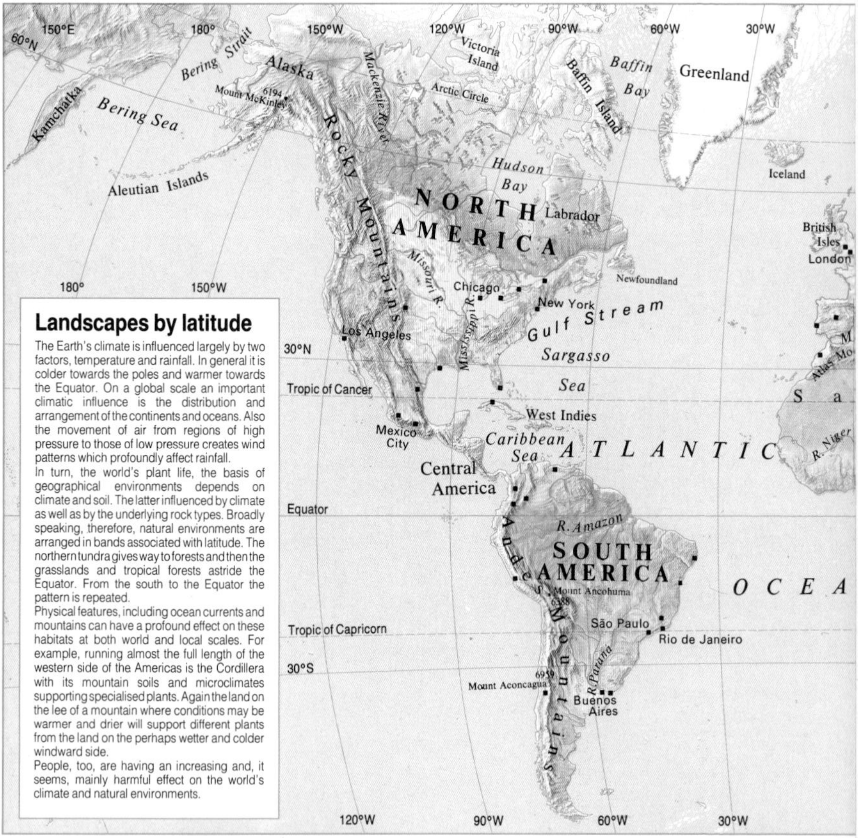

Landscapes by latitude

The Earth's climate is influenced largely by two factors, temperature and rainfall. In general it is colder towards the poles and warmer towards the Equator. On a global scale an important climatic influence is the distribution and arrangement of the continents and oceans. Also the movement of air from regions of high pressure to those of low pressure creates wind patterns which profoundly affect rainfall.

In turn, the world's plant life, the basis of geographical environments depends on climate and soil. The latter influenced by climate as well as by the underlying rock types. Broadly speaking, therefore, natural environments are arranged in bands associated with latitude. The northern tundra gives way to forests and then the grasslands and tropical forests astride the Equator. From the south to the Equator the pattern is repeated.

Physical features, including ocean currents and mountains can have a profound effect on these habitats at both world and local scales. For example, running almost the full length of the western side of the Americas is the Cordillera with its mountain soils and microclimates supporting specialised plants. Again the land on the lee of a mountain where conditions may be warmer and drier will support different plants from the land on the perhaps wetter and colder windward side.

People, too, are having an increasing and, it seems, mainly harmful effect on the world's climate and natural environments.

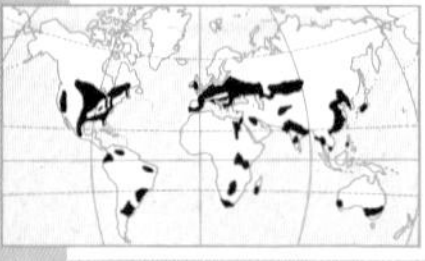

Cultivated land

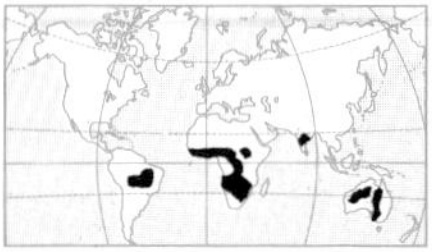

Savanna

. . .and trees, grass and deserts were created

Coniferous forest

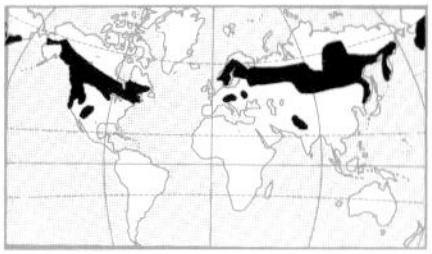

Rain forest

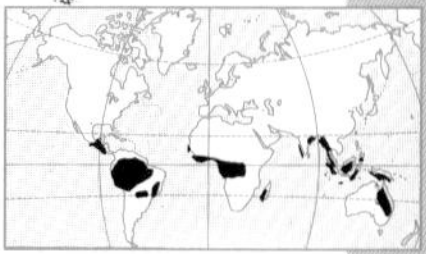

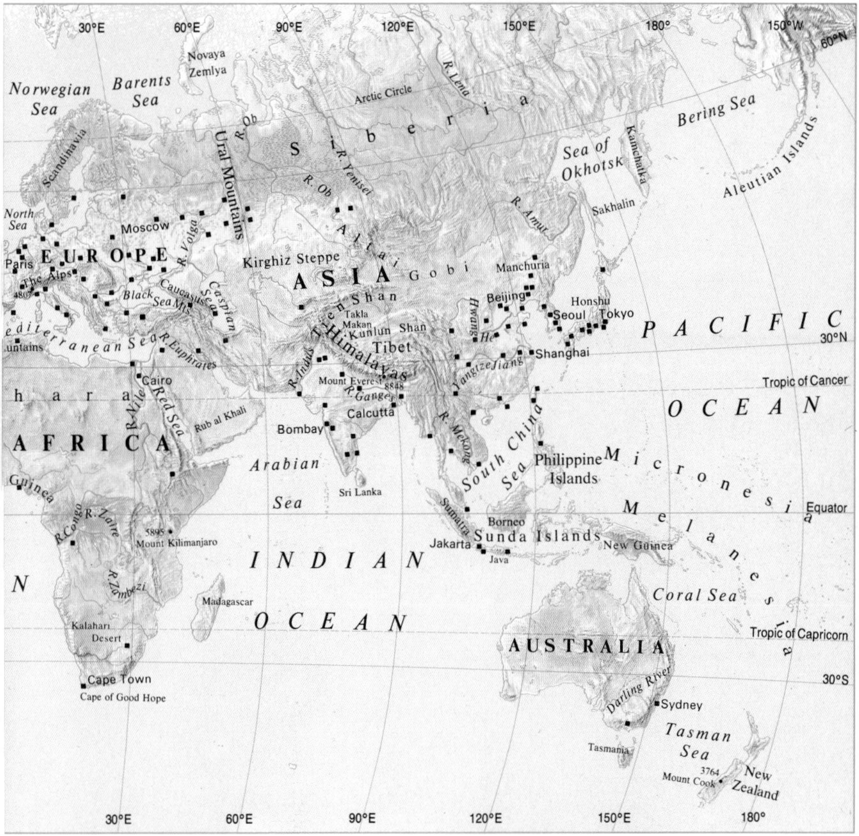

As Savanna gives way to **Steppe** the vegetation becomes lower and sparser with small thorny-leaved plants. **Deserts** are found on 8% of the world's surface. Little vegetation can survive but plants such as cacti, which conserve water may thrive. Much of the world's temperate forest and grassland has already been cleared and is now **cultivated land.** In some areas human activity is sometimes destroying the natural vegetation.

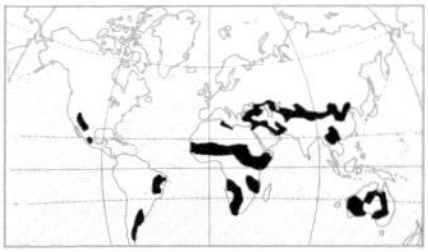

Steppe

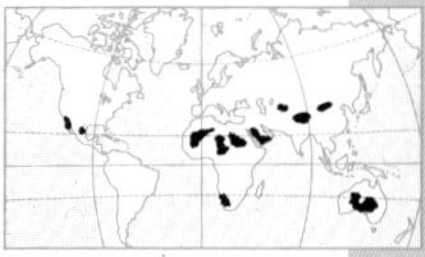

Desert

III Australasian Region. Zoogeographical regions are defined by the animals that occur only in that area. Present day boundaries are partly determined by climatic and physical features. The distribution of many animals is explained in terms of movements of the earth's crust (see page 12). Because this region has had no connection with any other land mass for millions of years many of the animals, including many marsupials, are unique.

I Palearctic Region. Contains most of the world's temperate lands. The deer family are characteristic, and there are many insect-eating animals.

Some animals such as the beaver occur almost only in North America, while hedgehogs and wild pigs occur only in Europe and Scandinavia.

II Oriental Region. The Thar desert and the Himalayas form a climatic boundary in the west.

The eastern limit is defined by Wallace's Line along the Macassar strait. Birds, reptiles, frogs and toads are well represented.

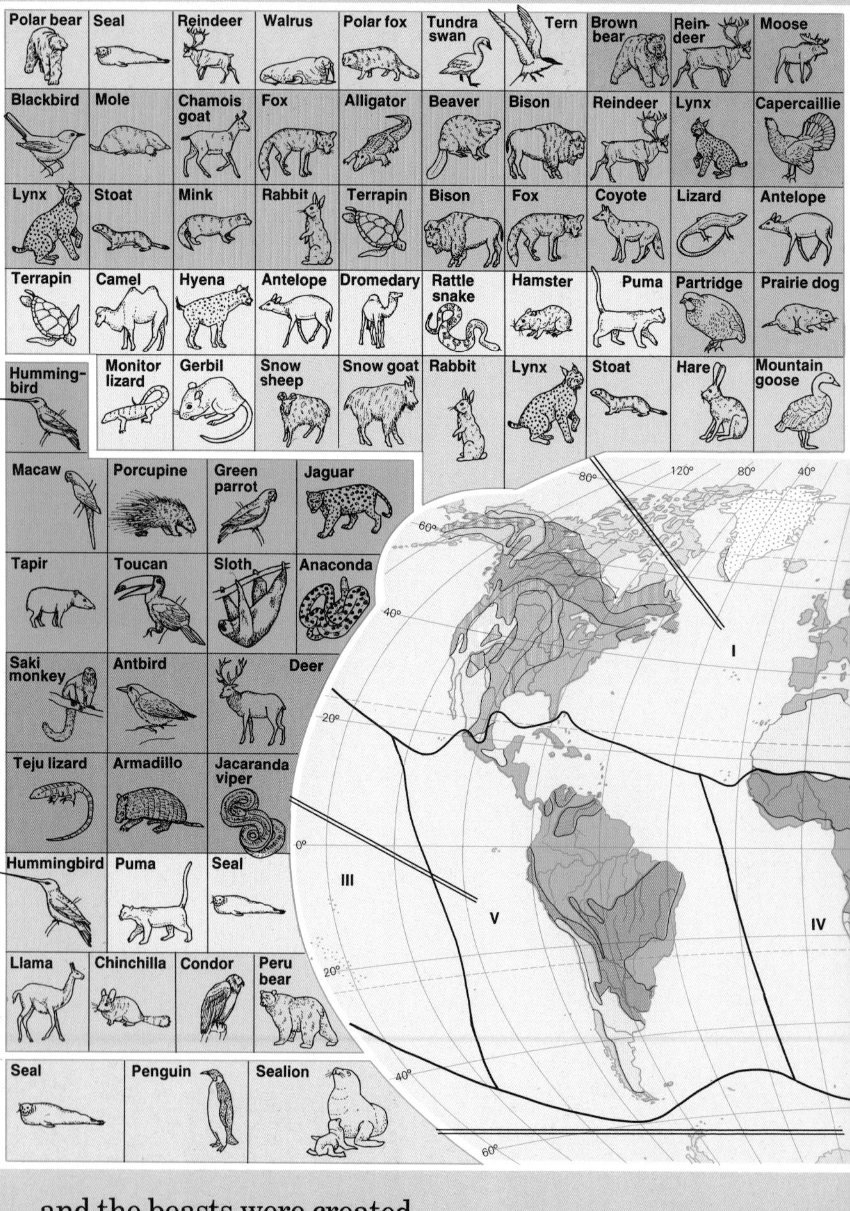

. . .and the beasts were created

adaptation of the species

IV Ethiopian Region. The Sahara and Arabian deserts form natural boundaries. There are many different mammals in this area. Thirty eight families of animals are represented and 12 are unique to the region. Giraffes and hippopotamuses are not found anywhere else. Some of the animals from this region have close relatives in the Oriental region. Others, particularly fish are related to South American species.

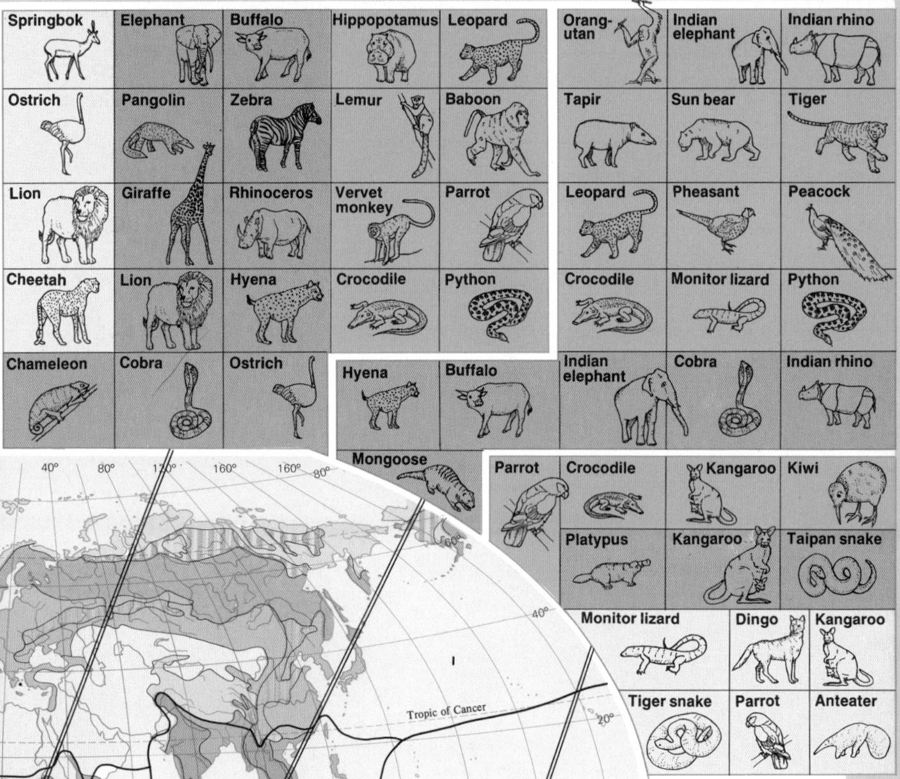

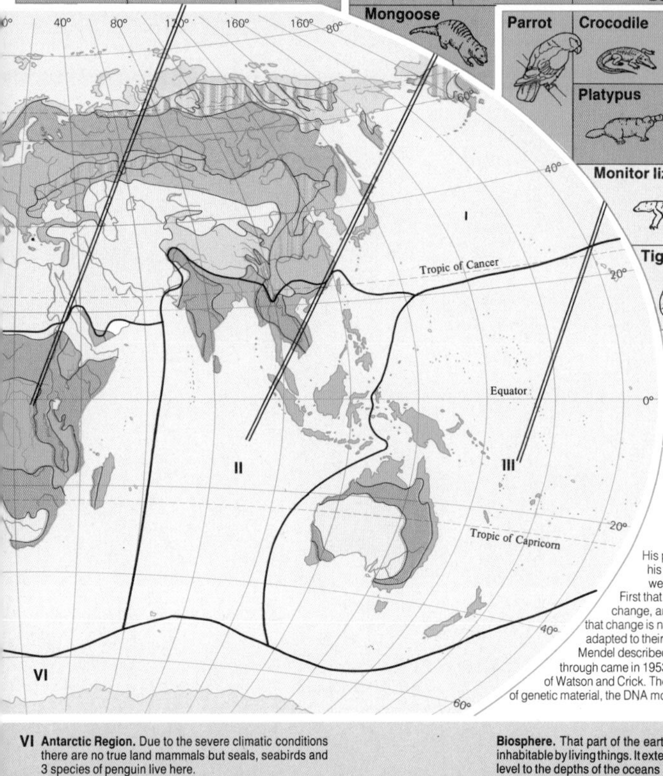

Adjusting to Nature

Much of the human race has believed that the enormous variety and complexity of living things must be the result of divine creation. Since 1859, the year Darwin's *Origin of Species* was published it has been difficult not to accept that man and all living things have evolved through different species and that all have their origin in very simple lifeforms. His painstaking work as a naturalist and his awareness of the geological record were the basis of his two major theses. First that species are not immutable and can change, and secondly that the mechanism for that change is natural selection. Those that are best adapted to their habitat flourish. The Austrian monk Mendel described patterns of heredity but the breakthrough came in 1953 with the Nobel prize-winning work of Watson and Crick. They explained the chemical structure of genetic material, the DNA molecule in the form of a double helix.

V Neotropical Region. Because of the isolation of this region for the last 60 million years there are many unique animals.

VI Antarctic Region. Due to the severe climatic conditions there are no true land mammals but seals, seabirds and 3 species of penguin live here.

Biosphere. That part of the earth, its atmosphere and oceans, that is inhabitable by living things. It extends from about 9 km, 6 miles above sea level to the depths of the oceans 11 km, 7 miles.

Pre-Cambrian	Million years ago 600	Cambrian	500	Ordovician	440	Silurian	400	Devonian	350

Creation or development. . .

In 1650 Bishop Ussher dated the creation to the 23rd October 4004 BC, and early fossil discoveries were seen as evidence of Noah's Flood. By the 1800's, geologists disputed this timescale. In 1859 Darwin's '*Origin of Species*' argued for a long gradual evolution for all species.

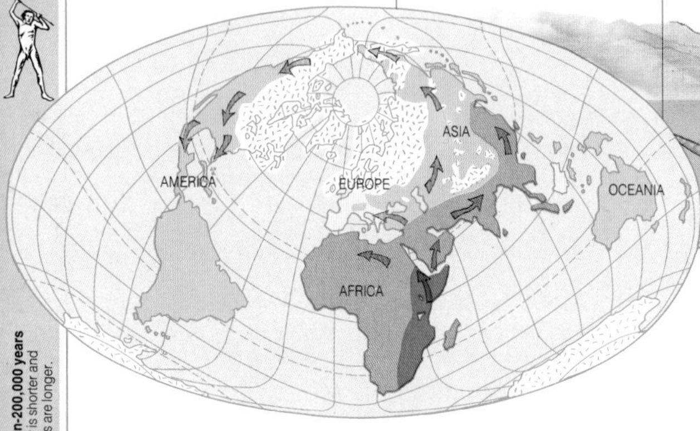

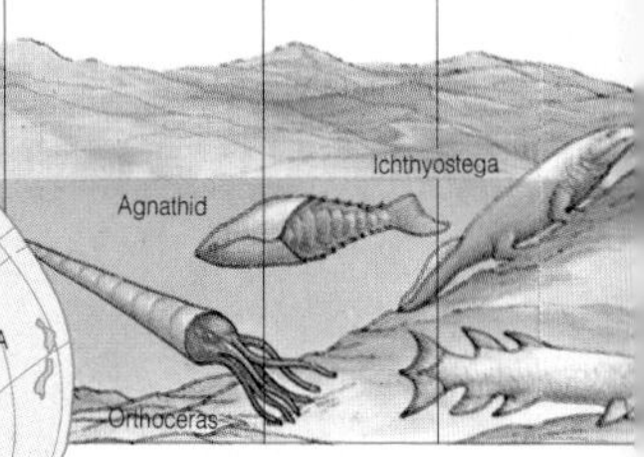

We populated the Earth. . .

4M years ago our ancestors evolved in east and south Africa. 1·5 million (M) years ago Homo erectus evolved. Their behaviour differed, in that they hunted and shared food collectively. This cooperative behaviour made migrations possible. 1M years ago, or earlier, H.erectus is found in Asia and Europe, in highly organised social groups, with knowledge of fire and hunting large animals.

Area inhabited already 3·5-1·5 million years ago

Areas where people lived 1·5 million years ago

Areas inhabited 500 000-100 000 years ago

Areas inhabited as late as 100 000-10 000 years ago

Glaciation 200 000 years ago

Man's movement from his original home in Africa to Europe, Asia, Oceania and America

Australopithecines and Hominids

Australopithecines lived in Africa 4-1·4 million years ago, and differed from great apes by walking upright (about 1·3 m, 4½ ft tall), and having slightly larger brains (700cc, c.f gorilla 585cc). It is still unclear how they were all related, or how the later Hominids developed from them. A.afarensis, named after remains from the Afar valley, Ethiopia, is the oldest. In 1974 40% of a young female skeleton was discovered, named 'Lucy', 3·5-3 million years old, and footprints 3·7 million years old from Laetani, Tanzania, show a short stride, but upright posture. Two other early types are A.africanus and A.robustus (A.boisei is an east African version of A.robustus), who may both have evolved from A.afarensis. A.africanus was a more lightweight, 'gracile' species, a meat-eater and perhaps a tool-user, and dates from 3-2 million years ago. Finds range from Koro Toro in Chad, to Taung in S.Africa, and the more recent fossils lived alongside A.robustus, a heavyweight and vegetarian who survived until about 1·4 million years ago. Hominids succeeded australopithecines, and fossils are found outside Africa.

a. Unmanufactured stone tools, more than 2 million years old. Used by Homo habilis.

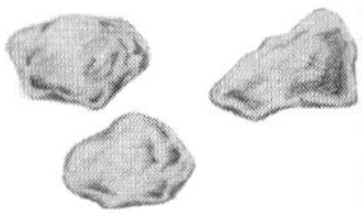

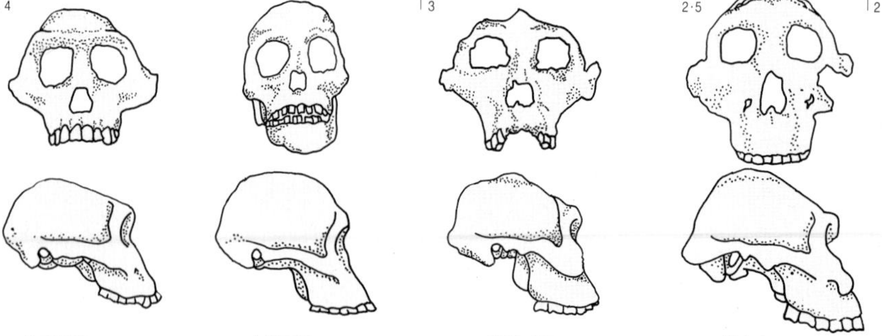

6-4 million years ago. Australopithecus learns to balance and walk a way on hind legs.

4-1·5 million years ago. The pelvis is now adjusted to an upright posture.

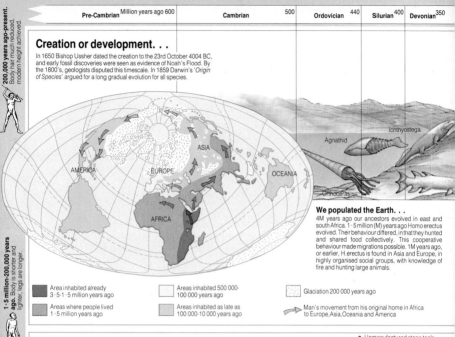

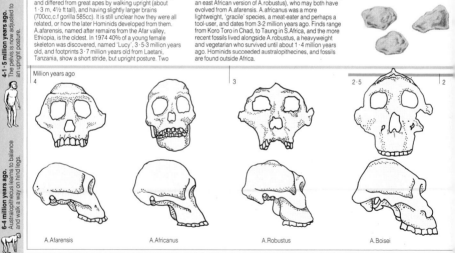

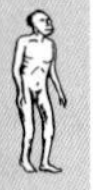

1·5 million-200,000 years ago. Body is shorter and lighter, legs are longer.

200,000 years ago-present. Body hair much reduced, modern height achieved.

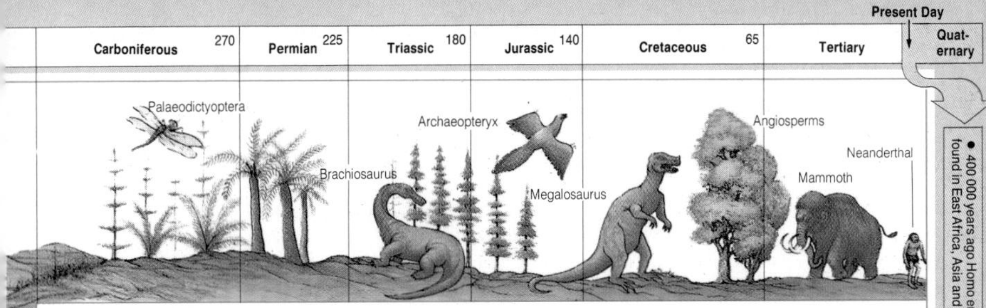

From single-cell to human being

The earth is over 4,500M years old, but the oldest signs of life are reefs (called 'stromatolites'), found in rocks in Zimbabwe and Australia, 3500M years ago. Protozoa first appear 1700M years ago; jellyfish and worms developed 670M years ago as the earliest animals. Life emerged onto land 425M years ago, and dinosaurs ruled by 225M years ago. The oldest fossil of a true placental mammal is 114M years old, from Mongolia, and early primates appear 69M years ago. The earliest known primate, Purgatorous, was living in Montana 69M years ago, watching tyrannosaurs pass by.

Our cradles

Homo sapiens dates from about 250 000 years ago, based on fossils from Swanscombe and Steinheim, 'Neanderthal' remains, named from a skull discovered in 1856, are now considered to be H.sapiens. It is Neanderthal types that first live beyond the frost-free zones and bury their dead with ceremony; at Toshik-Tash a burial is surrounded by goat horns, at Shanidar the grave contains garlands of medicinal flowers.

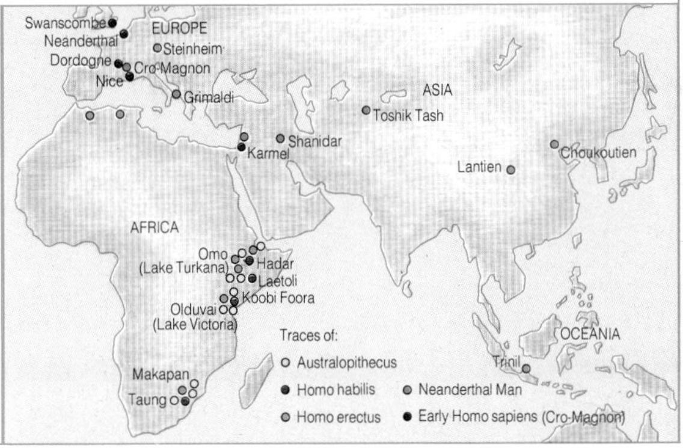

b. Side-scrapers, 1·5 million years old. One or two edges are chipped.

c. Small flake, used as a cutter by H.habilis.

d. Rounded pebble chopper, which could also be used as a hammer for making stone tools.

e. This early hand-axe was first used by H.erectus.

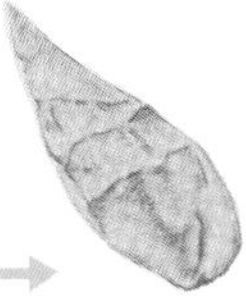

Homo sapiens probably evolved in west Asia, in temperate, not tropical, climates, about 230,000 years ago; the oldest fossils are of the neanderthal type. H.sapiens neanderthalensis extended from west Europe to Siberia, migrating as the ice-sheets retreated, also south as far as Zimbabwe. They had a brain slightly larger than ours, about 1650cc, and lived in larger social groups. The most recent neanderthal remains are 34,000 years old, from Cognac in France, but H.sapiens was living in caves in Cromagnon, Dordogne, 40,000 years ago. Apart from having a lighter, taller skeleton, H.sapiens probably had command of a fully developed language, which led to a very rapid colonisation of the world.

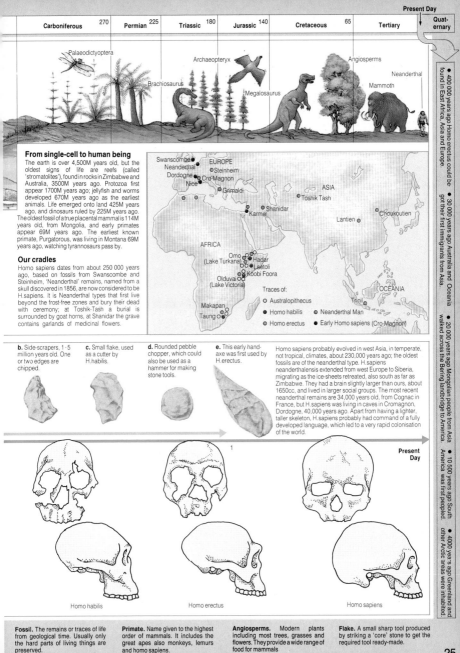

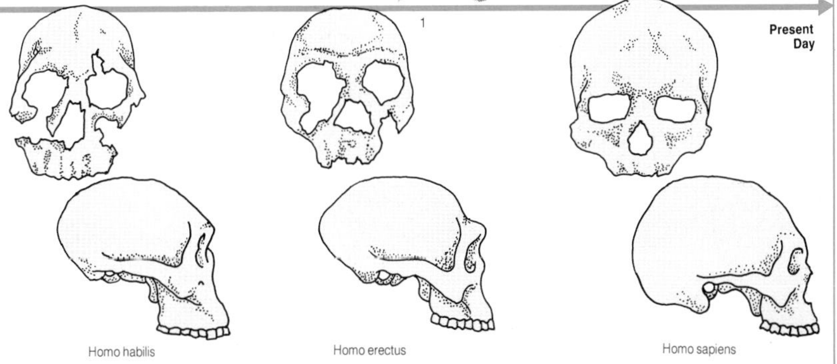

Fossil. The remains or traces of life from geological time. Usually only the hard parts of living things are preserved.

Primate. Name given to the highest order of mammals. It includes the great apes also monkeys, lemurs and homo sapiens.

Angiosperms. Modern plants including most trees, grasses and flowers. They provide a wide range of food for mammals.

Flake. A small sharp tool produced by striking a 'core' stone to get the required tool ready-made.

- 400 000 years ago Homo erectus could be found in East Africa, Asia and Europe.
- 30 000 years ago Australia and Oceania got their first immigrants from Asia.
- 20 000 years ago Mongolian people from Asia walked across the Bering landbridge to America.
- 10 500 years ago South America was first peopled.
- 4000 years ago Greenland and other Arctic areas were inhabited.

● World ● Africa ● Europe	26 births 45 births 14 births	10 deaths 15 deaths 11 deaths
● India ● UK ● Kenya	28 births 13 births 54 births	11 deaths 12 deaths 12 deaths
● USA ● El Salvador ● Australia	16 births 38 births 16 births	9 deaths 7 deaths 8 deaths

The birth and death rates are the numbers of these events in a year per 1000 of the population.

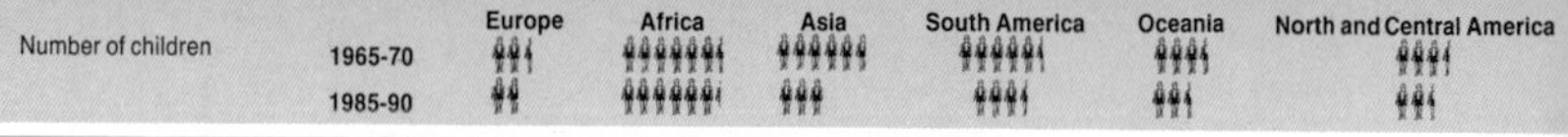

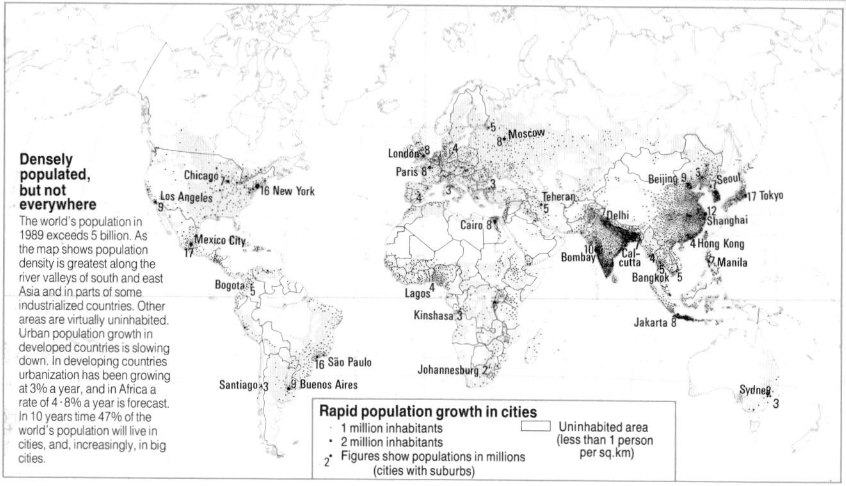

Densely populated, but not everywhere

The world's population in 1989 exceeds 5 billion. As the map shows population density is greatest along the river valleys of south and east Asia and in parts of some industrialized countries. Other areas are virtually uninhabited. Urban population growth in developed countries is slowing down. In developing countries urbanization has been growing at 3% a year, and in Africa a rate of 4·8% a year is forecast. In 10 years time 47% of the world's population will live in cities, and, increasingly, in big cities.

Rapid population growth in cities

- · 1 million inhabitants
- • 2 million inhabitants
- 2• Figures show populations in millions (cities with suburbs)
- ☐ Uninhabited area (less than 1 person per sq.km)

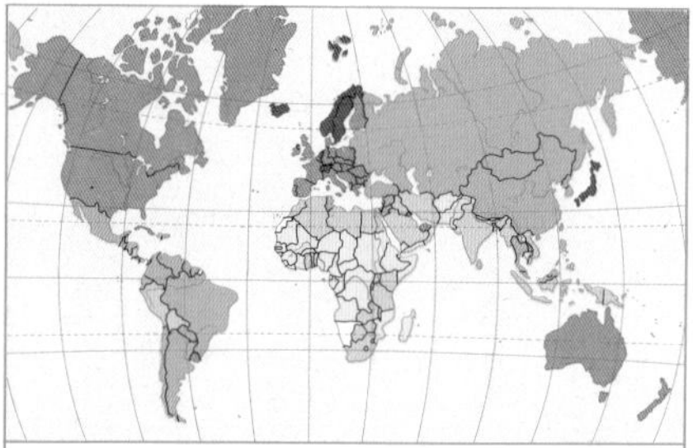

Life expectancy at birth

The age a newborn baby can expect to reach

Years old: No data, 40, 50, 60, 70, 75

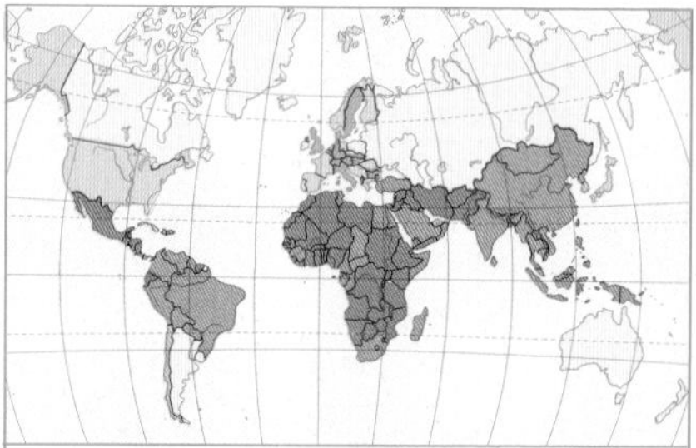

Median age in 1980

Median age splits the population into halves. If the median age is 22 then half the population is younger than 22.

Years old: 14, 18, 22, 26, 30, 34, 38

Comparison of population with area

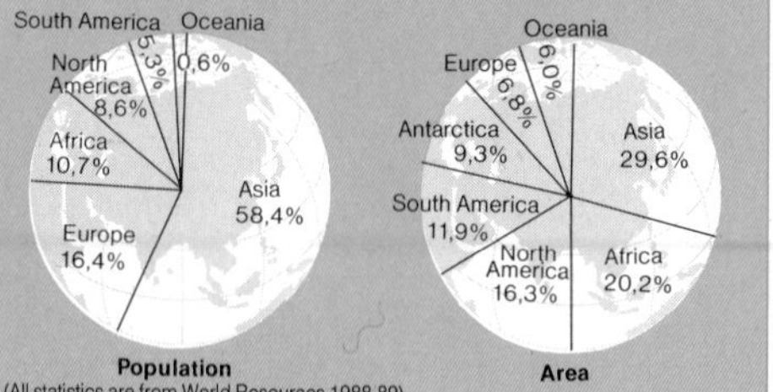

(All statistics are from World Resources 1988-89) (IIED and UNEP)

Predicting the future

Rapid population growth is a feature of the second half of the 20th century. Better medicine has resulted in fewer children dying and the old living longer. There were about 1·6 billion people in 1900 but the UN predicts 6 billion by the year 2000, with population stabilizing at 10 billion, twice the current number by the end of the next century.

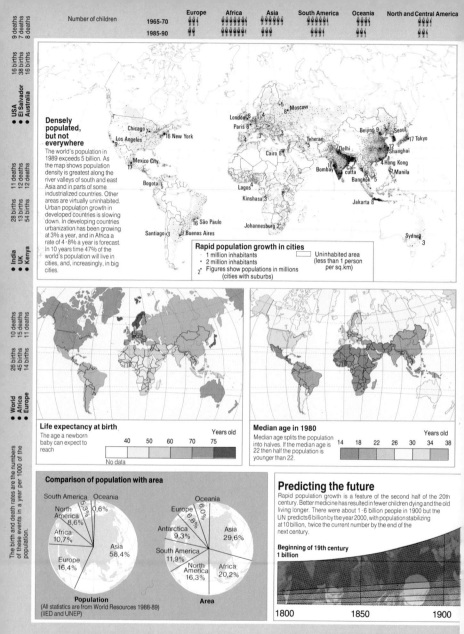

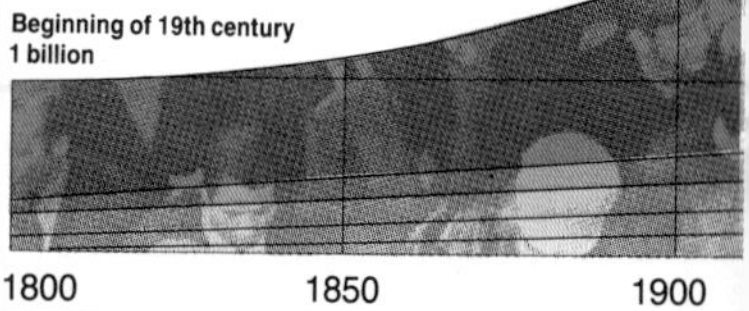

. . .and we populated the Earth

distribution and growth of world population

Population 1975 and 2000

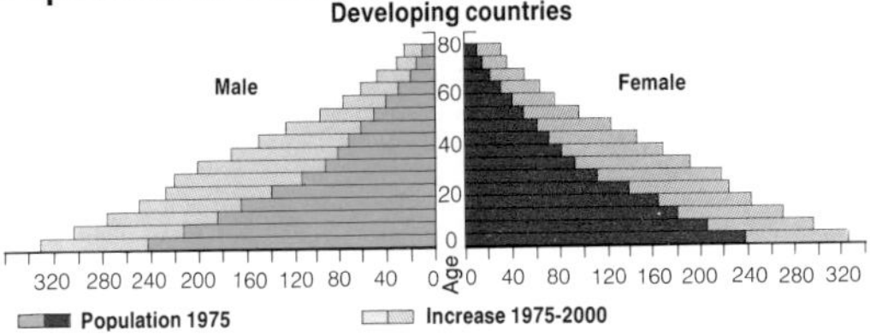

The broad base of the pyramid diagram (left) is typical of developing countries. The pyramid indicates that population is growing fast and child mortality is high. About 37% of the population is under 15 years of age so there will continue to be many more children even if family size is limited. Only 4% of the people are over 65 years. The pyramid for industrial countries is in sharp contrast. In Japan life expectancy is 77 years while in Afghanistan it is only 39 years.

The birth rate is usually the dominant factor in population growth. Fertility rates vary widely, with those for some African countries being the highest. The number of children born to the average woman has fallen sharply in some Asian and South American countries.
There are a number of measures that governments may choose to use to limit population growth; these may range from basic family planning to incentives and disincentives which affect family size.

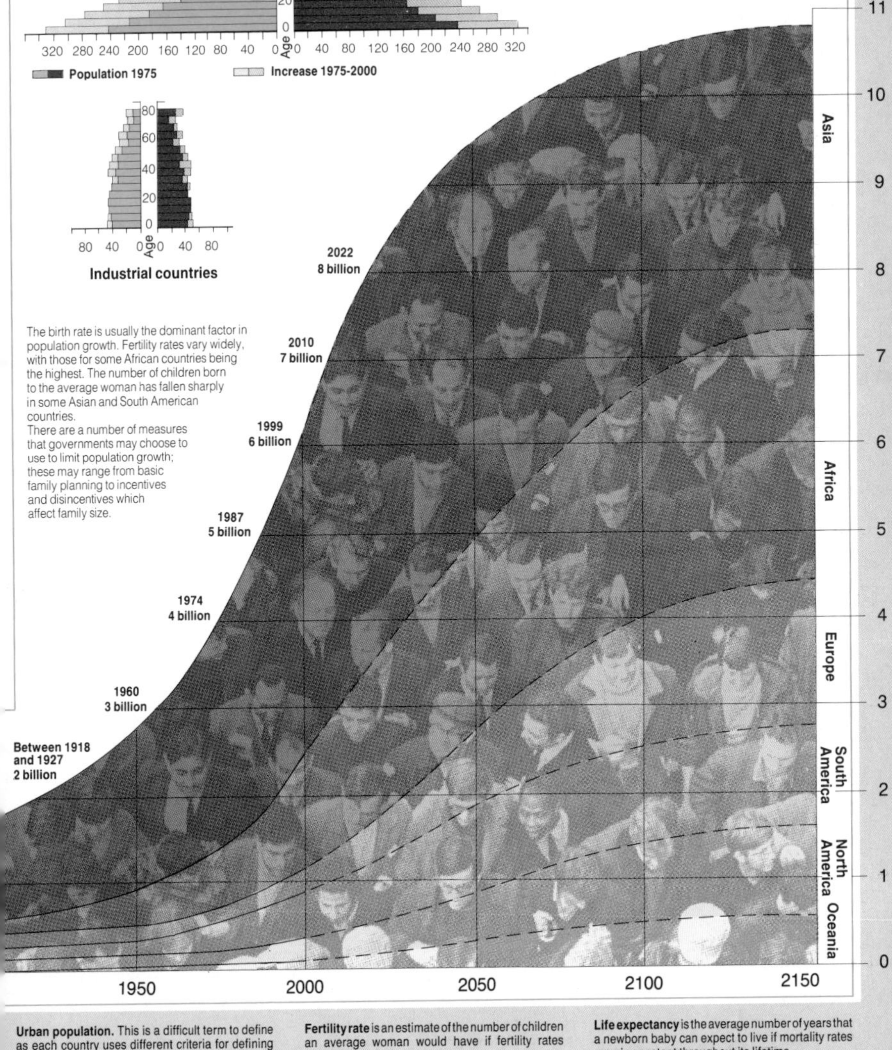

Urban population. This is a difficult term to define as each country uses different criteria for defining cities. A UN urban agglomeration includes a central city and surrounding urbanized localities.

Fertility rate is an estimate of the number of children an average woman would have if fertility rates remain constant.

Life expectancy is the average number of years that a newborn baby can expect to live if mortality rates remain constant throughout its lifetime.

1. America

2. Far East

3. Iran – India

4. Northern Europe

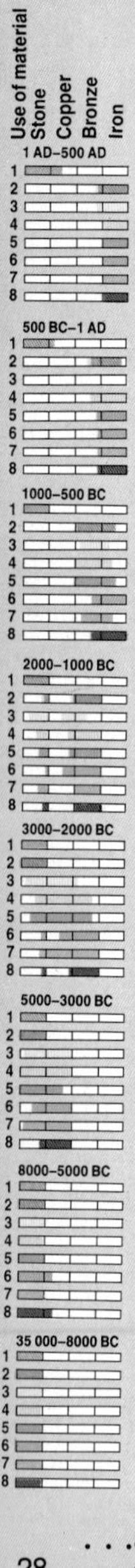

Technological advances

1 AD–500 AD

The Romans harnessed water power in the waterwheel. Underfloor heating systems and surgical instruments were also developed. In the Chinese seismograph, a trigger released a ball which fell into the frog's mouth.

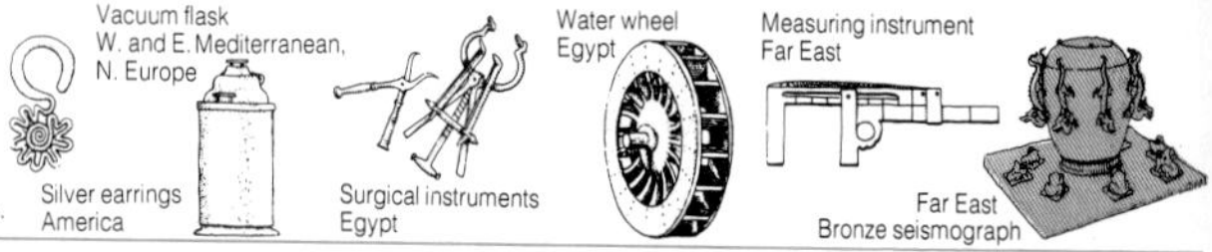

500 BC–1 AD

Many mechanical inventions were made; they include the pulley, lever, screw and gears. Hero made a 'steam engine' for opening temple doors. In N. Europe a wagon with 14 spoke wheels was made.

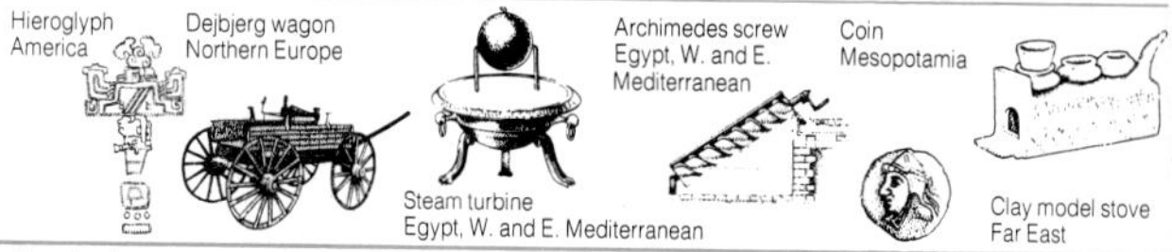

1000–500 BC

Iron, now widely available, was used to make high quality weapons. Helmets, horse bits and domestic goods such as candelabra and scales were made from bronze. Trade stimulated writing and coinage.

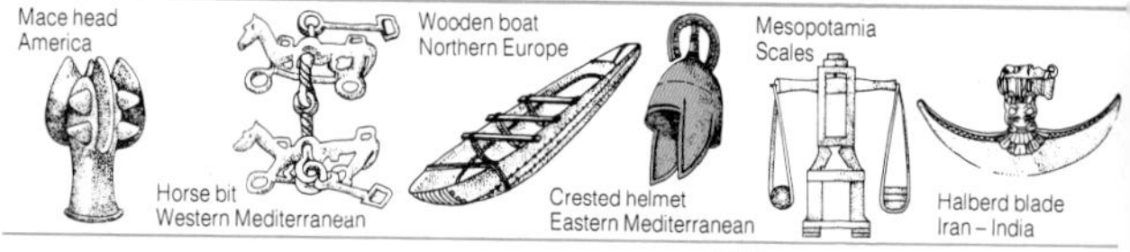

2000–1000 BC

Bronze was used to make better tools for working metal and wood. Decorated items such as the razor with a boat inscribed on it were created. Glass, a new material, was used in moulded form.

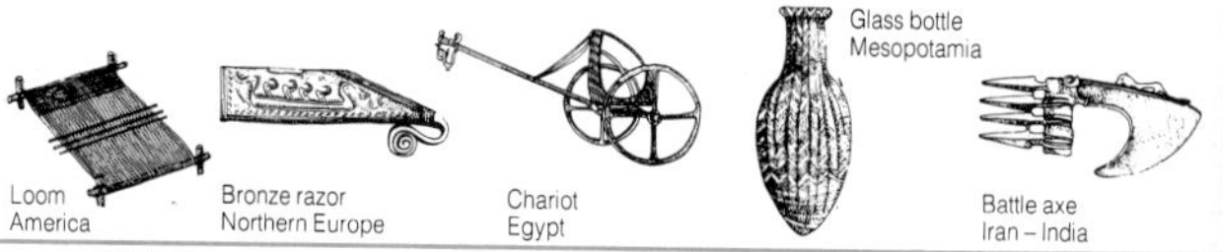

3000–2000 BC

In the Eastern Mediterranean experiments led to the discovery of bronze, a mix of 10% tin with 90% copper. Trade stimulated improved ship building.

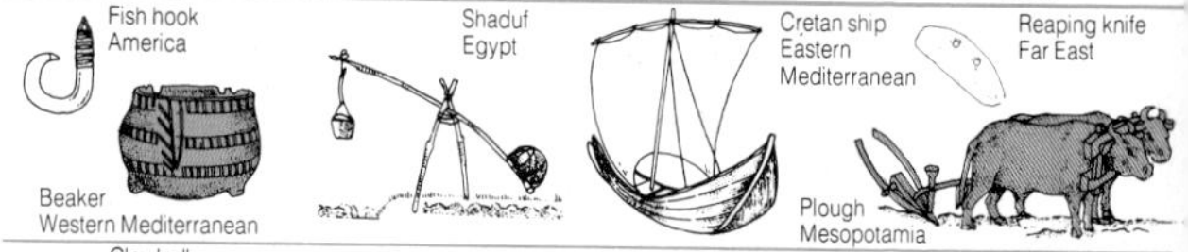

5000–3000 BC

In the cities of southwest Asia technological developments included wheeled vehicles and sea-going ships. The potters wheel and kiln were developed and copper tools and weapons were made.

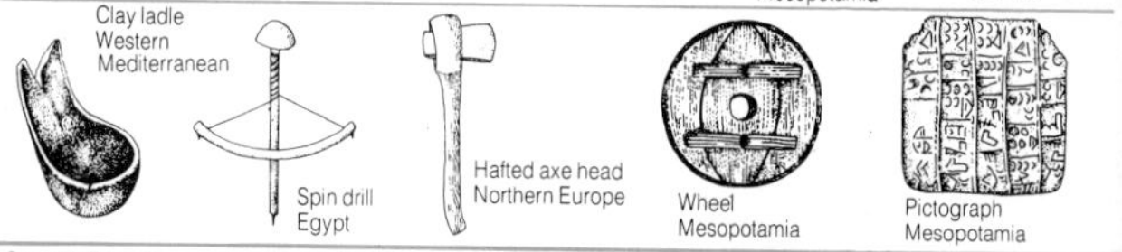

8000–5000 BC

The beginning of farming and the keeping of domesticated animals dates from this period. Stone querns were used for grinding corn, and clay stamps to mark property suggests the concept of personal ownership.

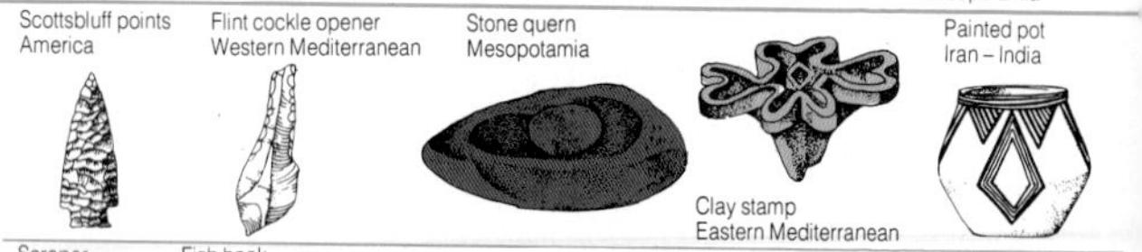

35 000–8000 BC

Natural materials including stone, bone and antler were used to produce hunting and fishing equipment. Decorative items and combs were also made from bone. Technology developed slowly.

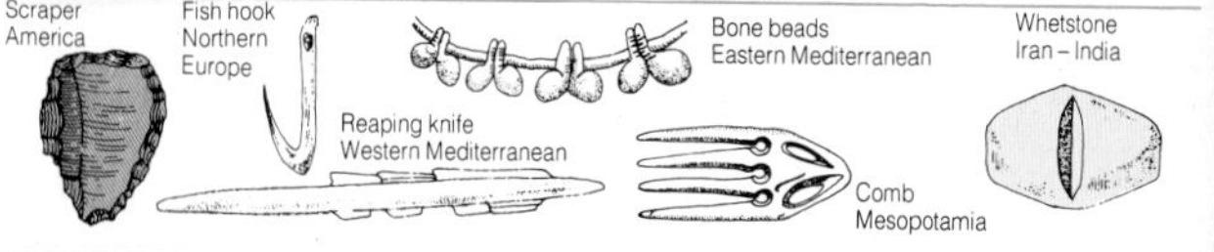

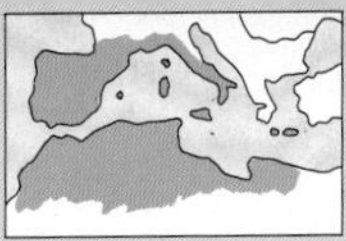

5. Western Mediterranean

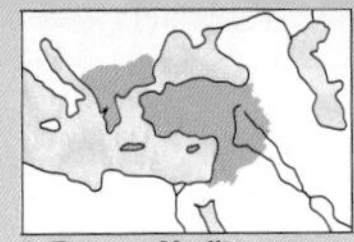

6. Eastern Mediterranean

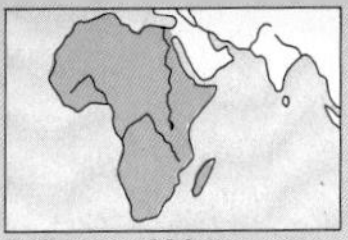

7. Egypt – Africa

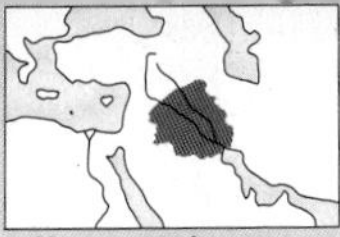

8. Mesopotamia

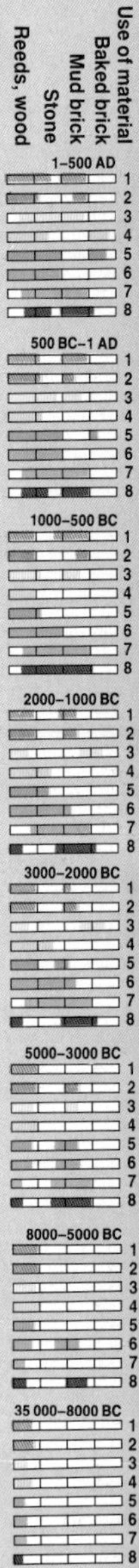

Architectural advances

Artwork courtesy Dorling Kindersley Ltd.

1 AD–500 AD

Roman civilization demanded a wide range of civic buildings including theatres, baths, bridges and aqueducts. The arch and barrel vaulting were developed together with the use of brick faced concrete.

500 BC–1 AD

The arch was developed for Greek theatres which were built on level ground. The Stupa at Sanchi in India was built as a hemispherical mound with elaborate carving. In Shetland the broch served as a refuge.

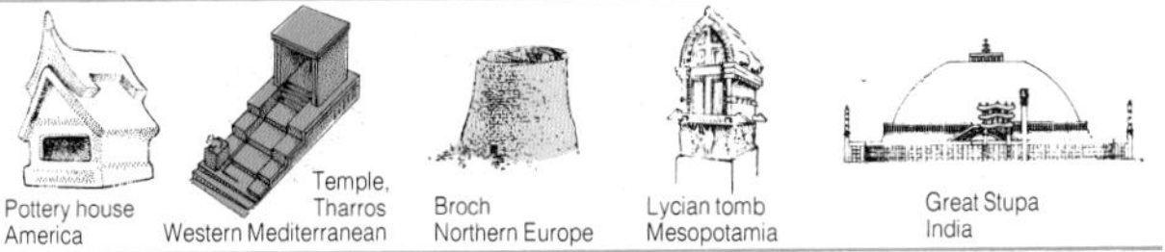

1000–500 BC

In western Europe tribes built hillforts with earth and timber ramparts. The Greek temples had colonnades with columns decorated in Doric, Ionic and later, Corinthian style.

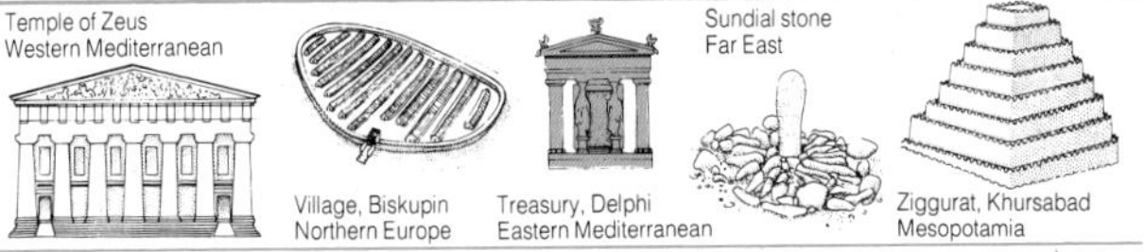

2000–1000 BC

Temples for ceremonial use were built in Peru using natural stone blocks. The Cretan palace of King Minos combined administrative space with living quarters. Stonehenge was being built in England.

3000–2000 BC

Large complex temples were built in Mesopotamia. In Egypt the Pyramids were constructed and in western Europe huge blocks of stone formed dolmens.

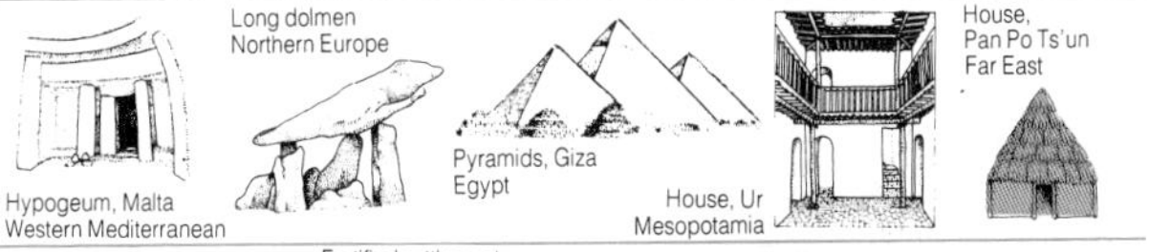

5000–3000 BC

The long gabled thatched houses in the Danube valley had timber frames and wooden lattice walls coated with clay. In the Eastern Mediterranean walled towns with a decorated palace were built.

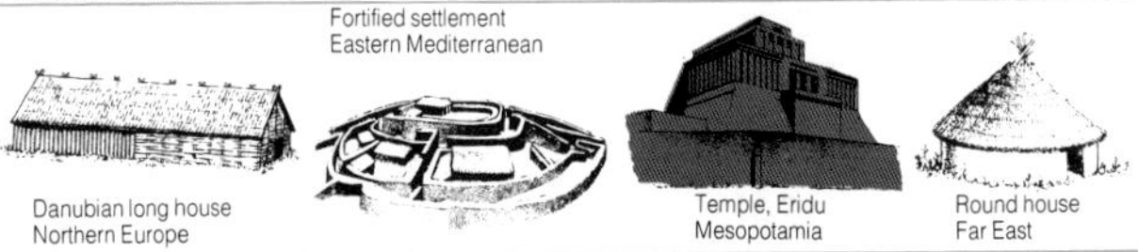

8000–5000 BC

Throughout most of the world temporary structures continued to be used. In southwest Asia the settled way of life associated with farming demanded more permanent buildings. Houses were built from mudbricks.

35 000–8000 BC

In the limestone areas of Western Europe caves were used. In addition the hunters dwelt in shelters made with wood, reeds, mud, mammoth bones or whatever was available.

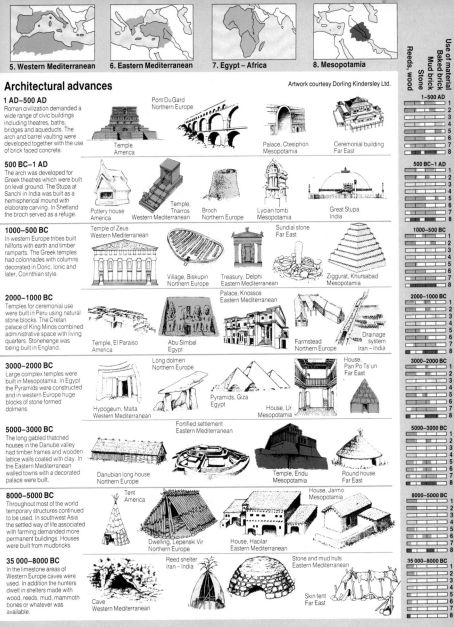

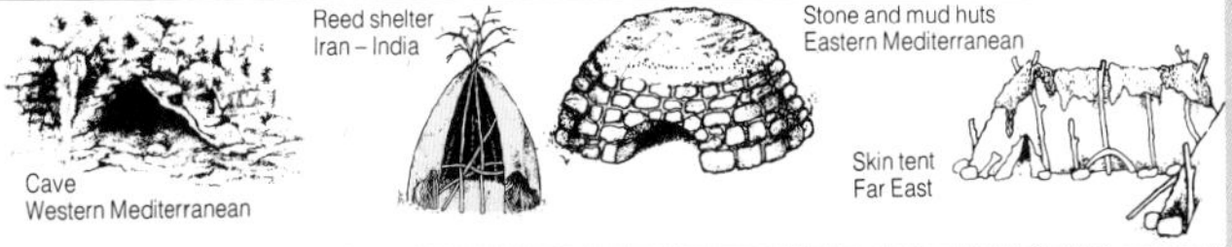

British sign language

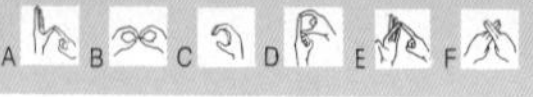

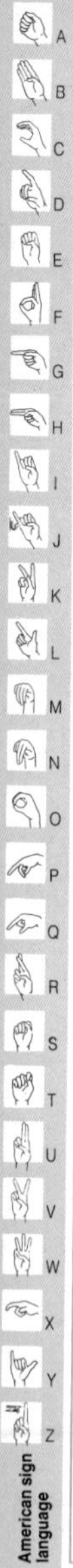

A babble of languages

Many mythologies show the importance of divine intervention in the creation of language. The Old Testament story of the Tower of Babel tries to explain why there are so many languages in the world and why nearby communities can have such trouble understanding each other. Sometime after the Deluge, the Babylonians wanted to impress the world by building a vast city, and a tower 'with its top in the heavens'. Building became impossible and was abandoned, when God confused the languages of the workers so that they couldn't work together. The people scattered and formed their own communities over the earth. Today, the number of languages and dialects spoken throughout the world could be as high as 6000: there are many languages which have no written form and are only spoken by a few elderly survivors of a particular tribe. Some languages have been saved, but others are disappearing as remote areas are being opened up and brought into the modern world. Missionaries in the 19th century invented scripts for some of the American Indian languages so that they could make translations of the Bible, but many others have been lost. Now attempts are being made among the new nations of Asia and Africa to safeguard their traditions and languages through writing, and the development of new alphabet.

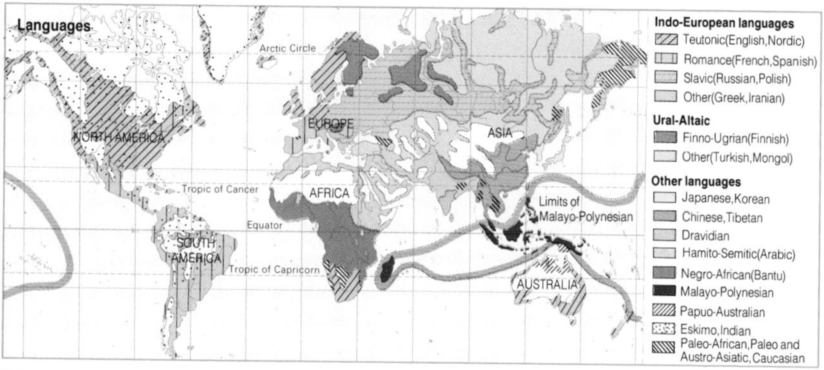

The seven largest language families of the world

	Speakers in millions	
1 Indo-European		2 000
Hindi, Urdu	400	
English	350	
Russian	200	
Spanish	200	
Portuguese	100	
2 Sino-Tibetan		1 000
Chinese	800	
3 Ural-Altaic		250
Japanese	100	
Korean	50	
Turkish	60	
Finno-Ugrian	20	
4 Malayo-Polynesian		150
Javanese	100	
5 Semitic		150
Arabic	100	
6 Dravidian		100
Tamil	50	
7 Bantu		100
Swahili		3 750

The existence of language families has been recognised for a long time. But the seven families shown in the chart were only discovered in the 19th century, when linguists formulated a methology of comparing resemblances and stating systematic relationships between languages. The new linguistic science shattered the belief of Christian Europe that, since God had spoken to Adam in Hebrew, this must be the parent language of the world. The resemblances between certain languages indicated several parent languages which no longer exist.

4 or 1687 newspapers to choose from!

There has been a world-wide attempt to promote literacy, and the right to free primary education has been embodied in the Universal Declaration of Rights. The advent of machine printing created the base for mass education and radio, television and newspapers now reach a wider audience than ever before. Yet these charts show the gap between the ideal and the reality. In many developing countries there is no local primary school for children to attend.

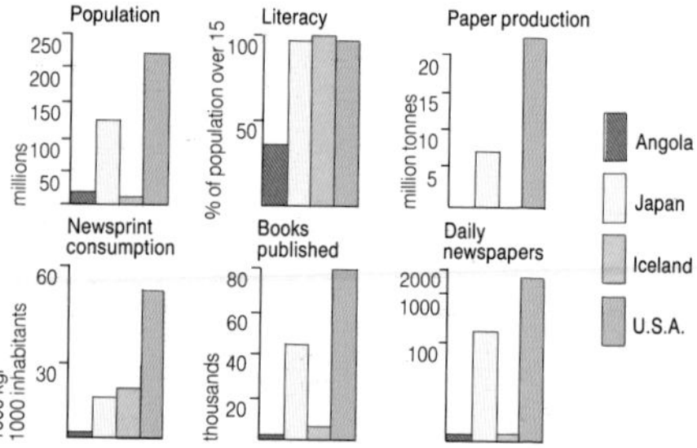

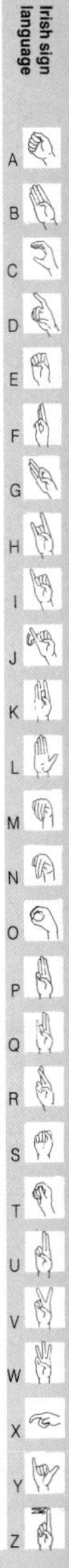

Schools, teachers and books

In the highly centralised, developed countries of the world, universal literacy has been all but achieved. In other parts of the world, a great deal of money and effort is being put into making this a reality, but literacy programmes are encountering many problems. In Africa, for example, indigenous languages are numerous. It is impossible to put the necessary investment in terms of teacher training, teacher time and resources, publications and printing to achieve literacy in every one of these languages – some of which are spoken by less than 50 people. Instead, resources have tended to be put into a Western-style education system which is often inappropriate to local needs. In Africa and India, this problem is being tackled and efforts are being made to use home-trained graduates to deal with the needs of the indigenous culture. But limited resources have to be shared between the training of specialists who are needed to lead the newly politically independent countries into economic independence and a broad-based primary education system which will be of long term value.

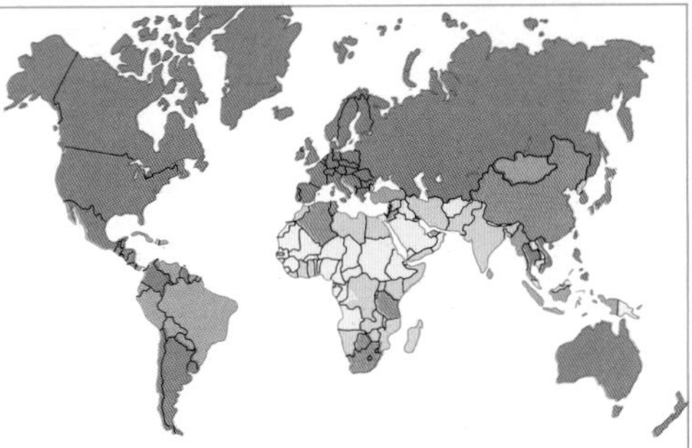

Literacy, 1980

< 20% | 20-40% | 40-60% | 60-80% | >80%

Business and pleasure

In many countries of the world, the language of officialdom and the commonly used language have become very close. But this wasn't always true, and there are many parts of the world where it still isn't true. Africa is the most multilingual area of the world, but the few colonial languages are used for government administration.

In Medieval Europe, Latin was the international language of educated people but it meant little to most people who had no schooling. Since World War II the dominance of English-speaking people in science, technology and international commerce led to English becoming the major international language. In the 19th century Dr. Zamenhof tried to create a simple international language, Esperanto, but it was never used as an official language. Now people are trying to solve the problems of international communications with computers. A translation program exists which can produce simple, reasonably accurate text at lightning speed, which can be checked for inaccuracies.

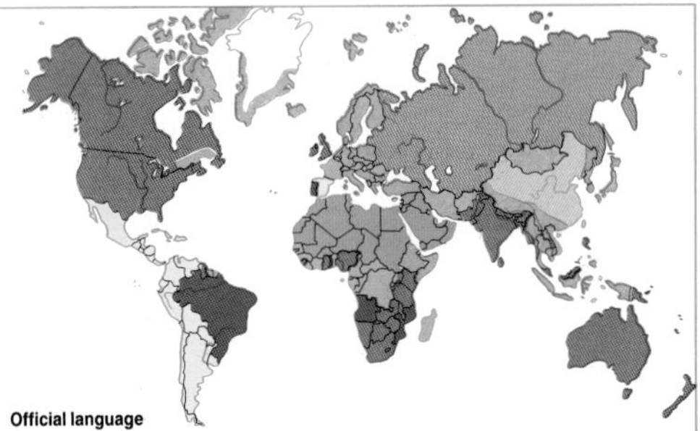

Official language

English | French | Russian | Chinese
Spanish | Portuguese | Arabic | Other

Families of Indian languages in California

California – the world's largest language cemetery

Anyone who comes to California today, for whatever reason, will expect to communicate in English or possibly Spanish. But there was a time when at least 75 very different languages were spoken, each with its own dialects.

At this time – before the mid-18th century – there were about 300,000 Indians of many different tribes living in California. Their peaceful life pattern was broken up after 1769 by the arrival of Spanish settlers in the south, who put the local people in missionary stations. Eighty years later, gold diggers crowded into the north, massacring the natives and driving them off their lands. Many of the survivors moved into the cities and most of them abandoned their native tongues. A few languages have been documented by talking to the last few surviving native speakers.

A success story! A group of Cahuilla Indians gained a legal right to a very lucrative piece of land that they had previously lost. They became very wealthy and they managed to keep their language alive by building schools where their language was studied.

The numerals, one, two and three are shown in some of the different Indian languages. Although they are from neighbouring areas the languages may be very different.

	Wiyot	Yurok	Hupa	Karok	Chimariko	Shasta	Wintun
1	kuts	korah!	lah!	yisa	pun	tsehla	ketem
2	dit	nihliyet	nax	axak	xoku	xuk'a	palet
3	dikh	nahkseyl	tak	kwirak	xotai	xatski	panol

Braille alphabet

a
b
c
d
e
f
g
h
i
j
k
l
m
n
o
p
q
r
s
t
u
v
w
x
y
z

- **Lower case letters** appeared in 300 AD and by 1500 AD had assumed their present form.

Egyptian hieroglyphs

From about 3000 BC the ancient Egyptians developed a form of picture writing using symbols for ideas and for sounds. They had 24 single consonants sufficient to write their languages. Hieroglyphs were carved on stone for royal or funeral inscriptions; professional scribes developed a cursive style more suited to writing with a rush pen on papyrus.

Cuneiform writing

For the riparian traders of Mesopotamia in the 4th century BC a form of registering property and recording commercial transactions was imperative. The Sumerian cuneiform signs were abstract and wedge-shaped, a form partly dictacted by the writing materials. Clay was abundant and soft clay tablets were inscribed with a reed stylus.

Chinese characters

The earliest Chinese inscriptions on bone or bronze date from ca 2000 BC. The modern script retains many of the characteristics with each character representing a concept. As writing materials changed, calligraphy became an art form and the number and complexity of the symbols increased. The characters are set in vertical columns from the top right.

Semitic scripts

A characteristic of Semitic scripts which developed in the Eastern Mediterranean between 1800-1300 BC, was the use of signs to represent single sound units. Writing was relatively simple as only 22 symbols were required. From branches of this script developed the alphabet as well as Arabic, Hebrew and Indian scripts. This enabled three of the major religions, Christianity, Islam and Hinduism to spread and Judaism to consolidate.

Greek alphabet

The importance of the Greek alphabet was that vowels and consonants were each represented by a symbol, thus enabling the script to be used for many different languages. Early Greek inscriptions date from the 8th century BC and by the 6th century AD the familiar Roman alphabet was developing. Writing was originally from right to left and attempts to use the boustrophedon fashion 'the way an ox plough moves' failed as the left to right direction became established.

Runic script

Runes were a remarkable form of writing probably derived from an Etruscan alphabet in the 1st century AD. They existed mainly in inscriptions on bone, stone or wood which could explain the angularity of the script. Only 24 letters existed in the Germanic form while the Scandinavian version made do with 16 signs. The script was associated with cult and secret writing and was discouraged by 7th century missionaries. After the 13th century Runes declined.

Tactual alphabet

In the early 19th century Louis Braille proposed a six dot code which could be embossed on card thus enabling a blind person to read. The Braille code was adopted in 1932 and has the advantage that it can be used for any language using the Roman alphabet. There are 63 possible dot combinations which represent the alphabet, punctuation and common words. A special braille typewriter has been developed.

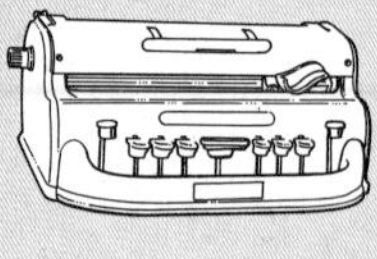

Up, down, left, right

Nowadays writing may go from left to right, right to left or top to bottom, but in earlier cultures it could be in a circle in either direction or even in a spiral. In Aztec manuscripts a red line indicated the continuation of the text.

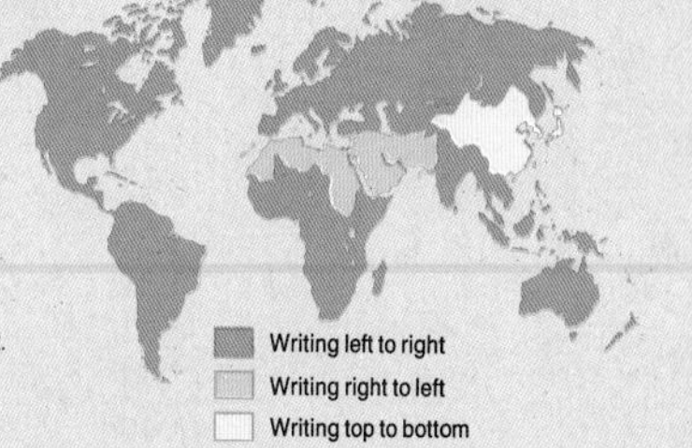

- **The Greeks** adapted some letters from the Phoenicians, but the Romans developed the final form.

Egyptian scribe

From snake to N

Arabic	ا	ب	د	ه			م	ن	ع	ف	ص	ت
Cyrillic	Aa	Вв	Дд	Ее	Ии	I	Мм	Нн	Оо	Рр	Сс	Тт
Roman	A	B	D	E	H	I	M	N	O	P	S	T

From the writing of Mesopotamia and the Ancient Mediterranean cultures the Semitic scripts seem to have evolved. Contemporary European scripts are one of the forms of writing which have developed from this source.

Peace on earth

Roman alphabet

বিশ্ব শান্তি

Bengali

ἐπὶ γῆς εἰρήνη

Ancient Greek

உலகின் மீது சாந்தம்

Tamil – southern Indian subcontinent

ειρήνη στη γη

Modern Greek

ကမ္ဘာ တွင်း ငြိမ်း ချမ်း ပါစေ

Burmese

विश्व शान्ति

Hindi – Indian subcontinent

สันติธรรมในโลก

Thai

Да здравствуй мир на свете!

Cyrillic – Russian

Gujerati – western Indian subcontinent

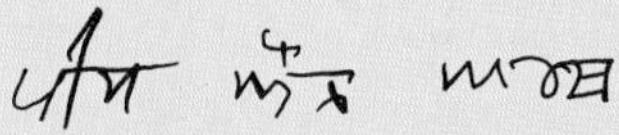

Gurmurki

Primary source; A History of Writing A.Gaur. The British Library.

Peace on earth written in most of the world's major scripts shows the variety of writing. There are many modern Indian scripts and most are derived from Brahmi, a script also used for stone cutting. The stone cutter's tool left a wedge shape at the top, which has developed into a horizontal line. In Hindi all characters are joined by a line. The 'circles' style of Burmese writing evolved because a linear script caused the pointed stylus to split the leaves which were the original material used.

Chinese characters may have as many as 17 different strokes which must be executed in a certain order. There are 50 000 characters in a modern dictionary. Japanese writing consists of two components, the Kanji, the third and fourth characters, which represent a sound and a meaning, and the Hiragana which represents a sound.

Arabic and Hebrew are written from right to left. They are characterized by fine calligraphy.

Chinese

世界和平

Japanese

地に平和がありますように

وعلى الأرض السلام

Arabic

שָׁלוֹם עַל הָאָרֶץ

Hebrew

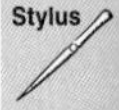

Stylus

Reed pen

Roman bronze pen

Quill

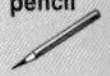

Lead pencil

Fountain pen

Ballpoint pen

Felt tip pen

We dropped the pen

In 1808 the Italian Pellegrino Turri invented a machine to enable a blind friend to write her own letters. This is thought to be the first typewriter.

Seventy years later the American Remington Company produced the first commercial machine. The design has been refined but mechnical typewriters are still remarkably similar.

In 1902 the Blickensderfer Company sold the first electrically assisted machine, but it was not until 1961 when the so-called 'golf-ball' machine replaced letter bars and the travelling carriage with a moving plastic sphere.

Just three years later IBM produced the first word processor and it became possible to store and retrieve huge amounts of text on electronic disks or magnetic tapes.

An early typewriter

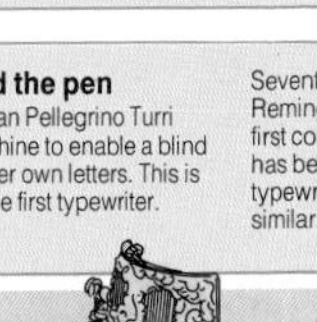

An electric machine

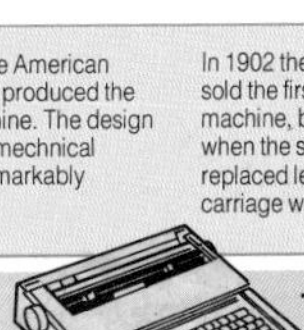

The 'golf-ball'

Lap top computer

strological signs

♈ Aries

♊ Gemini

♌ Leo

♎ Libra

♐ Sagittarius

♒ Aquarius

♉ Taurus

♋ Cancer

♍ Virgo

♏ Scorpio

♑ Capricorn

♓ Pisces

Astrology

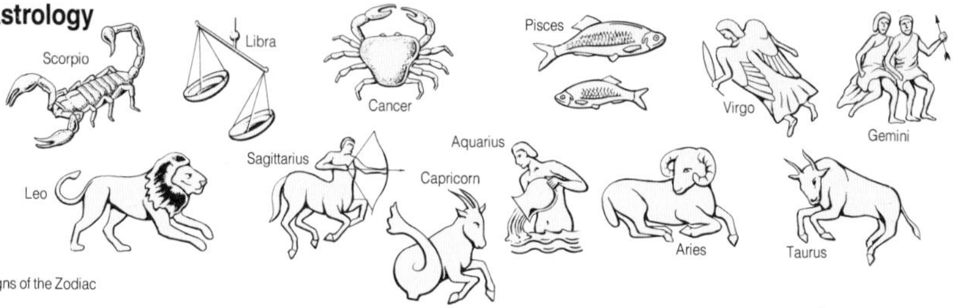

Signs of the Zodiac

History of an ancient science

Astrology is the study of the planets, including the sun and the moon, together with the stars, and how they relate to life on earth and to future events. Interest in the stars is as old as civilization itself and the study of them could be called the oldest science.

The study of the stars first became important in Mesopotamia and Ancient Egypt. Astrologers developed a system linking changes in the seasons with groups of stars. For example, heavy rainfall frequently occurred in Babylonia when the sun was in a particular constellation. Hence astrologers called the constellation Aquarius, the water carrier. The Nile also flooded at certain regular times and to predict these events the Egyptians developed a calender based on the movements of the heavenly bodies.

Both the Greeks and the Romans practised astrology and the classical names for the signs of the Zodiac and the planets are in use today. From the 14th century the movement of the heavenly bodies was shown on clocks, such as the one in Strasbourg Cathedral or the ornate one at Hampton Court, England.

During the 15th and 16th centuries the discoveries of Tycho Brahe, Copernicus and others conflicted with the established theory that the sun went round the earth. From then on astronomy developed as the scientific study of objects in space while astrology went along other lines.

The basic principle of astrology is that events on earth are related to the heavenly bodies. Astrologers can understand these relationships by casting a horoscope which shows the position of the planets, the stars and the earth at the time of a person's birth. Astrologers place the earth at the centre of the system with the other bodies revolving round. They believe that the planets, including the 'new' ones Uranus, Neptune and Pluto, strongly influence life on earth; each one has a particular quality, for instance Mars is aggressive and Venus is loving. The Zodiac is a band of stars divided into 12 equal parts called signs. Each of these have certain characteristics. Finally the earth's surface is divided into 12 Houses which determine how the planets affect a person's life. Interpretation can be made in different ways. In the Greek way the human life (the microcosmos) mirrors the greater system (the macrocosmos). Interpretations can be made from the greater to the smaller system. The Babylonian principle is that the heavenly bodies directly affect life on earth.

Many people throughout the world believe in astrology and base important decisions in their lives on the advice of an astrologer. Others believe that it is a superstitious belief. Nevertheless the ubiquitous newspaper horoscope suggests that many people enjoy astrology.

The astrologer is using a model showing the movements of the heavenly bodies to predict events on earth. Though the science of astronomy became separate from astrology, the academic study of astrology continued into the 19th century.

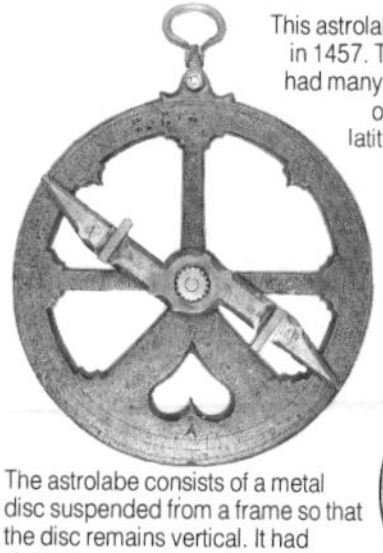

This astrolabe was made in Vienna in 1457. The beautiful instrument had many functions, the principal ones being to determine latitude and to tell the time.

The astrolabe consists of a metal disc suspended from a frame so that the disc remains vertical. It had sights for observing a star, and gradations for measuring its elevation.

Astronomy

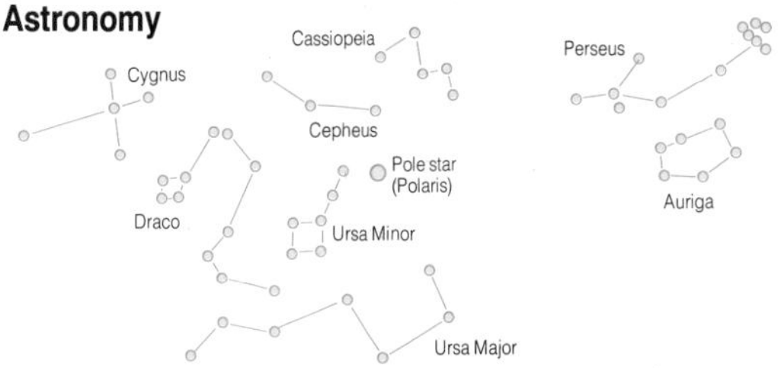

Using polaris

Astronomy is the scientific study of celestial bodies, and in the modern science knowledge reaches to outer space (see page 10).
The early Mediterranean navigators used the sun to give direction. Another important guide was Polaris (the Pole star) which can be seen in the Northern hemisphere near the constellation Ursa Major. This star remains stationary above the North Pole while the other stars revolve round it. A particular star may just touch the horizon as it moves round, and at the same latitude the same star will graze the horizon. Early sailors could sail to any destination if they knew which star touched the horizon at that latitude. They could then sail along the latitude until they reached port.
Polaris was not always in its present position and about 1000 B.C. the star Kochab in Ursa Minor appeared stationary and was doubtless used as a guide.

Looking into space

Since the earliest days people have used the stars both to tell the time and as a guide to their destination. The vast sky visible on dark clear nights enabled people to acquire a detailed knowledge of the stars and their movements. Although only stars visible to the naked eye could be observed, much useful information about the earth could be deduced. Not surprisingly, many people who live in cultures without modern aids such as watches and road maps, have a better practical knowledge of astronomical time and celestial navigation than the average city dweller.

Not only astronomers, but all who travelled had to understand the signs in the night sky. As well as being a signpost the stars were thought to foretell great events.

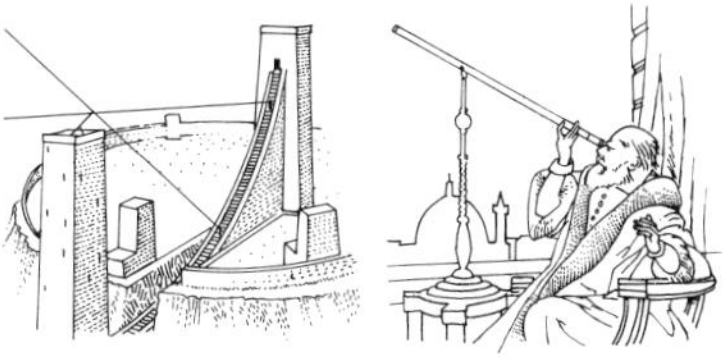

Astronomical instruments had to be very large to maintain accuracy. The illustration shows a quadrant, built in 1440, in an observatory in Samarkand. Galileo constructed his first telescope in 1609. He discovered the moons of Jupiter and the rings around Saturn.

The Chinese were probably the first to discover magnetism and to use it in such instruments as this mariner's compass.

The quadrant or backstaff enabled observations of the altitude of the sun and stars to be made at sea, using the principle of double reflection.

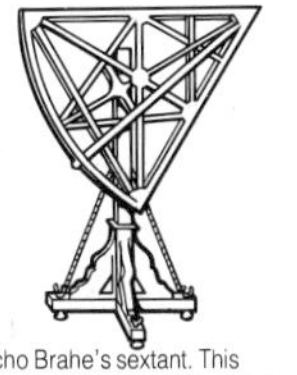

Tycho Brahe's sextant. This instrument from the 16th century enabled angles between the stars to be measured.

Find your way

The properties of magnetic ore were known in ancient times but it was not until the 12th century that the compass was used by Mediterranean navigators. The sextant was developed from the astrolabe and greatly assisted celestial navigation.
A big step forward in position finding was the invention of the chronometer in 1760 by John Harrison. This made it possible for the first time to establish longitude correctly.

Modern navigation

Using a modern sextant astronomical observations can be used to establish latitude. Longitude can be determined from the time of the reading. Horizontal angles (bearings) can be made on two known places to find out the observer's position. These angles can be read off a bearing plate.

The bearings of two radio transmitters can be fixed by means of a receiver with a directional antenna. The ship's course (shown by the compass) and the distance covered (shown by the log) gives a good idea of the ship's position.

Using radar, radio waves give bearing and distance to coast, islands and buoys. Phase shifting in signals from three specialized transmitters gives a very accurate position.

Satellites have revolutionized international communications and navigation satellites are already in use. Signals can be beamed from a transmitter and can be relayed back from the satellite. A highly accurate global positioning system using high orbital satellites enables vessels to locate within 10m of accuracy.

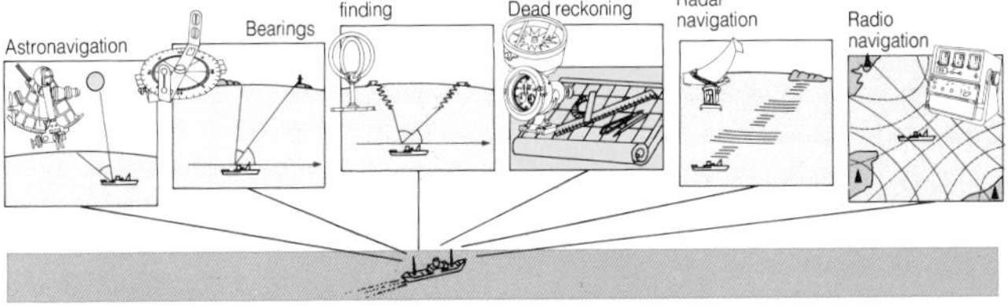

Chronometer. A very accurate clock which can be used at sea. Harrison won a competition to find the best timekeeper.

Geosynchronous satellite. A satellite which circles the Earth in 24 hours will remain stationary at the same point over the Earth.

Log. Originally a log-and-line was used to determine a ship's speed. Some modern logs compare the water pressure as the ship moves, with still water.

Planetary signs

- **The longest distance** flown by a ringed bird is 22 530 km, 14 000 miles, by an Arctic Tern flying from the White Sea to Western Australia.
- **The highest altitude** that has been recorded for a bird is 11 277 m, 37 000 ft. The bird was a Ruppell's Vulture which hit a plane over West Africa.
- **The fastest animal** over a short distance is the cheetah. Its probable maximum speed is 100 km per hour, 63 mph over level ground.

Why do they move?

Most people associate migration with the seasonal movements of certain birds. In the spring birds arrive to occupy their summer breeding areas and then in the autumn when food supplies become scarce they migrate to their warmer winter resting places. At the same time winter visitors including many ducks and geese arrive from their their summer breeding ground in the Arctic. This is called return migration.

Many other animals also move immense distances by land, sea or air. Whales for instance swim thousands of kilometres returning to their traditional areas to mate and raise young.

Somewhat different are the migrating patterns of some insects such as locusts. Swarms of locusts tend to migrate to areas of low pressure which are likely to be wet. Rain stops migration and the insects breed. As the larva hatches overcrowding and food shortage trigger another migration. Although the locusts seem to move in seasonal circuits the life cycle of the locust is too short for any one individual to complete the circuit.

A different sort of migration occurs when a whole species of animal spreads out from its original area to new ranges. An interesting example is the Collared Dove *(Streptopelia decaocto).* Before 1930 this bird was found chiefly in Asia, but by 1955 it was found in the Netherlands and Denmark and by the 70's it had extended its range to include Britain and southern Scandinavia. It appears that the species has taken up an underused ecological niche.

With the Barn Swallow from Europe to Port Elizabeth

When the young Barn Swallows *(Hirundo rustica)* are fledged ringing reveals that they may fly as much as 200 km, 125 miles in any direction from the nest. This exploratory migration familiarises them with their birth area and perhaps enables them to locate nesting sites for the future.

In the autumn when food becomes scarce all the swallows will migrate from Europe to various parts of Africa. There is evidence that they winter regularly in the same area before returning in the spring. Some say that the same pair of birds will return each year to the same nest site.

The birds perform complex feats of navigation. Studies suggest that for most of their journey the birds fly on a bearing which is particular to that species. It has been suggested that they may have some substance in their brains which acts as a compass.

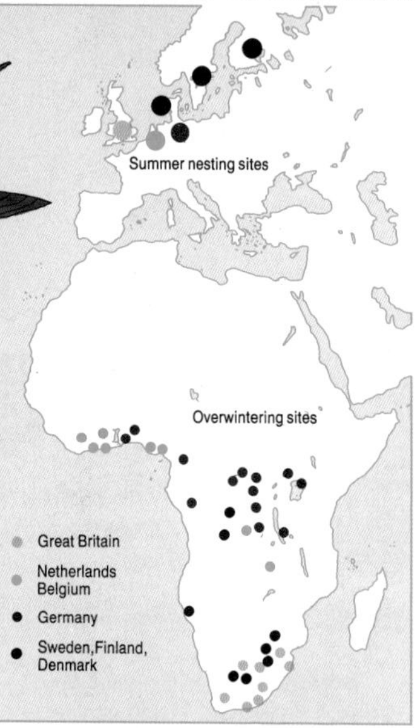

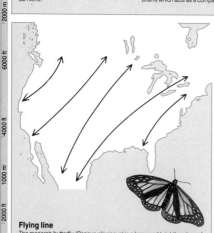

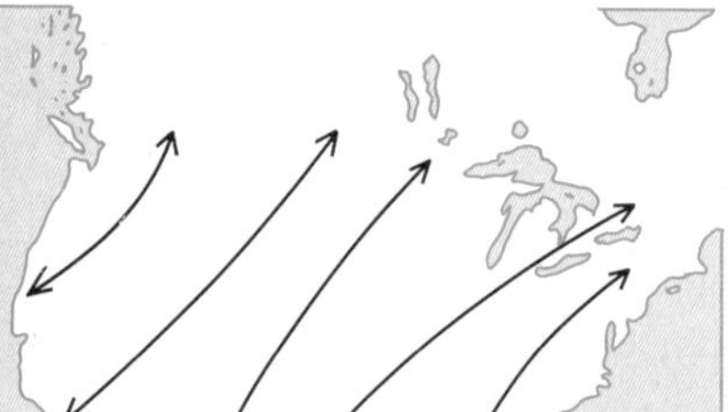

Flying line

The monarch butterfly *(Danaus plexippus)* is a large and brightly coloured butterfly which lives in the USA and Canada. A large proportion of the population migrates to warmer areas in the autumn. On their flight to Mexico and California they may fly up to 120 km, 75 miles a day. They will over-winter there and on a cold day may be seen clustering in dense and spectacular roosts. During their slower return in spring, mating and egg laying take place. The hatching young may also produce another generation before they reach their summer quarters. It is uncertain whether any of the original migrants return. Moreover about 30% of the population hibernate and their young are the first seen in the spring.

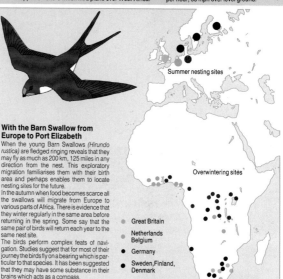

Follow my leader

The Norway lemming *(Lemmus lemmus)* is a small rodent, looking not unlike a hamster. Lemmings' normal habitat is above the treeline in the alpine zone of the Scandinavian mountains. When climatic conditions lead to an abundance of food the young mature more rapidly. They breed at a much earlier age than normal and numbers increase very quickly. With the growth of the lemming population there is an increase in the number of predators which are on the next level of the food chain.

The population explosion triggers off a migration with immature males predominating. Each animal moves in a straight line and the whole group heads down the mountainside into the valley. Although they are strong swimmers lemmings do not usually go into water unless they can see the other side. However a concentration of animals on the edge of open water may force those on the shore to start swimming. Despite popular belief the lemming is not bent on suicide but rather on finding and colonizing a new habitat. Scientifically this is known as non-calculated removal migration.

- **The fastest fish** in the sea is the Blue fin tuna with a speed of 70 km per hour, 43 mph over a short distance.

A yearlong walk for food and water

The Serengeti national park in northwest Tanzania also borders on to Kenya. With an area of 15 000 km², 5800 square miles, it is home to 25 species of the larger mammals. As well as zebra and wildebeest *(Connochaetes taurinus)* there are several species of gazelle, impala and other large herbivores. Altogether there are about 2 million of the larger animals.

In their search for food and water the animals complete an annual circuit of about 800 km, 500 miles. At the beginning of the year the beasts are to be found in the savanna grasslands to the southeast of the park. The dry season begins in about May and then the wildebeest gather in great herds, sometimes as many as 400 000 strong. The herds migrate north and west in lines and columns up to 40 km, 25 miles long. They use their sense of sight and sound but not, apparently, smell to guide them to the storms and the fresh grass. At each halt the bulls try to establish their own territories to enclose the passing females. There are many minor skirmishes as small adjoining territories are set up.

Most of the animals have reached the northern limits of the park and the Masai Mara by the end of the dry season in July. A short rainy season in November and December brings fresh grass. By the end of the year the animals will have migrated back to towards the south again.

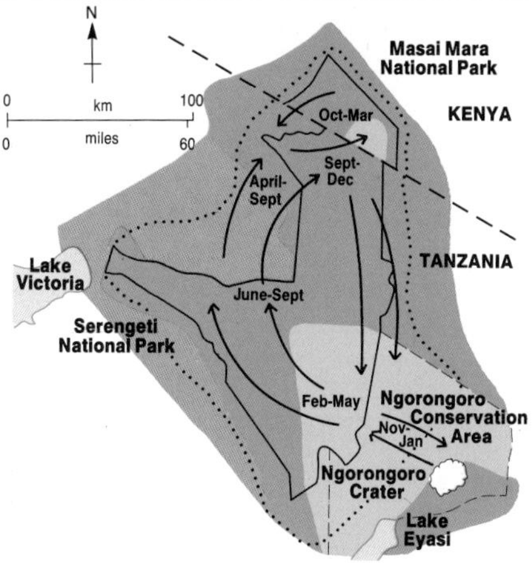

Across the Atlantic

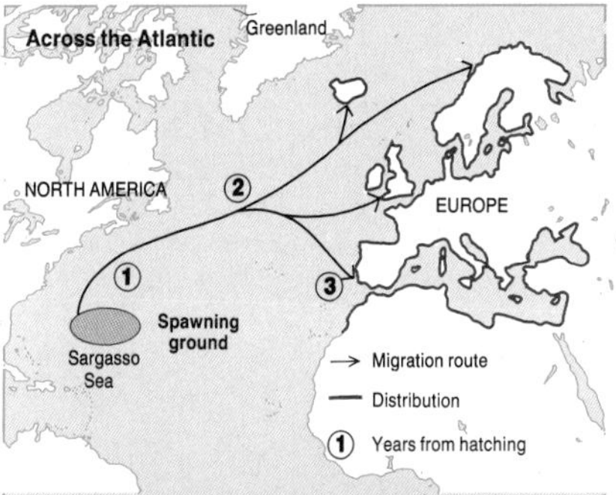

The European eel *(Anguilla anguilla)* is, for most of its life a freshwater animal although it is born at sea. The eels hatch out in the Western Atlantic in the warm waters of the Sargasso Sea. The larva is called leptocephalus (Greek for slender head). They spend three years journeying to Europe carried along near the surface by the current known as the North Atlantic Drift. During this time they actually shrink in size, though many of them enter the food chain. When they reach the rivers of Europe they metamorphose into elvers and swim upstream. Their adult life is spent in freshwater. Up to 20 years later the adults swim to the sea and move at great depths thus avoiding the surface currents. On reaching the Sargasso Sea again the eels spawn and then die.

Run for home

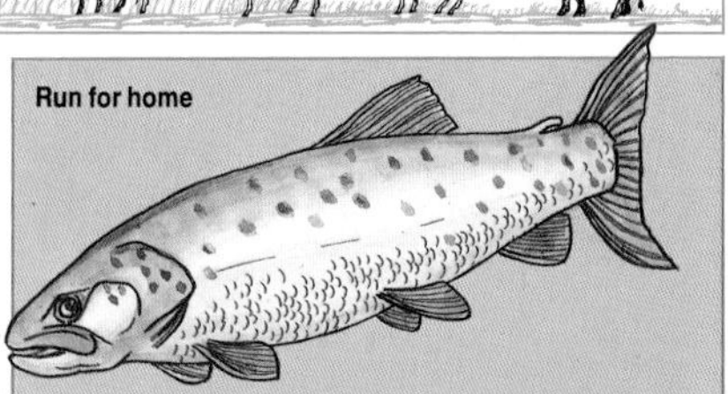

The Atlantic Salmon *(Salmo salar)* migrates from its birthplace in freshwater into the sea to exploit the richer feeding available there. These migrations involve swimming thousands of kilometres to waters off Greenland and the Faroe Islands. A remarkable feature of its migration is the salmon's ability to return to the same river it left several years earlier, and in so doing, to leap up waterfalls and negotiate strong rapids. When it reaches coastal waters it can identify the river by the chemical characteristics of the water, by its taste and smell. How the salmon navigates during the long ocean journey is still a mystery.

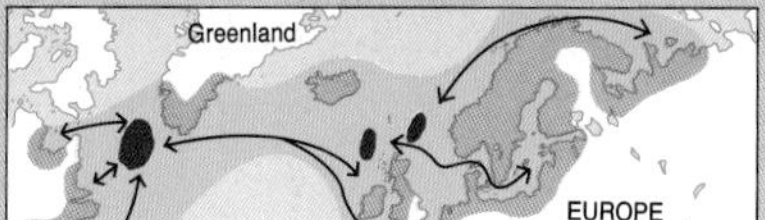

Ecological niche. A species will live where the supply of food is adequate and there are not too many predators.

Habitat. All the conditions that affect plants and animals. These include the physical environment and the inter-relationship between species.

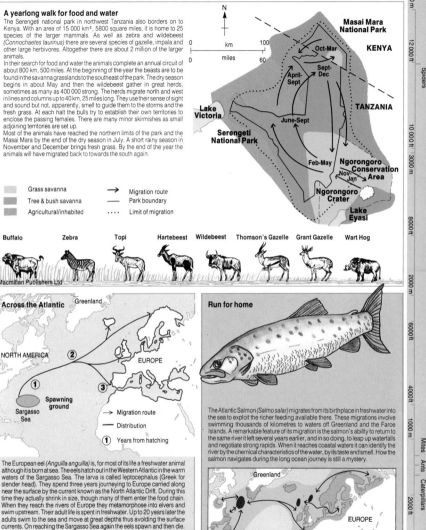

- **Rene Caillie.** In 1828 this French explorer travelled through West Africa to Timbuctu and then crossed the Sahara.
- **Charles Sturt.** In 1828-29 he explored southeast Australia, but in 1844 failed to reach the centre, defeated by the desert.
- **David Livingstone.** Between 1840 and 1872 he explored central Africa, discovering the Victoria Falls and crossing the continent.
- **Lewis and Clark.** 1804-08. The expedition travelled up the Missouri, crossed the Rockies and down the Columbia to the Pacific.
- **Mungo Park.** In 1795 he was the first European to explore the Niger river in West Africa.
- **A.E. Nordenskiold** headed the first expedition to sail the North East passage between the Atlantic and Pacific Oceans in 1878-79.
- **Alexander Mackenzie.** In 1792-93 he travelled by canoe from the St. Lawrence up the Peace river to the Pacific Ocean.

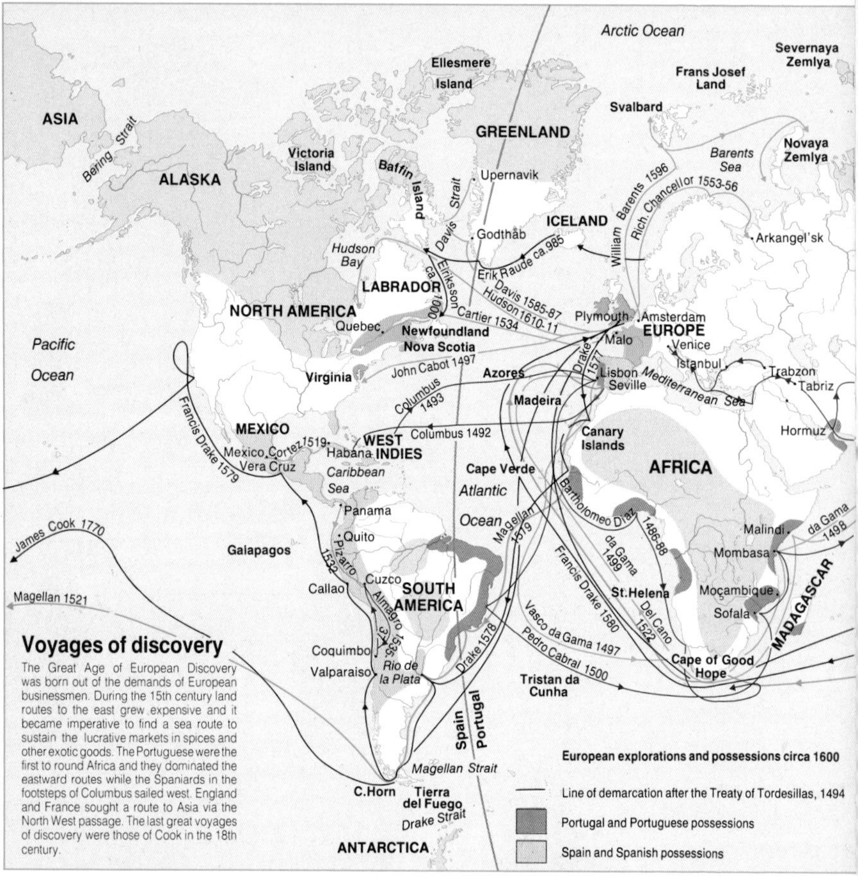

Voyages of discovery

The Great Age of European Discovery was born out of the demands of European businessmen. During the 15th century land routes to the east grew expensive and it became imperative to find a sea route to sustain the lucrative markets in spices and other exotic goods. The Portuguese were the first to round Africa and they dominated the eastward routes while the Spaniards in the footsteps of Columbus sailed west. England and France sought a route to Asia via the North West passage. The last great voyages of discovery were those of Cook in the 18th century.

Christopher Columbus (1451-1506)
In 1492 Columbus sailed west from Spain hoping to find a sea route to India. Relying on the Ptolemy map (see front endpaper) he probably underestimated the length of his journey; instead he discovered a new continent, America. His first landing was at Watling Island in the Bahamas. On subsequent voyages he discovered most of the islands in the West Indies but failed to find the gold his backers wanted.

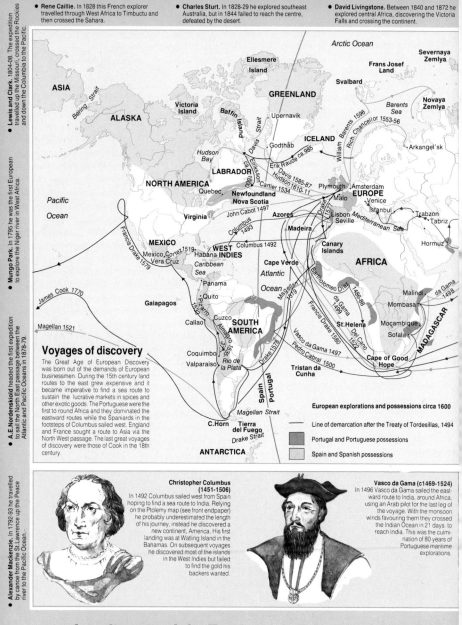

Vasco da Gama (c1469-1524)
In 1496 Vasco da Gama sailed the eastward route to India, around Africa, using an Arab pilot for the last leg of the voyage. With the monsoon winds favouring them they crossed the Indian Ocean in 21 days. to reach India. This was the culmination of 80 years of Portuguese maritime explorations.

. . .and we discovered the Earth
geographical exploration

- **Henry M.Stanley** met Livingstone in 1871. Afterwards Stanley continued exploring Africa and traced the Congo to its mouth.
- **Robert Peary.** In 1909 he was the first man to reach the North Pole. The last 210 km, 130 miles was covered in just 2 days.
- **Roald Amundsen** was the first man to sail through the North West Passage. He was also the first at the South Pole in 1911.

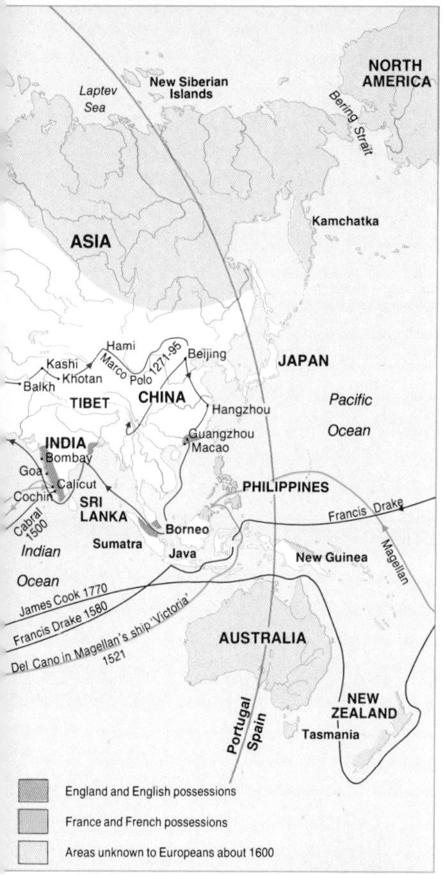

Remaining white spots

In Medieval times unknown places were thought to be peopled by monsters. The merchant traders dispelled these fantasies and opened up much of the world beyond the Mediterranean Sea. Explorer scientists such as Cook and Humboldt revealed more of the world and its ways. The last great continent to be explored was Antarctica in the early 20th century.

There are now few places left to discover although as recently as the 1930s tribes living in New Guinea were 'discovered'. Perhaps there are still communities untouched by 'civilization'.

Modern exploration whether on land or probing the ocean depths, seeks to combine specialist scientific study with a concern for the environment. Thanks to the National Geographic and other sponsors a film record is obligatory. We are all now armchair explorers.

The Polynesians
sailed vast distances during their voyages of colonization in the Pacific about 2000 years ago. Their craft were outrigger canoes with a triangular sail. They used charts made of sticks and shells. The arrangement of the Sticks representing the interference pattern set up by several wave masses.

The Vikings
reached the Caspian Sea in the east and 'Vinland' (presumably Newfoundland) in the west. In the 10th century they settled in Iceland and Greenland, although their voyages along the rivers of Russia were economically and historically more important.

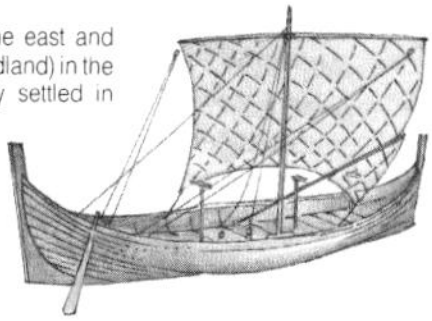

The Chinese
were probably the best shipbuilders of the Middle Ages. Their junks carried several masts. In the 15th century explorers reached India and East Africa.

The Arabs
were skilful sailors. In their lateen-rigged ships and aided by the steady monsoon winds they traded during the Middle Ages from East Africa ('Azania') in the west, to the Indonesian spice islands in the east.

Ferdinand Magellan (c1480-1521)
In 1519 Ferdinand Magellan set forth on the first circumnavigation of the globe. They sailed down the east coast of South America and in October 1520 they discovered the strait to the Pacific. In 1522 one ship of the three returned to Portugal with only 18 of the original 265 men. Magellan was not among them as he had been killed by natives in the Philippines.

James Cook (1728-1779)
In 1768 Cook sailed from England ostensibly to study the transit of the planet Venus in Tahiti. This accomplished he pursued the main purpose of his voyage, the search for the Southern Land. He in fact reached and charted the New Zealand coast and sailed west to Australia. In later voyages he charted many islands in the Pacific Ocean but was killed by natives in Hawaii in 1779. His fine seamanship, scientific interests, and humanitarian behaviour made him an exceptional explorer.

- **Robert Scott.** At the South Pole in January 1912 he found Amundsen's tent. On the return journey all the party died.
- **Joseph Rock** led National Geographic sponsored botanical expeditions in 1923-30 to the mountains between China and Tibet.
- **Nobile, Ellsworth and Amundsen** flew in a dirigible airship across the North Pole from Spitsbergen to Alaska in 1926.
- **Bradford Washburn.** Between 1935 and 1951 he surveyed Mt.McKinley, discovered many peaks and mapped the 'roof' of America.

- **Vivian Fuchs** crossed Antarctica from the Weddell Sea to the Ross Sea using Sno-cats. The journey between 1957-58 covered 3500 km, 2200 miles, and took 99 days.
- **Edmund Hillary and Sherpa Tensing** were the first to climb Everest in May 1953. News broke on June 2nd, the coronation day of Queen Elizabeth II.
- **Jacques Cousteau.** Since 1951 his underwater explorations have revealed much about the seabed, its biology and archaeology.

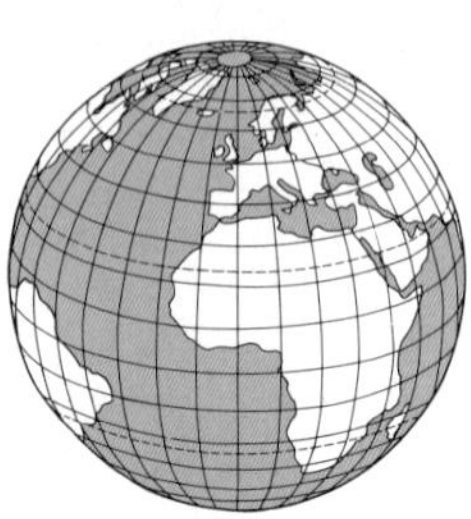

The globe: – a small image of our world

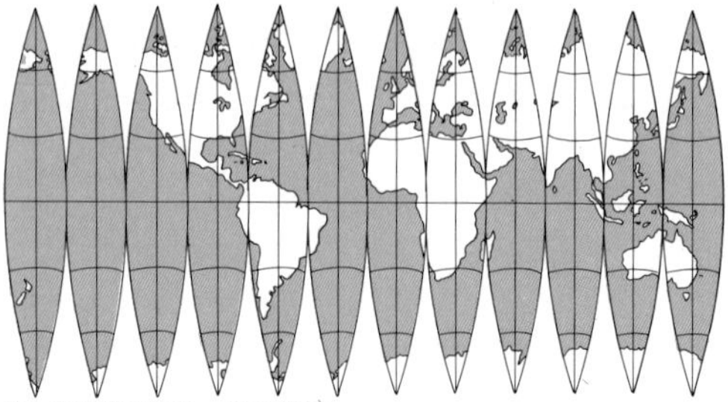

How to make a map. . .or a globe

If a globe were to be cut along lines of longitude from the pole nearly to the equator and then opened out and pressed flat it would be a map. A globe can be made by printing a map in gores and fixing it to a sphere, matching along the lines of longitude.

Polar projection

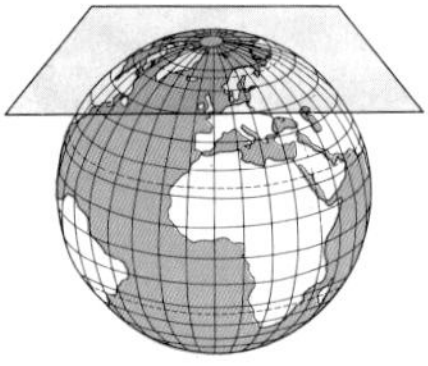

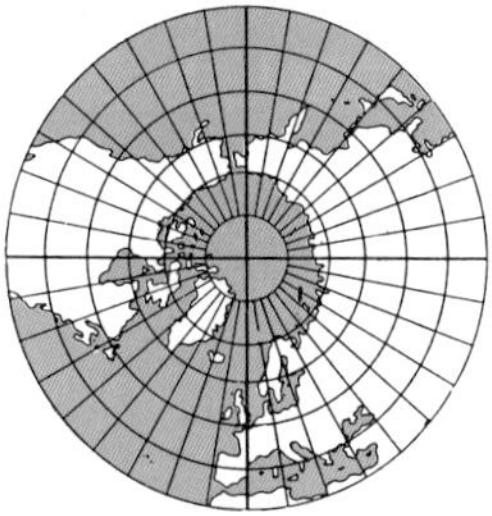

The earth from above

To map the earth requires, first of all the drawing of a framework. This framework is the network of the lines of longitude and latitude on the earth, and the pattern of lines is called the map projection of the earth. Once the graticule is drawn the countries may be superimposed. There are many ways of generating the pattern.
If a piece of paper is put over the pole of a glass globe at right angles to the polar axis and a light placed at the centre of the globe, the lines of longitude and latitude cast a shadow like a spider's web. This is the map projection. The distances along the lines of longitude can be made correct and such maps are of course useful for mapping polar regions.

Equatorial projection

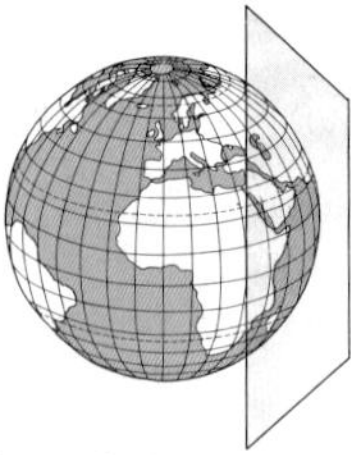

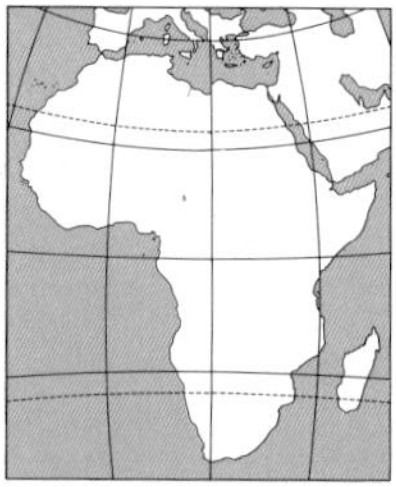

The earth sideways

If the viewpoint is from the side then the paper will be at right angles to the equator. The globe and the paper will touch at a point on the equator. If this point of contact is longitude 20° east then a very useful map of Africa can be drawn. In this illustration of Millers stereographic projection the lines of latitude and of longitude are both curved. The only straight graticule lines are the equator and the line of longitude that is in the centre of the map.
Both the shape of the continent and its area are not very different from the reality as seen on a globe at the same scale. However towards the edges of the map some distortion will occur.
Other maps projections such as Mercators and Peters which are shown on the opposite page are also equatorial projections of a different type.

Oblique projection

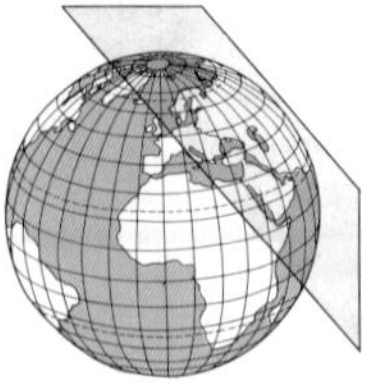

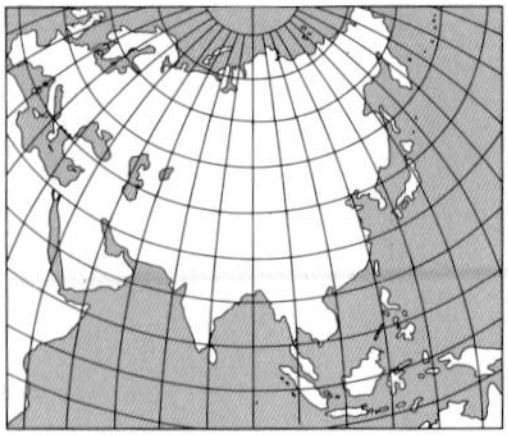

The earth from an angle

When the viewpoint is from an oblique angle, the paper may touch at any place on the globe. So the point of contact, where the map is most accurate can be chosen so that it is in the centre of the area to be mapped.
Away from the centre there will be distortions of shape or area but by adjusting the curvature and spacing of the graticule the errors can be minimised. The illustration shows a map of Asia on Lamberts Azimuthal projection. The point of contact is in the centre of the map, and at that point the map will show area and shape exactly as they would be on a globe at the same scale as the map. The bigger the area to be shown the more difficult it is to avoid distortion.

Projection – a flat picture of a globe

The only true 'picture' of the earth is a globe. Only on a globe are the shapes and areas of countries truly correct, but at a reduced scale of course. And only on a globe are angles and distances between two points truthfully shown. Because it is not possible for a curved surface to be shown on a flat piece of paper a number of ways of portraying the lines of longitude and latitude, the map's framework, are in use. These are different map projections. Once the graticule or framework of a map is established then the outlines of countries, the rivers, roads and all the other features can be drawn in correct relation to each other. The use of computer programs has made the computation and drawing of networks much quicker, and lines may also be transformed from one projection to another. To show the whole world, however, certain properties such as true area or true angles may be preserved but only at the expense of other properties, such as shape. For most world maps a compromise solution is adopted.

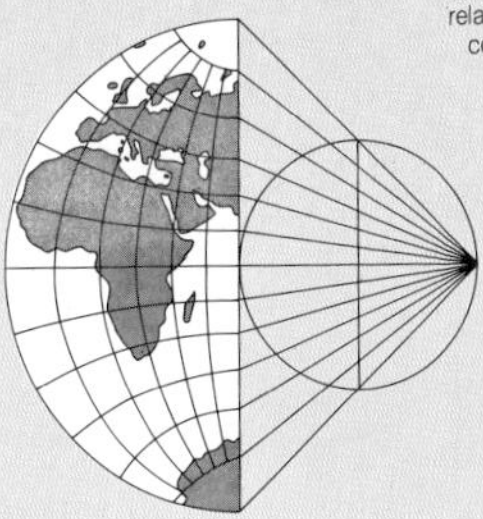

Transverse projection

Some graticules can be envisaged as though they had been projected on to a cylinder fitting round the globe. The cylinder then touches the globe along a chosen line of longitude.

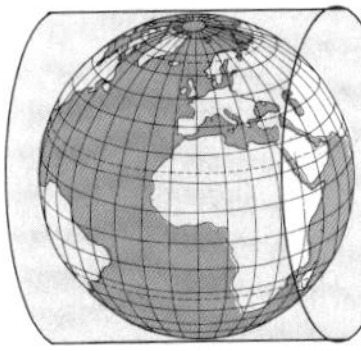

Normal projection

Where the line of contact between paper and globe is the equator, equatorial areas will be shown best. Many world maps use variations on this theme. The lines of longitude may be straight lines and the spacing of the lines of latitude varied to give either the correct area or the true shape of the countries.

Conical projection

The distortion of polar areas associated with cylindrical projections can be lessened by using a cone which is then cut and opened out. The paper will touch along a chosen line of latitude. In this case a band along the selected latitude will be well shown and errors will build up away from this strip.

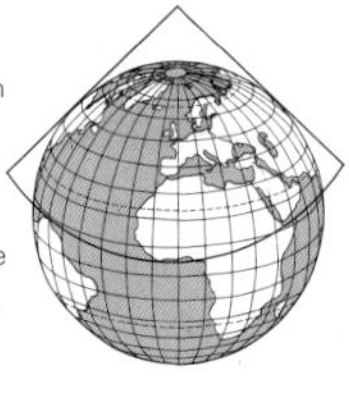

Robinson

This projection was devised in the 1960's by Professor Robinson. It was adopted by the National Geographic magazine for its world maps in 1988 because there are no marked distortions. Scale is correct at 38°N and 38°S. Three-quarters of the world has a distortion of less than 20%.

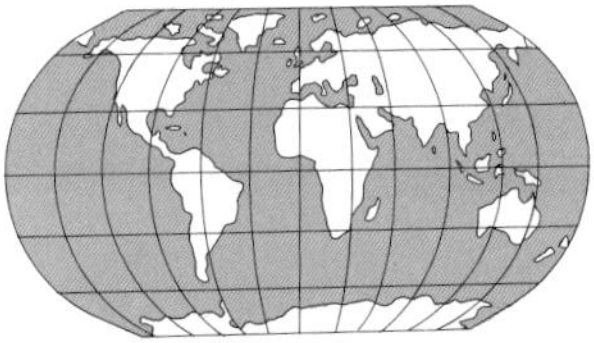

Van der Grinten

This projection shows the whole world within a circle. It has no particular properties but gives a good representation of all countries except those in the far north. For instance Canada is shown with 258% of its true area while Greenland is 5·5 times too big. This projection was used by the National Geographic from 1922 until it was replaced in 1988.

Mercator

This projection was first proposed in 1564 by Gerardus Mercator and is still used. The unique property of this projection is that the great circle or direct routes along which ships sail are shown as straight lines. Its main disadvantage is that countries furthest from the equator in high northern and southern latitudes appear to be very much bigger than they really are.

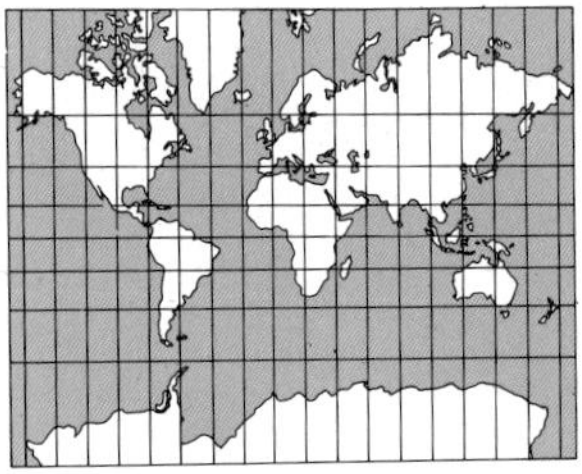

Peters

First published in 1973, this projection was developed by Professor Peters. All countries are shown with their true area, but inevitably shapes are altered, and in particular South America and Africa are somewhat elongated. The projection is used in some United Nations maps of the world and shows developing countries clearly.

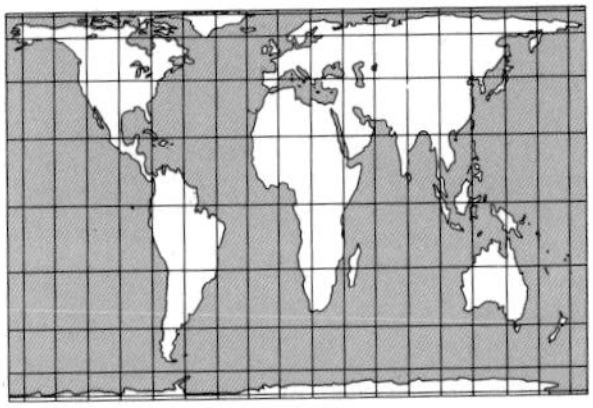

Graticule. The imaginary lines of latitude and longitude on the earth as shown on the map.

Longitude. In 1884 the line of longitude through Greenwich was selected as the Prime Meridian. At the Equator meridians are 112 km, 70 miles apart.

Latitude. 1 degree of latitude is about 100 km, 63 miles on the earth's surface.

Urbanization of the landscape

The sites on which our great cities stand were once virgin land on to which people gradually encroached. As population grew towns started to develop. Initially there was a clear demarcation between town and countryside with an exchange and interaction between these two environments.

The pioneers cleared new ground in the forests because of the shortage of land in the earliest farming areas. Increasing population necessitated a settled life and with it improved farming methods and the development of a more complex social structure.

Villages evolved as the clearings merged and formed farming districts with the farmhouses clustering together. Woodland remained where it was difficult to work the soil and marginal land was used for grazing. Both the woodland and the pasture was held in common by the villagers.

Vertical and horizontal growth

The once compact and centralized town has become an urban sprawl. Rising land prices have led to the building of tall buildings and skyscrapers, a typical feature of New York (right). Large areas are also covered by motorways and complex traffic systems. Big cities in less developed countries have, superficially a modern appearance; Mexico City (below) is characteristic. However the infrastructure is often poorly developed and unemployment is high. The attraction of temporary work or the possibility of setting up a small business contribute to the drift to the towns. Social mobility is a feature of these cities, but too often expectations go unfulfilled and the landless jobless poor live in conditions of extreme squalor.

In the central area of Mexico City skyscrapers and other high-rise office buildings bear witness to a busy commercial life. However in the shanty towns on the outskirts the poor have a very different lifestyle with inadequate housing and few services.

Micro-megalopolis

With their tunnels and chambers the anthill and the termite nest (shown here in cross-section) are not unlike the cities that people build. Insect communities are not, however, political systems. They have an elaborate social organization with the colonies structured round a male and female. This pair is attended by workers, each group having a specific function.

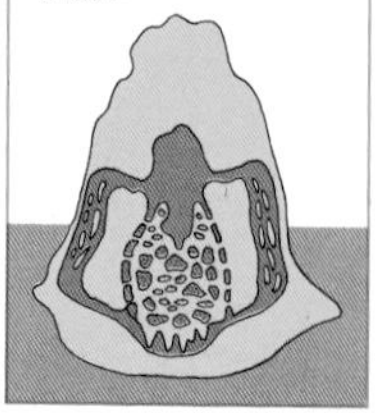

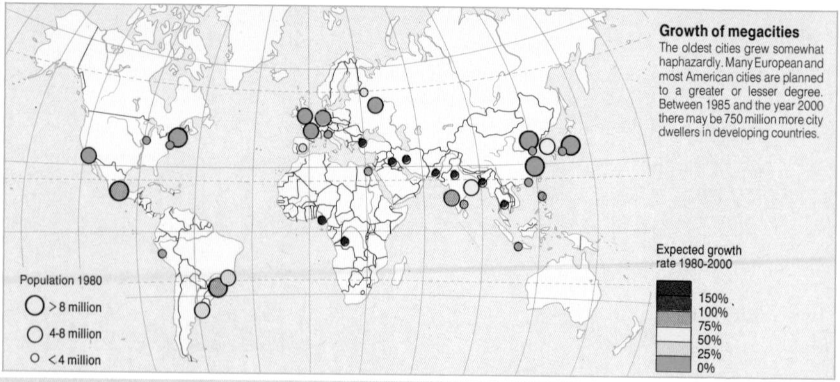

Growth of megacities

The oldest cities grew somewhat haphazardly. Many European and most American cities are planned to a greater or lesser degree. Between 1985 and the year 2000 there may be 750 million more city dwellers in developing countries.

Population in some big cities in thousands

Year:	1900	1930	1988
New York	3 437	6 930	9 120
Calcutta	1 322	1 197	9 166
Cairo	570	1 064	5 074
Sydney	112	1 238	3 281
Moscow	1 173	2 412	8 396
Paris	2 714	2 891	8 510
Lagos	42	100	2 100

Medieval towns could develop where a village was at the intersection of trade routes or had a harbour for overseas shipping. The setting up of political or eccesiastical authorities may have stimulated growth. Walls protected the citizens who still maintained a semi-agricultural way of life.

Industrial towns developed a more regular layout with paved streets, piped water and drainage systems. Factory chimneys competed with church towers and steeples to form the skyline. A structure evolved with administrative and commercial functions at the centre while industry and residential developments were located on the outskirts.

Mexico City – a bad prognosis

Mexico City is a great spawling agglomeration. The population is growing fast and is probably already 19 million; it could rise to 26 million by the year 2000. Since the earthquake in 1985 concerted efforts have provided new low cost housing. The rebuilding is the largest postwar reconstruction effort but nevertheless only 1 in 7 families in the old part are rehoused.

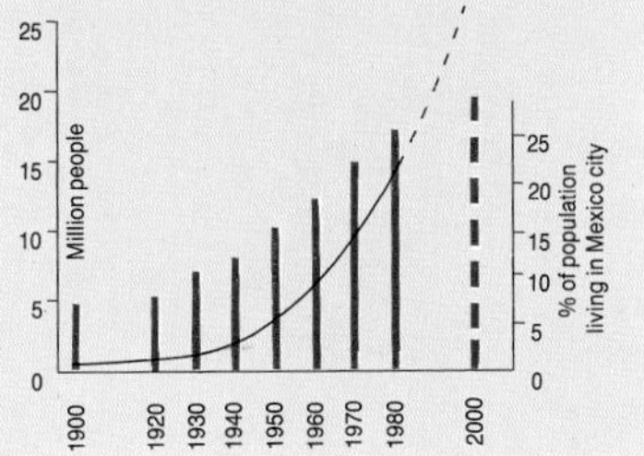

Town versus country

The vast shift of people from the countryside to the towns has been a feature of the past decades. For the richer countries there are the problems of congestion and of maintaining an overloaded infrastructure. In the poorer countries the situation is infinitely worse. In addition to inadequate housing there are many problems relating to health, water supply and sanitation.

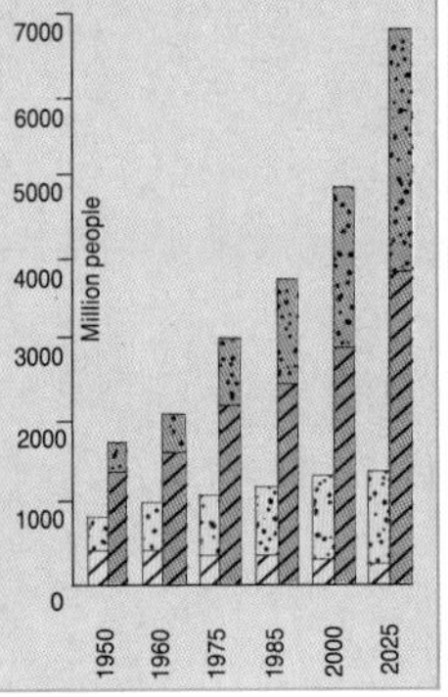

Year:	1900	1930	1988
• Tokyo	1 839	2 070	11 635
• Toronto	208	631	2 999
• London	4 536	4 396	6 765
• Shanghai	600	2 674	11 860
• Buenos Aires	821	2 149	9 710

Organization and expansion

The growth of a city creates a system of differentiated zones. The historical nucleus, comprising administrative and commercial functions is surrounded by densely built older housing. Beyond this there is a belt of industry and beyond that again more recent housing areas.

Pressure on the city centre is relieved by satellite towns which are smaller editions of the city itself, with their own centres and industrialized areas. The system of radial transport routes reveals that their independence is illusionary.

The city plan (right) is based on Paris. For historical reasons other cities, especially those outside Europe may have quite a different structure.

In developing countries attempts are often made to expand city areas systematically. However in Cairo, for example most of the new growth has taken place, unplanned, along valuable farmland adjacent to the Nile. The area near the Giza Pyramids has seen a rise in population of 50% in the last 10 years. Planned desert communities some distance from the city have failed to attract residents.

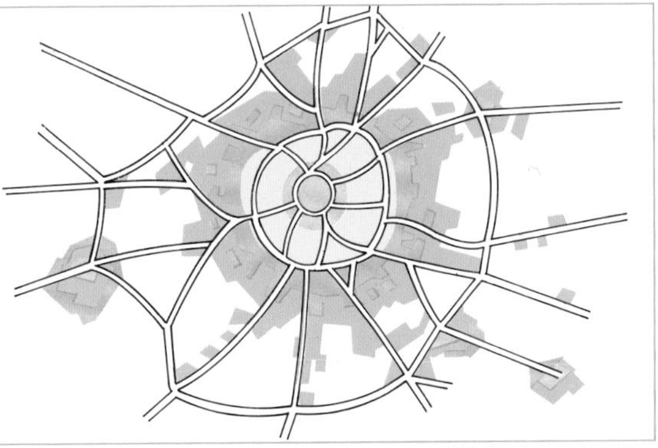

Satellite town

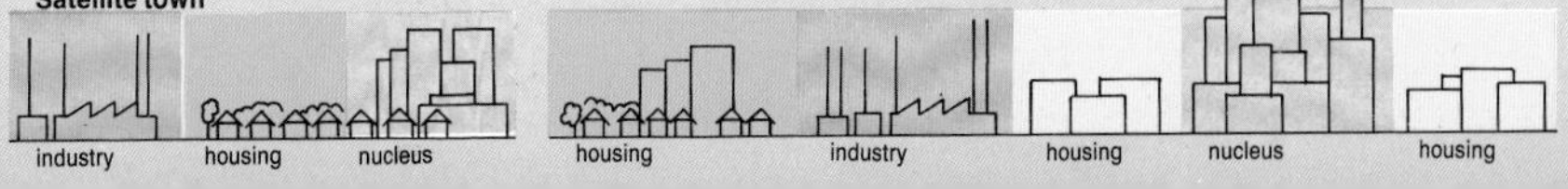

Percentage of people with access to:
- clean water
- sanitation

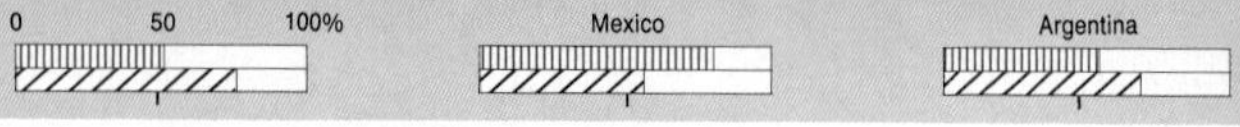

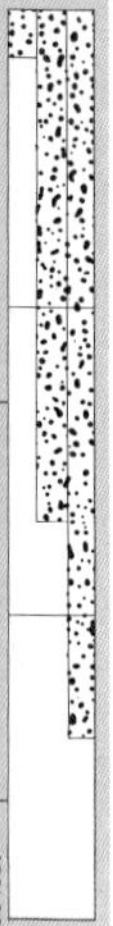

Depth to water

- Afghanistan 60
- UK 175
- USA 635

The amount of water used by a household varies greatly from one country to another. The figures are for litres for one person for one day.

15 September 1988

On one day in September some people are waiting for water while others are splashing in it. In many parts of Africa animals are dying and thirsty people are queueing to fill their pots. Where there is a suitable underground water reservoir wells may be drilled to provide clean drinking water.

The damming of a river provides water storage in the reservoir created. Hydroelectric power may also be generated. In countries where there is plenty of water and the people are relatively well-off swimming pools are popular, and pleasant fountains are enjoyed. Water a luxury for some a plaything for others.

If no change, then disaster

Although the world's rainfall would be ample for our needs if it fell evenly in time and space, its distribution over the globe is uneven and variations from year to year, and from season to season are great. Therefore it is extremely difficult to provide a constant supply to all who need water, as the population increases and demands for water per individual rise. Recent droughts in West Africa show how vulnerable certain areas are to natural fluctuations.

The social and health aspects of lack of water are clear and the provision of adequate supplies of safe water is one of the world's most pressing problems. Agriculture and livestock rearing require water and economic development too will suffer without sufficient water.

Some areas are short of water for human use and agriculture, but the development of groundwater potential and reservoir storage can help with fluctuations in supply. There are important advantages in using groundwater, these include its widespread occurrence, relatively easy development and natural protection against surface pollution and evaporation losses.

2000

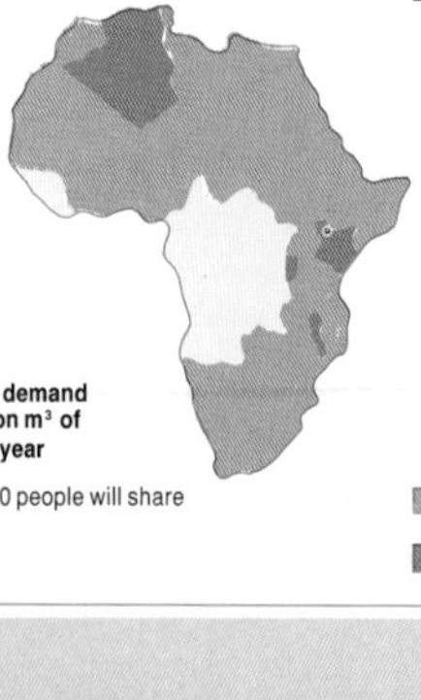

2025

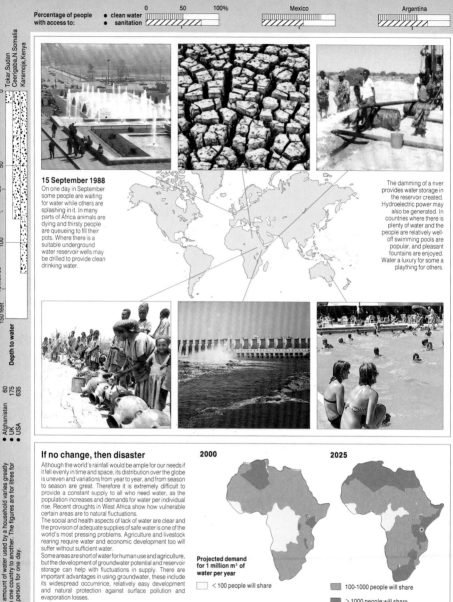

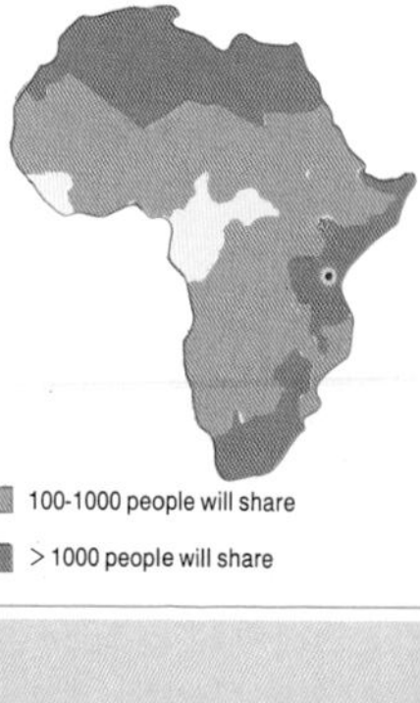

Projected demand for 1 million m³ of water per year

- < 100 people will share
- 100-1000 people will share
- > 1000 people will share

. . .and we found water

distribution and consumption of water resources

Walking all day to survive

In many developing areas a large proportion of the work of maintaining a household is the carrying of water from rivers or wells. Women walk for several hours to and from a source of water which may provide just enough for all her family's needs. Dry periods make the work even more onerous as rivers dry up and well levels fall. At the same time the chemical and biological quality of the water deteriorates.

Aid has concentrated on providing hand-operated wells which are cheap to install and maintain, and provide uncontaminated water. For towns the aim is to use stored river flows or large well-fields as local conditions dictate.

Inequality, even in water supply

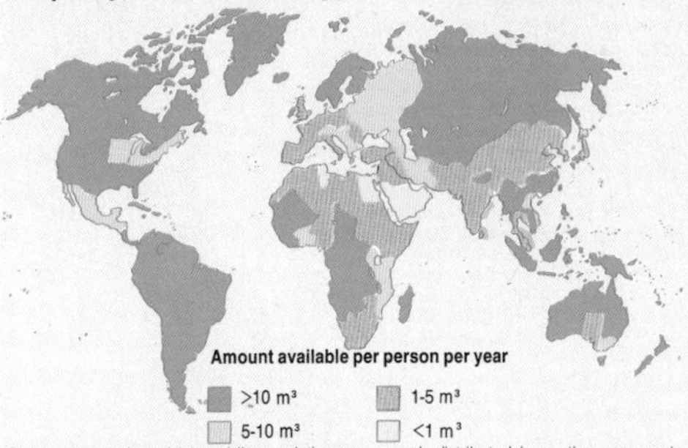

Both water supply and the world's population are unevenly distributed, hence there are great variations in the amount of water available per head. Regions in the low categories are prone to drought and by the year 2000 at least 30 countries may be short of water.

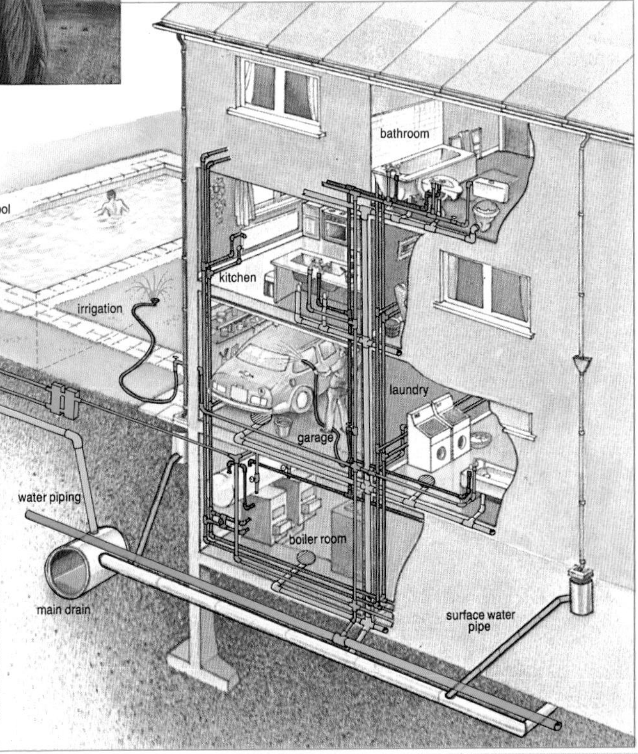

The height of luxury

Water consumption has increased greatly during the last 100 years. The UN recommends a minimum daily supply of 80 litres, 16 gallons per person. In countries with piped water consumption is greater.

The illustration shows the profligate way we use water. In an affluent water-rich society a lot of water can be used for the garden, the swimming pool and cleaning the car, as well as for cooking, washing and flushing the toilet. In the picture the hot and cold running water are shown in red and blue respectively, the water in the central heating in orange and the drainage in green.

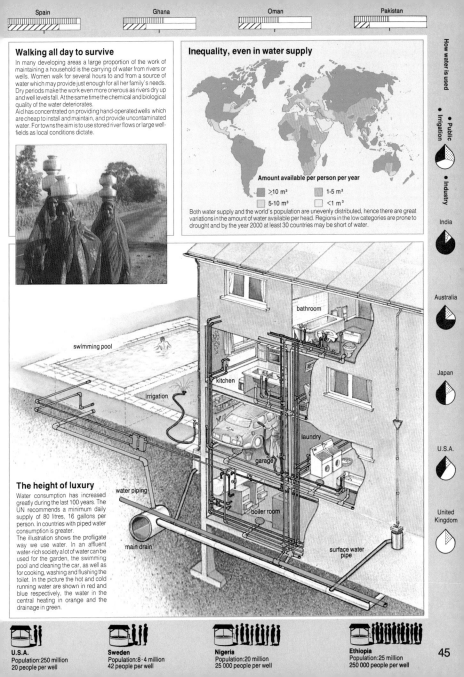

U.S.A.
Population:250 million
20 people per well

Sweden
Population:8·4 million
42 people per well

Nigeria
Population:20 million
25 000 people per well

Ethiopia
Population:25 million
250 000 people per well

Kansai Airport. The international airport for Osaka in Japan is being built on an artificial island 85 km², 33 sq miles in area. The terminal building will be 1.5 km, 1 mile long.

Stockholm. Land can be reclaimed naturally. Isostatic readjustment has lifted the city 40 cm per century. The shoreline is now 2.5 m lower than in the 13th century, and uplift and infilling have combined to push the coast 100 m back into the sea.

Fenlands in England. The Romans started to drain this lowlying area round the Wash in eastern England. In the 17th century the Duke of Bedford with the help of the Dutch engineer Vermuyden completed the reclamation. The rich farmland of the Bedford Level was created by a complex of drains and sluices. As the land was drained the peat shrunk and the surface level rapidly became lower. This very fertile area now specialises in vegetables and bulbs.

Reclaiming land

From the earliest times defences were built to protect lowlying areas and settlements against flooding from rivers or from the sea. Land was also reclaimed by enclosing areas that had silted up naturally. By the 17th century the demand for new farmland led to the drainage of lakes and wetlands in the Low Countries and the Fens in England. Increasingly sophisticated windmills and later steam pumps were used to control the water level.

After 1900 the Dutch began the ambitious plan to enclose the Zuyder Zee with a 30 km, 19 mile long barrier dam. The first polder was drained in 1930 and the freshwater Ijsselmeer was created.

During the second half of the 19th century land has largely been reclaimed from the sea for residential and industrial purposes.

Inland, the process of reshaping the earth also has a long history. In England the Saxons cut platforms in the hillside for farming. They are called strip lynchets. And in parts of Asia terracing the hillside is a long established practice.

Land reclaimed 1200-1600
Land reclaimed 1600-1900
Land reclaimed since 1900

1. Wieringermeer 1927-1933
2. North east polder 1937-1942
3. Eastern Flevoland 1950-1957
4. Southern Flevoland 1959-1968

Using the reclaimed land

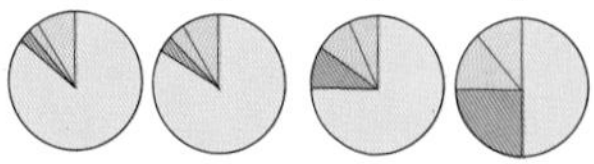

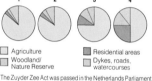

Agriculture
Woodland/ Nature Reserve
Residential areas
Dykes, roads, watercourses

The Zuyder Zee Act was passed in the Netherlands Parliament in 1918. It was the start of a plan to make a large inland sea and 5 huge polders, which would increase the surface area of the Netherlands by 6%. Polders are totally enclosed by dykes. Water levels within the polder are rigorously controlled and each polder has a specific groundwater level.

The building of the polder dyke is a complex business. The core of the dyke consists of sand which is surrounded by impervious clay. Piles provide a retaining wall, and to help protect and stabilise the shoreline osier 'mattresses' filled with stones are used. A facing of stone blocks adds further protection and the dyke is complete when a road is built and the structure grassed over.

Building for generations to come

Once the land has been enclosed and drained, which may take several years, the polders are laid out to provide good communications, focal centres and farms of an appropriate size. The land is excellent for farming and after the first few years of state management is sold or leased to private individuals. Initially most of the land was used for agricultural purposes. However only half of the most recent polder is farming land and a quarter is devoted to woodland and an important nature reserve. There are also plans to build modest hills to create a more varied landscape.

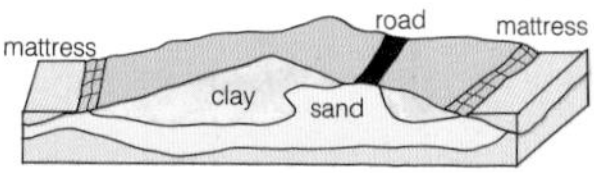

Hong Kong – The New Territories. Already a lot of land has been reclaimed for housing. Hill slopes are terraced and the levelled hills provide fill for reclamation. Parts of the New Towns are built on this land.

Kenya. 'Fanya-juu' is a farming technique designed to reduce erosion. It is used on more than 1.5 million hill farms. A trench is dug and a bank made. Soil builds up against the bank to level the terrace.

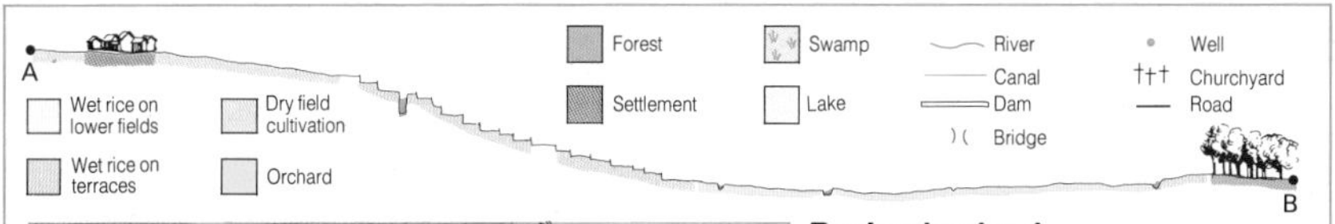

Reshaping land

In Asia population pressure on the land has, for many centuries, necessitated the cultivation of hillsides. To facilitate cultivation and prevent erosion terracing is carried out. Fields are cut from the slopes and low walls built to support and protect them. This practice creates a distinctive landscape with quite narrow field strips following the contour of the land.

The system as practised in China, the Philippines and much of south Asia is characterized by intensive cultivation methods and high yields. The size and shape of the fields preclude the use of machinery. A wide variety of crops can be grown and where rice is cultivated, the fields, known as paddies, may be flooded.

In Bali and elsewhere communities must cooperate to control water supplies, so that planting, harvesting and meeting the requirements for duck rearing or fishing can be coordinated.

Venice. To prevent flooding Venice requires 4 flood gates to protect the lagoon. The scheme being tested envisages 'chocolate block' steel canisters hinged on the sea bed. It is essential that shipping is not hampered when the barrier is not in use. The scouring by the sea of the lagoon must not be impaired or the ecology of the region damaged. Currently maintenance and reliability are being tested at the Lido. The whole construction is due to be completed by 1995.

City in the sea
The first Venetians found refuge from the invaders of the 5th century on the islands of the lagoon. Subsequently the capital moved to the Rialto and from the 11th century shipping and trade expanded. Venice has long ceased to be a commercial centre, but the palaces, churches, and canals make it a unique historic city. Now, despite a sea wall high tides invade the centre, flooding San Marco ever more often. Despite international concern no satisfactory solution has yet been found.

Land hungry island
Singapore has no natural resources and very limited land space. Its prosperity is built on entrepot trade. Oil refining and ship-repairing are important and Singapore has recently developed its manufacturing base.
Reclaiming of land has allowed the development of one of the world's largest best-equipped ports. Changi International Airport, with a runway length of 4 km opened in 1981. It is on reclaimed land, as is Jurong Town, the largest industrial estate in Southeast Asia.

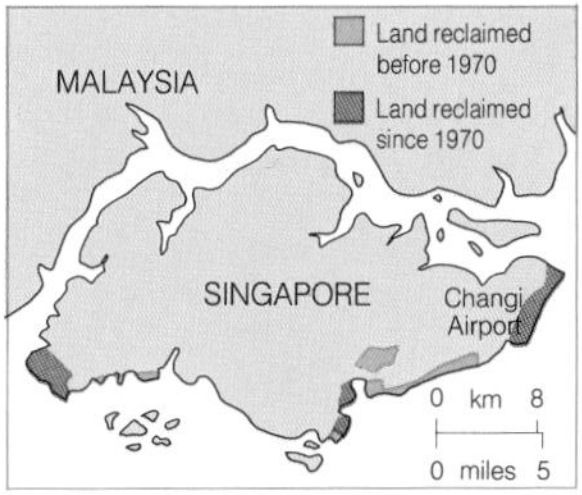

Isostatic readjustment. After the melting of the icecap at the end of the Ice Age, the earth's crust began to rise because it was no longer bearing the weight of the ice.

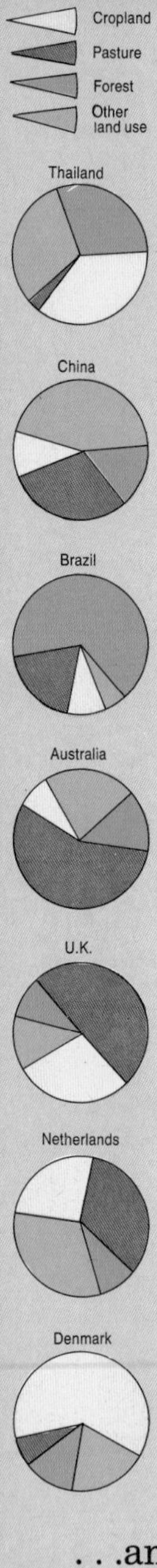

Subsistence farming

The most important consideration for the subsistence farmer is to provide enough food for the family throughout the year. The system has evolved over many centuries and is still the dominant one in most tropical regions. Where rainfall is sparse or unreliable, pastoralism is common. As long as the number of animals, which may be cattle, sheep or goats, is not greater than the land can sustain, the system is satisfactory. Cultivation is practised where the climate is more favourable. A lot of effort is required as it may be necessary to walk long distances to the fields, tools are simple and bringing water to the crops may be arduous. Only when the needs of the family are satisfied, can the farmer cultivate cash crops. Income from these crops is used to buy salt, pots, cloth and other simple necessities.

In some areas cultivation has been continuous for many centuries, and with careful husbandry the soil fertility is maintained.

Commercial agriculture

When the farmer can grow more than the family requires, then crops and livestock can be produced for market. With mechanization the pattern of agriculture changed dramatically, as a smaller number of men using machinery could then produce food for a large number of people. A characteristic of modern farming in a country such as the United States is the high level of efficiency in producing crops. Commercial livestock-raising involves careful breeding to obtain the best animals. On huge ranches the motorcycle and the helicopter are as much part of the scene as the horse.

The small mixed farm which includes both livestock and crops is still common. But there are now many big agricultural businesses which concentrate on a few carefully bred crops, and use substantial amounts of fertilizers and pesticides.

In contrast, intensive use of a small area for the production of vegetables, fruit or flowers can also yield good returns for the farmer.

Forestry

The function of forests is to conserve the soil, to protect river basins, to provide a habitat for wildlife and perhaps a recreation area or a home for people, and to regulate the global climate; all this in addition to providing timber and fuelwood.

In temperate areas both hardwoods and softwoods are grown. The latter is used for building and also in the pulp and paper industry. Generally the forests have been well managed with replanting programmes. The tropical forests have been subjected to a massive programme of deforestation to provide hardwood for Europe and Japan. The United Nations reckons that 100 000 km², 62 000 sq miles are lost each year, some authorities say twice as much is lost. In the Amazon basin an area of forest the size of West Germany was destroyed by fire in just one month recently. The purpose of the clearance is to provide land for cattle rearing, but the clearance not only destroys homes and habitats, it probably interferes with global climate patterns.

Living in cities

In some towns there still remains a medieval core often close by a religious building or an ancient marketplace. More often the original street pattern remains but the houses along them have been rebuilt, perhaps several times. The history and changing importance of a town can often be traced through its buildings.

Many large cities now have multi-storey office blocks and perhaps skyscrapers. These may be clustered in the commercial heart of the city, or built in a separate development area such as Défense in Paris.

The outskirts of towns are the main residential areas and there are stark contrasts to be seen. In many Third World countries the poorest people often live closely packed, and in very basic accommodation. Services may be minimal. For some, in the shanty towns, the shelter provided by a sheet of corrugated iron may be all they have. In contrast there is the large well-appointed house with every luxury.

Working areas

The sort of work people do varies enormously. There are differences between those who work indoors and those who work outside, between manual workers and white-collar workers, between those who work as individuals and those who work as a team, between paid and voluntary workers, and there are many other variations. A convenient simplification, for those in paid employment, is between agricultural workers, those who work in industry, and people in the service industries.

Those in agriculture will include both the farmer and the market gardener in his nursery. The industrial sector ranges from the self-employed blacksmith, to the workers in a huge industrial complex such as the Ruhr. Service industries include a wide range of employments; examples are work in banks and insurance, schools and hospitals, and shops and hotels. The proportion of people who work in the service industries tends to be highest in the wealthiest countries.

Traffic and recreation

During this century there has been a great increase in travel and the number of cars on the roads grows each year.

In medieval days a track might link two settlements and along each side of this road houses would be built. In time there would be a straggling linear village or strassendorf. In contrast to the straight road, are the loops of the slip roads in the multi-level junctions needed in a modern road network. In some countries the traffic problem is acute, even new constructions, such as the London Orbital road, can soon become choked with traffic.

Despite the demands of housing, agriculture, industry and transport, there is still land left for recreation. While for some the open countryside provides space for recreation, there are also many facilities for the active or the spectator sportsman. Golf courses and sportsfields are popular, while for those who enjoy horse-racing, speedway or Grand Prix events, there are also plenty of racetracks.

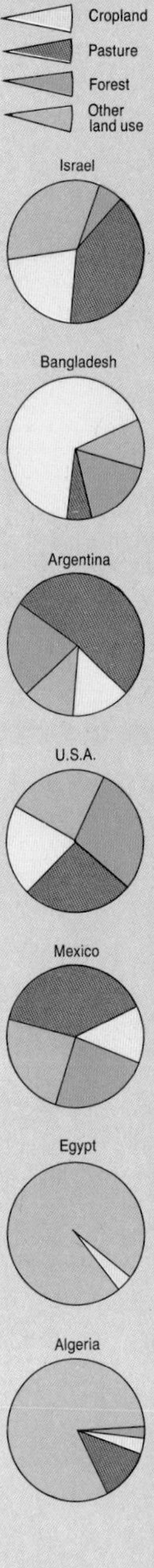

• The tropical forest is cut down and the branches and tree trunks burned. The wood ash enriches the soil, and the remaining stumps inhibit erosion. The fields are enclosed to keep out animals. Crops of varying heights are grown to intercept rain and prevent damage from runoff. New trees are allowed to grow. After a few years the land becomes exhausted and the forest is left to regenerate.

Shifting cultivation

• A soil profile shows the different bands of soil from the surface down to the bedrock.

Luvisol. A clay layer gives a badly drained soil.

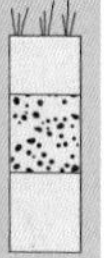

Podsol. Sandy soil above a layer of iron oxide.

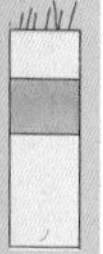

Chernozem. A thick black organic surface layer.

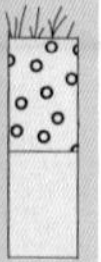

Farming our Earth

Agricultural patterns vary greatly from one part of the world to another. Climate and soil are the most important influences. Uplands and valleys support different types of farming. Wealth or poverty, and population density are reflected in methods used. A reliable supply of water permitting irrigation can transform farming practices. There are also economic factors, including markets, incentives and pricing systems that can modify farming. Above all the landscape had been shaped by history.

In Western Europe the small intensively cultivated holding is common, while in North America farms tend to be much larger. Most of the world's farmers are peasants trying to produce enough food for their own needs; holdings are small and methods are simple.

Birthplaces of farming

By 3000 BC people in the Middle East were domesticating animals and growing plants for food. Asia and Central America were other centres from which the skills of agriculture spread.

How people use the land

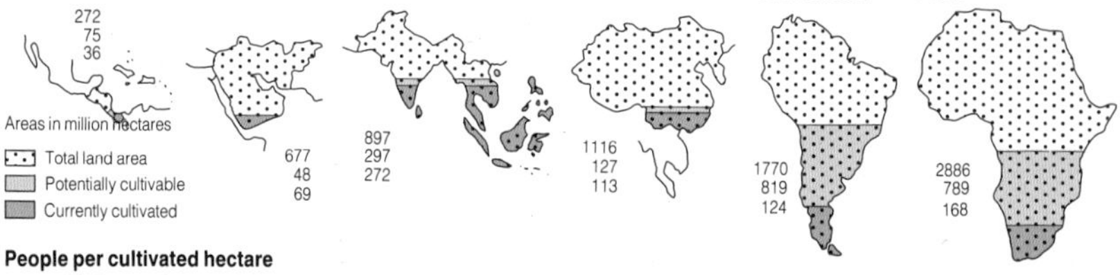

People per cultivated hectare

In some parts of Asia all the suitable land that exists is already being used for farming. In Africa and South America there is still land that could be cultivated. However it is largely marginal land with unreliable rainfall. Unless carefully farmed this fragile land can easily be destroyed.

Slash and burn

The traditional method of subsistence farming in tropical forests involves making a clearing in the forest. After a few years cultivation the land is abandoned and the farmer moves on to new ground. At low population densities this is a sustainable practice.

Rice growing

The Green Revolution. Since the 1960s there have been record crops of cereals including rice. High-yielding varieties can provide several crops a year, but need irrigation and fertilizers. Rice is grown on 80% of the farmed land in Bangladesh, Thailand and Malaysia.

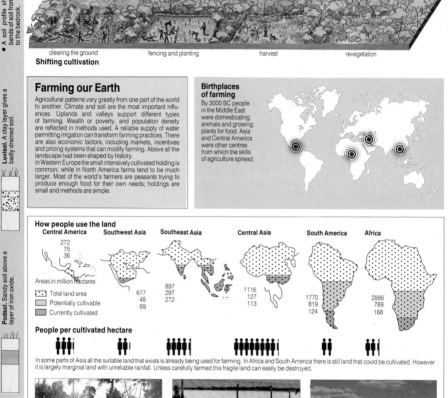

Mechanized farming

In North America farming has become an industry. Holdings are large, the labour force small but heavily reliant on machinery. Hign yields are obtained by using selected strains, pesticides and fertilizers. In the future genetically-engineered plants may be tailored to particular environments.

The field is flooded and the water retained by banks which also serve as paths. A layer of mud is produced by cultivating with hoes or buffalo-drawn ploughs. The seedlings are grown in seedtrays and transplanted after 20-30 days. Rice growing is labour intensive. At harvest time the field is drained and the rice cut. The hoe, the sickle and the carrying pole are the tools of Asian rice growing.

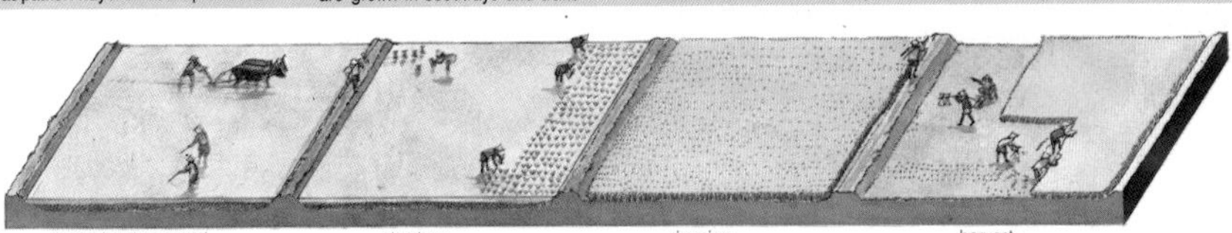

Rice growing

Soil curtails farming. Only 11% of the world's land is at present suitable for agriculture.

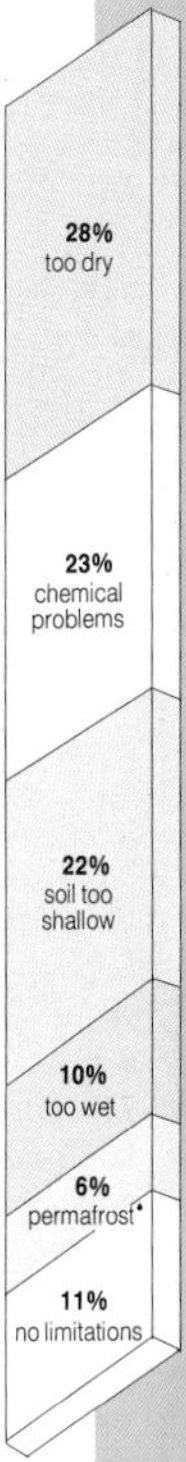

N, P_2O_5 and K_2O

The success of new crop varieties which respond to fertilizers and pesticides has alleviated hunger in many Third World countries. The most important artificial fertilizers in terms of nutrients are nitrogen, phosphate and potash. World consumption of fertilizers increased 2.5 times between 1965 and 1985, and in Asia fertilizer use quadrupled. The results, as demonstrated in crop yields have been impressive.

Many countries, however, are now concerned about the pollution of water caused by nitrates and other chemicals. These wash out of the soil into rivers or penetrate underground water reservoirs. Valuable humus and natural fertilizer can be provided by ploughing in stubble and waste plant material after harvesting.

Nevertheless supplying adequate fertilizers and the correct techniques for their use could be better for poor countries than supplying food, except in a crisis.

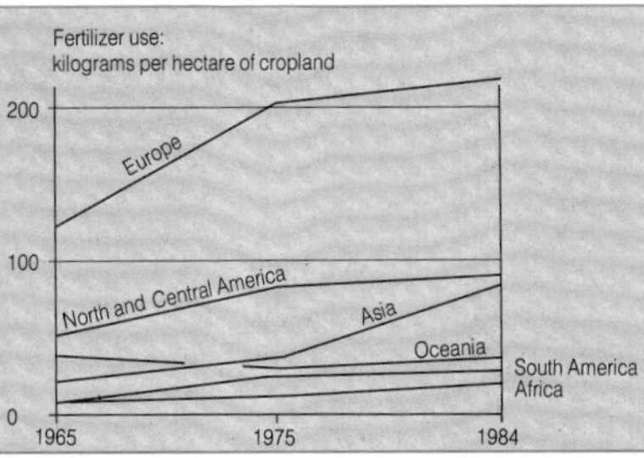

Irrigation – some old and new ways

Primitive water wheel used to irrigate land beside the Nile in Egypt.

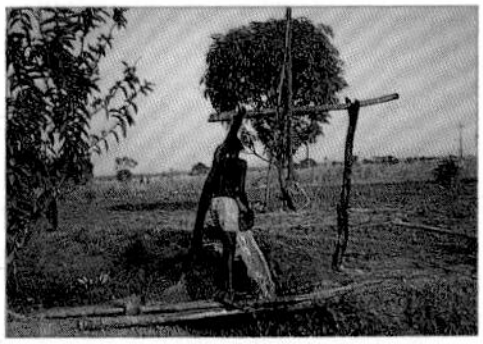

A shaduf in Sudan. Water is raised using a bucket with a counterweight.

Paddy fields in Indonesia. Growing wet rice involves flooding the field with water.

Drip irrigation. Water is fed through a plastic hose and reaches the roots without wastage.

Huge mobile overhead irrigation spray in use in Saudi Arabia.

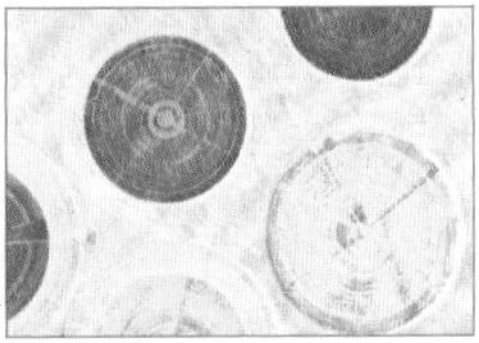

Rotary irrigation needs expensive equipment and a good water supply.

Mechanized farming

The land is first ploughed and harrowed. The seeds, which may be treated to assist germination and prevent disease are sown by machine. In Europe planting may take place in autumn or spring. In some countries if thick snow insulates the soil spring growth is better. Harvesting is critical and highly mechanized.

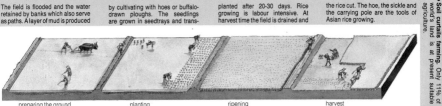

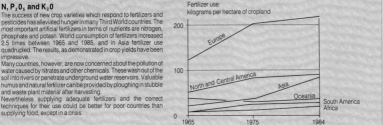

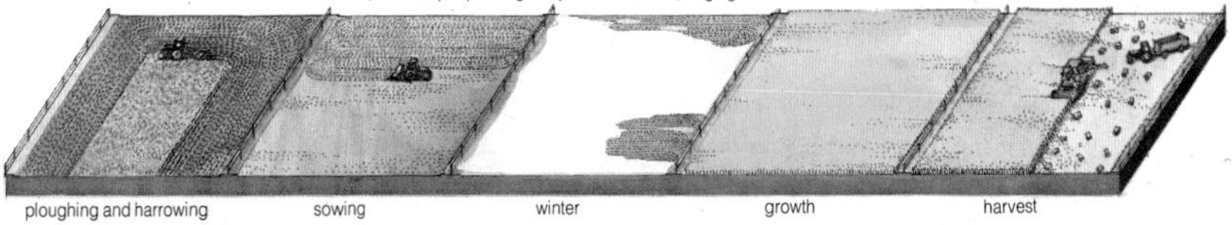

Genetic engineering. A single gene can be 'spliced' into another organism allowing scientists to create entirely new animals or plants.

Sustainable development aims to provide for our basic needs without jeopardizing the ability of our descendants to meet theirs.

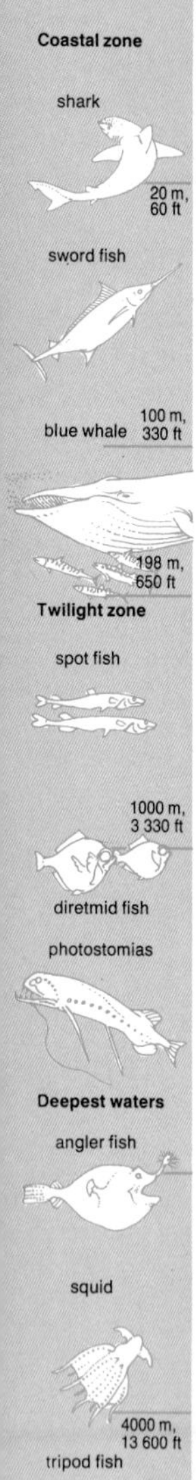

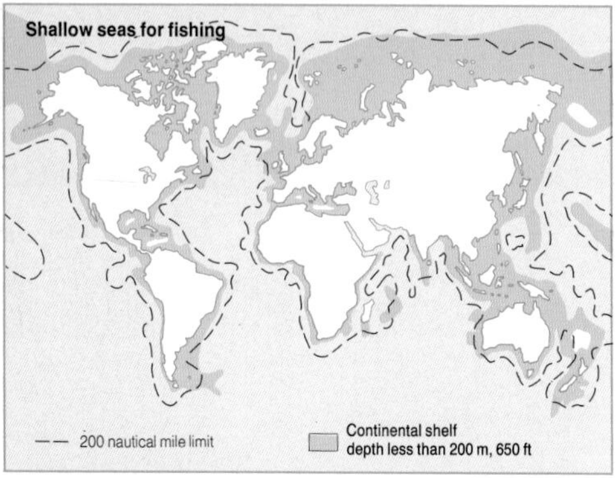

Polluting the seas

Pollution of the marine environment is the introduction of substances or energy which results in harm to living resources, hazards to health, hindrance to marine activities or impairment of water quality.

The practice of dumping chemicals and radioactive waste in the seas is not new (see page 74). But the concept that the oceans' immense size can dilute all waste to acceptable levels no longer holds true. With time waste containers can erode away and the poisonous contents be released into the seawater. Not a lot is known about the circulation of deep and surface water (arrows right) and this process may take place faster than was formerly realized, so that poisons spread to the biologically-active layer.

Heavy metals such as lead, mercury, zinc and cadmium when absorbed by marine organisms can affect animals higher up the food chain. Concentrations of certain metals can cause brain damage in communities dependent on fish. Dumping of dredged waste increases turbidity, reducing light penetration and thus inhibiting plankton growth.

From angling to trawling

There are two methods by which we can obtain protein from the seas: by fishing (hunting) and by aquaculture (cultivation). The farming of fish is still in its infancy but is becoming an important source of protein. Fishing emerged only 8000 years ago when the basic equipment of spear, hook and net were devised. Some 80% of the world's fishermen still use similar techniques, though these traditional methods account for a tiny percentage of the total world catch. The traditional fisheries are mainly concentrated in developing countries and rely on simple boat designs adapted to local conditions.

In the developed countries much of the fishing was also undertaken from open boats until the last century. The introduction of steam and diesel engines revolutionized commercial fishing. Fishermen were no longer so dependent on weather conditions, they could reach distant fishing grounds and the extra power enabled the use of larger nets.

Modern techniques include freezing the catch on board so that more time is spent at sea. Processing the catch increases the value for the effort expended, whilst the heads, bones and offal can be converted into fishmeal. Factory ships are floating processing plants which buy fish from the home water fleets.

It is difficult to assess stock levels and the increased level of effort has led to severe problems of overfishing in many regions. The most spectacular example is the history of whaling. In the North Atlantic it took only 20 years from the introduction of the explosive harpoon in the 1860's for the fishery to collapse. Subsequently, when new areas have been exploited the stocks have soon been decimated as the level of exploitation far exceeds the reproductive capacity of the mammals.

Scientists have tried to assess the maximum sustainable yield from a particular stock, and to this end have imposed quotas or bans.

Eutrophication

The over-nourishing of water results in algal blooms and lack of oxygen in the water which becomes foul smelling and virtually lifeless. This can be caused by phosphorus or nitrogen in sewage, or, as here by nitrogen draining off fertilized land. If the oxygen demand by algae and bacteria is high locally then fish and other aquatic life may suffocate.

Environmental poisons

are spread mainly by industry and agriculture. Poisonous substances such as DDT and PCBs cause serious damage to restricted waters like the Baltic where the water exchange rate is not very great. Accumulation of PCBs via the Baltic herring in the food chain has reduced the reproductive capacity of the grey seal.

Excessive fishing

does not necessarily lead to extermination of a fish species, but it does upset the balance between the different species. Unselective methods of fishing, in which all sizes and ages of fish are taken, will eventually reduce the spawning stock by reducing the survival rates of all age groups. When catches diminish dramatically, serious problems arise.

Sea fishing techniques

Cultivation of fish and shellfish usually takes place in sheltered coastal waters in floating cages or rafts which can be moved to ease cropping or to avoid local pollution. Coastal fishing provides mainly for local and household needs. The methods employed are diverse having developed over a long period and being subject to local conditions. Passive methods such as salmon trapping are still used today where migration routes are known, or where the fisherman can entice a lobster to a baited trap.

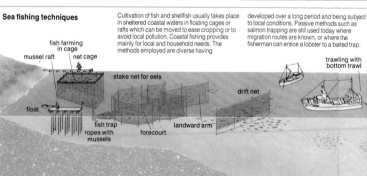

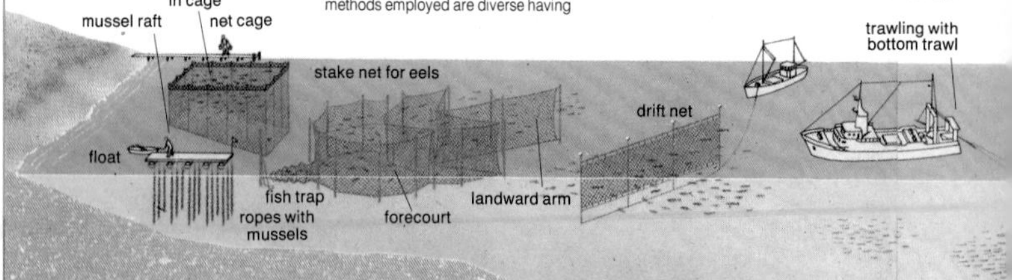

. . .and we fished the sea

catching shrimps and whales

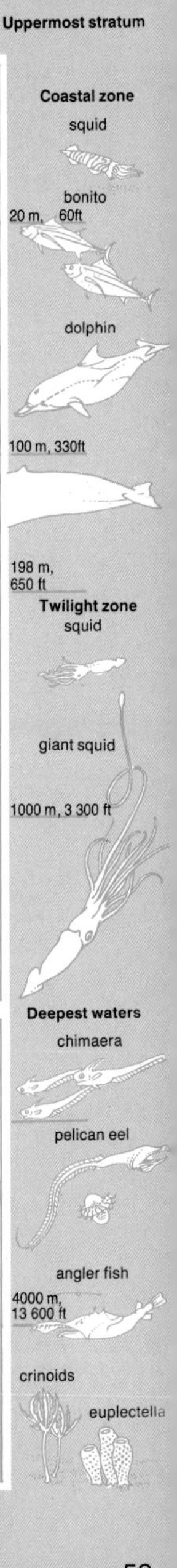

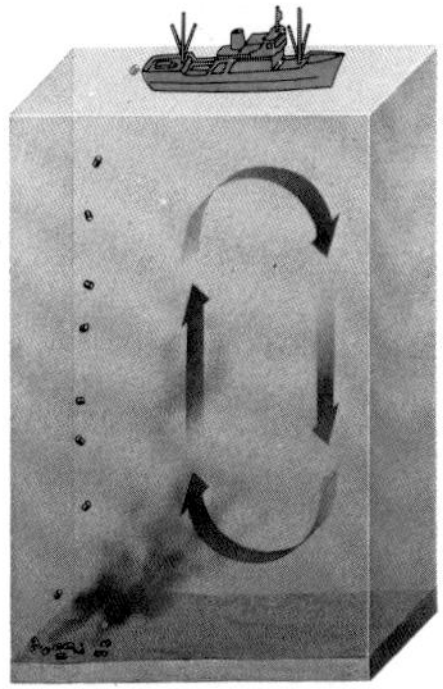

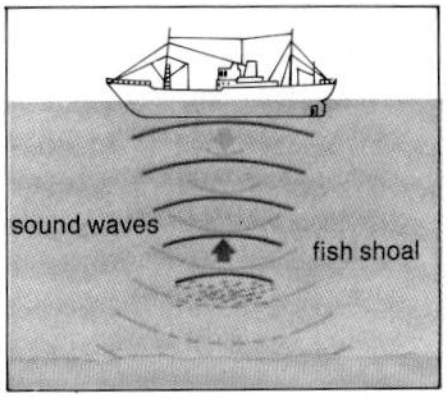

Deep-sea fishing

is conducted mainly by large trawlers. These are highly mechanized vessels which locate fish shoals by means of echo sounding (above). The fisherman also knows exactly how his nets are performing. Computer and satellite systems help him manoeuvre the vessel and advanced processing facilities on board preserve the catch to ensure maximum return for his effort.

Warm water stops fishing

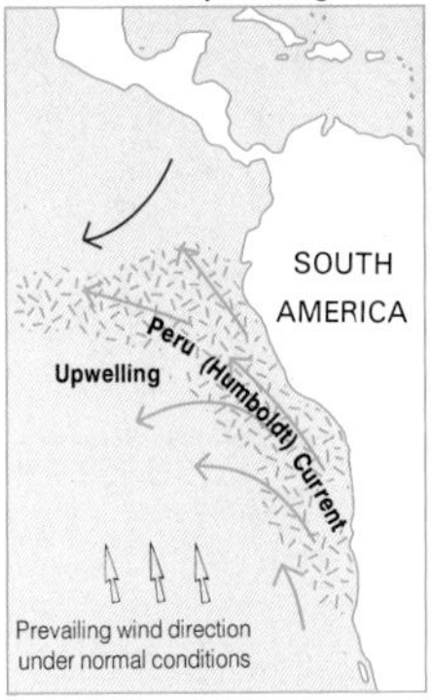

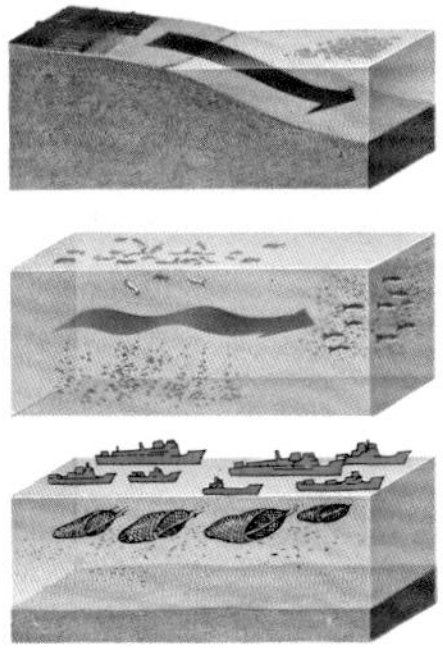

Peruvian anchovetta

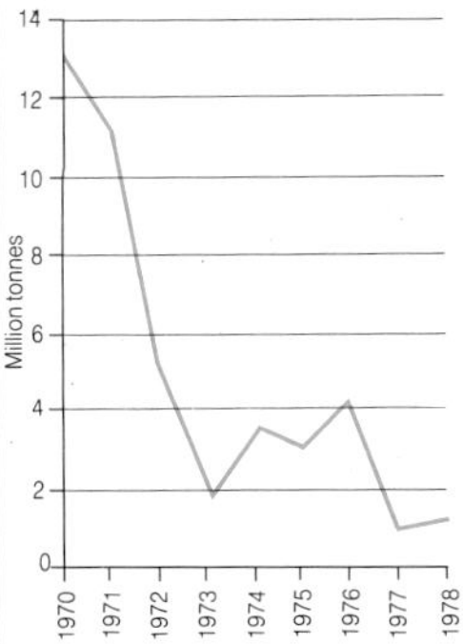

Climate and fisheries

The map above shows the normal climatic conditions upon which the Peruvian anchovetta fishery depends. Offshore winds drive water away from the coast causing colder nutrient-rich water to well up. This supply of nutrients supports vast quantities of marine life.

However the upwelling can cease when the prevailing winds are replaced by northerly winds which drive warm surface water shoreward. The nutrients are no longer available, fish die and the fishery collapses. This warm water incursion is called El Nino.

A particularly severe case of El Nino occurred in 1973 though catches were already declining due to overfishing. The mechanics of El Nino and similar aberrations have not been fully explained but they are known to be associated with anomalous behaviour of the atmosphere and oceans in other parts of the world.

Nets can be used to encircle fish or put out from a boat. The use of a boat enables a fisherman to follow shoals of fish and to deploy his nets to best effect. Gill nets are made of fine material which the fish cannot see, and they are set against the direction in which the fish usually swim thus entangling the fish. In contrast, trawl or seine nets envelop fish which either collect in a finer meshed cod end (trawling) or can be concentrated into so small an area that they can be pumped out. Lines with hooks tend to be used only for fish with a high individual commercial value.

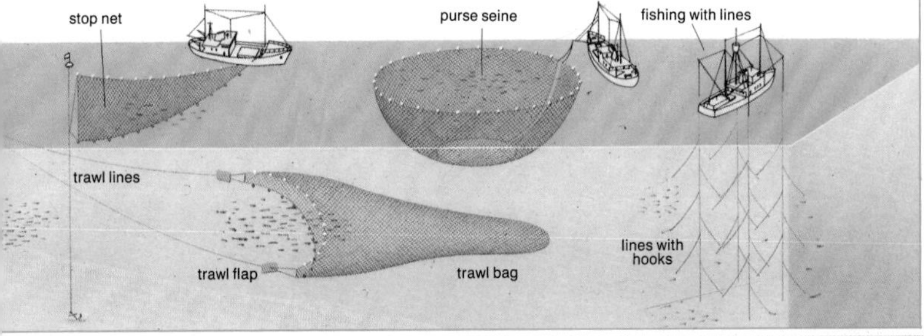

Purse seine. A fine-meshed net which encircles a shoal of fish. It is closed at the bottom to trap the fish and drawn in.

Trawling. Fishing method by which a net is towed on wires in midwater or along the sea floor.

PCBs. Poly chlorinated biphenyls are very stable and highly toxic chemicals which have been used for their heat resisting capacity and low chemical conductivity.

Food values. Calories per 25 g, 1 ounce

Cassava	30
Dates	27
Honey	72
Carrots	6
Chinese leaves	6
Herring	59
Bulrush millet	94
Cheese	40-130
Finger millet	94
Asparagus	5
Mixed beans	97
Butter	185
Apple (each)	44
Banana (each)	70
Bread	56
Almonds	139
Bacon	107
Beefburger	133
Ice cream	47
Maize	100
Oysters	13
Lamb	41
Muesli	165
Plantains	20
Lentils	25
Onions	10
Potatoes	22
Raisins	61
Sorghum	97
Yoghurt	13
Rice	101
Sugar	100
Sim sim	162
Sweet potatoes	27

Calorie consumption: daily consumption per person

> 2900 | 2500-2900 | 2100-2500 | < 2100

2350 calories = 9839 joules

In the cartograms at the top of these pages, each country is shown as a rectangle. The size of the rectangle is proportional to the population of each country. This diagram shows where the largest numbers of the under-nourished and the well-fed are living. It can be compared with the main map which shows similar information in a more conventional way, as well as showing where main foodstuff is produced.

The United Nations recommends a basic diet of 2350 calories a day for the 'average' person. However, while the average intake for developed countries is 3300 calories, for developing countries the daily calorie consumption on average is only 2200 per person. Many of the poorest consume less than 1500 calories daily, a dietary level at which hunger and many symptoms of undernutrition are evident. The average consumption of protein per person in developed countries is twice as much as in poorer countries, and of animal protein it is four times as much.

Too little. . .

Over the past few decades total food production for the world has increased greatly. The output of cereals, which provide half the world's calorie intake, has risen and since 1965 the amount produced per person for all continents except Africa has shown a marked increase.

There is, however, a great discrepancy between those countries with too much food and those with too little. In 1965 average calorie intake levels needed for healthy growth and productive work were below those needed in 46 countries, of which 29 were in Africa, 8 in Asia and 6 in Latin America and 3 elsewhere. In 1987 as many as 37 countries, mainly in Africa could not meet their food needs.

A further contrast between the hungry and the well-fed can occur within countries. In some apparently well-nourished countries, there may be many close to starvation, while in most developing countries the privileged may eat well while the poorest have a very low nutritional level.

. . .or too much

While many people in Third World countries go hungry, there is more than enough food in many developed countries. Indeed food is wasted because it is thrown away or allowed to go bad.

Malnutrition is a problem too; the malnutrition of obesity caused by overeating. The production of slimming aids and dietary products is a 'growth' industry. Other problems relating to nutrition may be heart disease and diabetes. Hypertension and some allergies have also been linked to certain foodstuffs in the diet of the affluent, though it is sometimes difficult to separate the effect of the food itself from the effect of the processing.

Still a hungry world

Efforts to combat hunger have had some effect and an average increase of 100 calories per day worldwide has been recorded over recent years. There has also been a drop in the percentage of the population with a diet too poor for an average day's work. However the decline is offset by the increase in population and because of this the number of hungry in the world is rising.

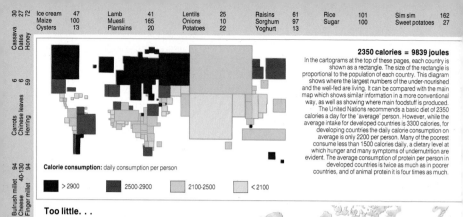

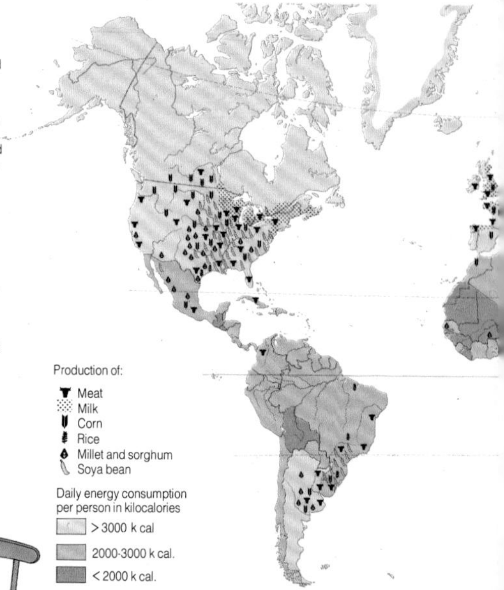

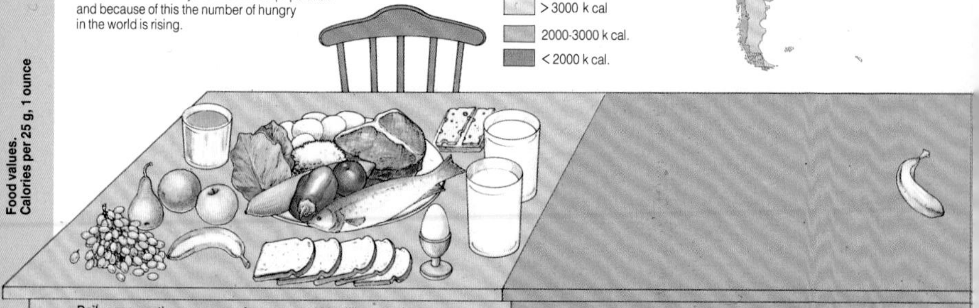

Daily consumption per person for approx. 1200 million people

Protein content of foods in g per 100 g

Chicken	20.8	Pork	12.0	Cheese	25.4	Dried milk	26.6	Lentils	23.8	Bread	8.3
Beef	14.8	Salmon	19.1	Egg	11.9	Dried egg	43.4	Peanuts	28.1	Rice	6.2
Lamb	13.0	Plaice	15.3	Fresh milk	3.3	Soya flour	40.3	Flour	11.6	Barley	7.7

Peas	5.8
Potato	2.1
Dates	2.0
Apple	0.3

What we eat matters

In developed countries where meat and milk products are an important part of the diet, it is believed that certain cancers and some cardiovascular diseases may relate to the consumption of animal products.

In the less developed countries protein-energy malnutrition (PEM) and other diseases caused by a lack of vitamins or minerals are widespread. PEM affects all age groups causing debility and a susceptibility to infection. In some countries it may cause as much as 60% of the high infant death rate. Yet kwashiorkor, a deficiency condition particularly affecting children, can be greatly reduced by education of the mother in the dietary needs of young children. Poorly nourished infants are also very susceptible to diarrhoeal infection which retards growth. The pregnant woman herself should eat good food, but she is unlikely to put her needs before those of her family. Protein can be obtained from many vegetables and pulses where meat and fish may be unacceptable, unavailable or beyond the means of the family.

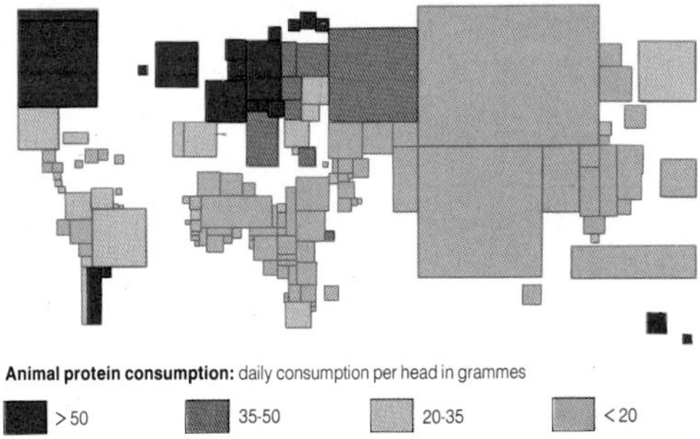

Animal protein consumption: daily consumption per head in grammes

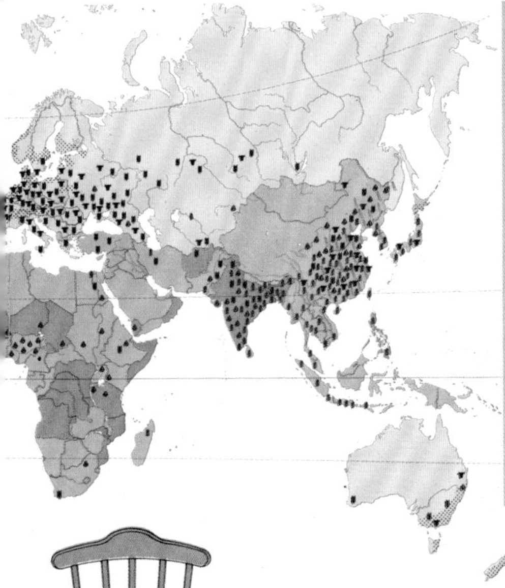

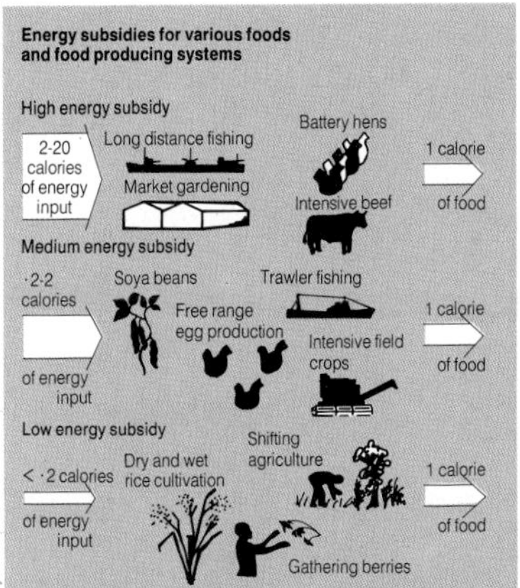

A low energy support system is one in which few calories of energy subsidy are expended to produce 1 calorie of food. Shifting cultivation is a low energy system while highly intensive rearing of cattle utilises 10 times more energy than the energy value of the food produced.

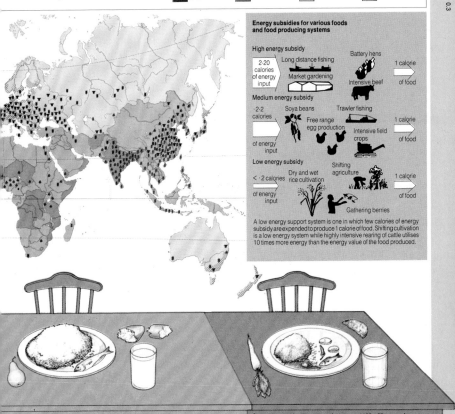

Daily consumption per person for approx. 2500 million people

Daily consumption per person for approx. 1100 million people

Vitamins are groups of nutrients which are needed for growth and health. There are 13 of them. A normal varied diet contains sufficient, but vitamin deficiency diseases are associated with poor diets. Insufficient vitamin B (vegetables and fish) leads to beriberi, vitamin C (citrus fruits) to scurvy and vitamin D (fish liver oil) to rickets.

Christmas pudding. A steamed pudding containing dried fruit. Of ancient Greek origin, currant is derived from Corinth.

Marmalade. Conserve made from bitter Seville oranges. Originated from a cargo of fruit landed in Dundee in the C 17th.

Pastrami. Salt beef, smoked and seasoned with peppercorns.

Cassoulet. Stew of dried beans and preserved goose and herbs.

Specialities from round the world which rely on preserved foods.

Agen prunes. Made from the dried red and purple plums of southwest France.

Smörgåsbord. Raw, pickled and smoked fish and meat are typical Swedish delicacies.

Hominy. American Indian name for dried white corn.

Kaestur hakarl. Fermented skate wing. An Icelandic winter delicacy.

Knickerbocker glory. A spectacular icecream with fruit, nuts and cream.

Gravad lax. Fresh salmon pickled with dill, sugar, salt and peppercorns.

Saving food

From the earliest times people have found it necessary to preserve food after harvest or when animals were killed in the autumn to provide winter stores. The natural methods of preservation, including using wind, sun, smoke and salt have been used over many centuries. Often food was dried for long storage, and to make it palatable later required a prolonged period of soaking; but at least a regular supply of food year round was ensured.

There are now a variety of new techniques for preserving food, and almost any foodstuff can be obtained at any time of the year in a nutritious, convenient and attractively packaged form. What was once a necessity has now become a method of food processing which we take for granted.

The principal cause of food going bad is the growth of a variety of organisms including yeasts, moulds and bacteria. The actions of agents called enzymes cause chemical reactions such as oxidation which also spoil food. The type of spoilage varies with the structure of the food and with the action of particular organisms.

Temperature and moisture are the principal factors which control the growth of the organisms, but oxygen content and the availability of nutrients are also important.

The successful preservation of food depends on destroying the micro-organisms usually by heat, or controlling the factors that allow them to grow. This slows down the rate at which the food will spoil.

Drying

The drying of meat or fish was one of the earliest methods of preserving food. Prehistoric man dried meat and fruits in the sun or by the fire. Drying prevents the growth of micro-organisms though some can grow on foods with as little as 16% moisture.

The first hot air dehydrator was developed in France in 1795. As well as air drying other techniques used now include superheated steam, vacuum drying or the use of inert gases.

A wide variety of foods can be dried; from milk and soup mixes to fruit and prepared foods such as pastas. As a powder or a solid, dried foods can be easily packaged, stored and transported with little deterioration. Dried meat, or pemmican stored on Scott's polar expedition in 1911 (see page 38) was brought back 40 years later and found to be quite edible. In recent times some food aid to Third World countries is in the form of dried milk or egg. Drying increases the concentration of nutrients, and although vitamins may be lost in some foods, dried food when reconstituted is generally of equal nutritive value to the fresh product.

Salting

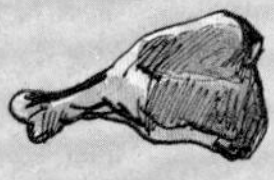

Soaking food in brine or dry-salting preserves food by preventing the growth of micro-organisms and inhibiting enzyme activity. Salt penetrates the tissues and extracts moisture.

The oldest road in Italy is the Salaria (Salt route). Along this route salt from Ostia was taken to other parts of Italy. It was an important item in medieval trade, particularly to the countries of northern Europe which formed the Hanseatic League.

Salting meat or fish often requires a brine to which spices, herbs or honey have been added to impart subtle flavours.

Conserving

Using fruits both in and out of season, by preserving them in sugar as jams or marmalade, has always been popular. The high concentration of sugar, 60% or more, and the removal by boiling of water, prevents the growth of spoilage organisms. Candied fruits are made by impregnating the fruit with increasingly stronger solutions of sugar. Another method of treating fruit is by preserving in alcohol.

Smoking

After salting or air drying some foods, particularly fish and meat, may be smoked; the cool fumes from burning oak, apple or hornbeam chips penetrate the tissue giving a particular flavour. In Scandinavia a method of hot smoking cooks as well as flavours fish and meat, particularly reindeer meat. Smoke houses are a characteristic in some parts.

Milk and butter and cheese and . . .

In many countries half the milk produced is sold as fresh milk, the rest is made into a wide variety of dairy products with better keeping qualities than milk.

Butter is produced by separating off the cream and churning to break up the fat globules. There are a great variety of cheeses; hard, soft, cream or flavoured with many things from walnuts to wine.

In recent years milk based desserts including ice creams, yoghurts and mousses have become very popular. Buttermilk, sour cream and goats cheese are among the wide selection of dairy products available today. These products however need careful storage and some do not keep for very long.

. . .and we preserved our food

from drying and smoking to irradiation

Arbroath smokies. Small haddock that are hot smoked. A Scottish delicacy.

Marrons glacé. Sweet chestnuts preserved with sugar.

Lime pickle. Can be very hot. A favourite in India eaten with curried food.

Feta cheese. This Greek cheese is often made from ewes milk and ripened in brine.

Freezing

The preservation of food by keeping it at low temperatures has a long history. In cold climates meat could be kept in natural ice and cool caves also provided some storage. Ice houses were sometimes made by constructing an underground chamber and lining it with cut blocks of ice. The ice melted slowly and the storage life of the food was prolonged.

Freezing does not destroy micro-organisms but inactivates them, however some enzymes may continue to act slowly so food will not keep indefinitely. Food must be frozen quickly to a temperature of 0°C – −4°C, 32°F – 25°F and stored at −18°C, 0 F for long periods. When food is thawed the processes of spoilage will immediately start up.

There are several commercial methods of freezing which involve cold air blast, direct immersion in a cooling medium or the use of liquid nitrogen and CO_2. The substances used in refrigeration are in part responsible for the CFC gases which are depleting the ozone layer (see page 106), so attempts are being made to modify the processes.

The frozen food cabinet now contains not only frozen peas, but a wide variety of vegetables and meats, as well as complete meals from the basic to the gourmet.

Pickling

Certain foods such as eggs and some vegetables can be preserved by pickling, which involves using a salt stock with an acid, often vinegar. The acid inhibits the action of enzymes. Herbs such as dill may be added to the acidified brine to flavour vegetables such as cucumbers or gherkins. Both green and black olives may be treated in this way and cabbage becomes sauerkraut. Pickling is particularly suited to oily fish such as herring, which can be flavoured with a variety of herbs. Pickles and chutneys are popular in many countries where they are a piquant addition to other foods.

Canning

High temperatures kill the micro-organisms which cause food to go bad, and this method of food preservation is one of the most widespread.

Canning involves putting food in the peak of condition into tins that are then sealed. The tins are then heat sterilized and cooled. In 1810 Nicolas Appart pioneered the preservation of food by heat treating food in sealed glass containers.

Aluminium, and steel with special coatings are now used for cans and plastic materials are increasingly being used to package heat treated foods.

Irradiation

Irradiation preserves food either by destroying the micro-organisms which make it go bad, or by slowing down the ripening process thus prolonging the storage life of certain foods. The process involves using a radioactive source and passing gamma rays through the food. At low doses it does not increase the natural level of radioactivity.

There are some concerns about its safety and effect on the nutrient content of the food. Because it is impossible to identify treated food in the shops, a clear label in addition to a symbol will be used on irradiated goods.

Smörgåsbord

From the delicatessen it is possible to buy food to prepare a typical Swedish smörgåsbord. The smörgåsbord includes a number of dishes which are examples of traditional methods of preservation which have been practised for centuries.

Such foods include salted and raw pickled herring, marinated, salted and smoked salmon, also smoked sausages and dried reindeer meat. The traditional foods which were once essential for survival have become expensive delicacies.

Freeze drying and acoustic drying

Freeze drying is a relatively new method of preserving food which involves a drying process while the food product remains frozen. The liquid in the food escapes as a vapour. It is particularly important to control temperature in meat processing to avoid destroying the protein content, but this is less critical for vegetables and fruit. The process has been particularly successful with instant coffee.

In America a new method of drying sticky substances such as fruit juices and tomato paste to a powder is being developed. Acoustic drying involves using sound waves to keep sticky substances moving whilst warm air can be used to dry them. The process, though noisy, is said to produce a better quality product more cheaply.

Salami Finocchioni. Pure pork meat flavoured with fennel seeds. One of many Italian salamis.

Stilton cheese. Penicillin moulds make the blue veins in this English cheese.

Jambon de Bayonne. Wine is used in the curing of this ham from the Basses-Pyrenees region of France.

Rumtopf. Rum is used to preserve summer fruits in a special jug.

Umeboshi. A Japanese dish of pickled sour plums.

Salary. In the Roman army a salt allowance (salarium) was made to each officer. Later this was converted to money. This is the origin of the word salary.

Micro-organisms. A collective name for the countless minute organisms, including moulds, yeasts and bacteria which are everywhere. Some are harmful, but those used for instance, in brewing and baking are useful.

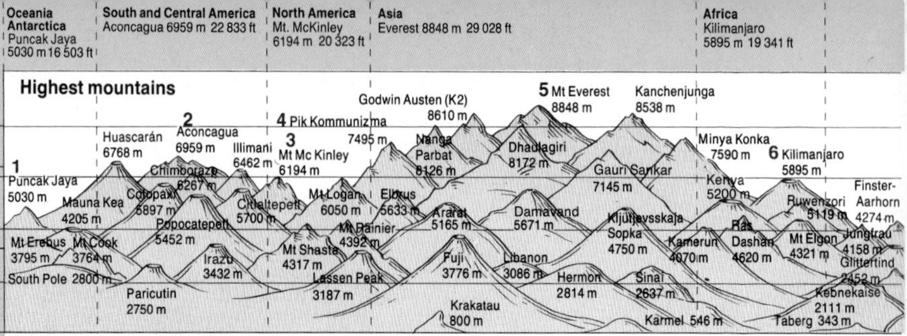

The world's highest peaks rise from the great mountain chains of the world. The Rocky Mountains with the Andes form the American Cordillera, and the Alps and Himalaya are features of the Eurasian mountain chain. More than 100 of the world's peaks over 7000 m, 23 000 feet rise from the Himalaya Karakoram ranges which extend 2500 km, 1500 miles across Asia. The volcanic peaks of Africa are part of the unstable region called the Rift Valley. The word Himalaya is derived from 'snow' and 'home' in Sanskrit. (See page 30.)

Islands and lakes

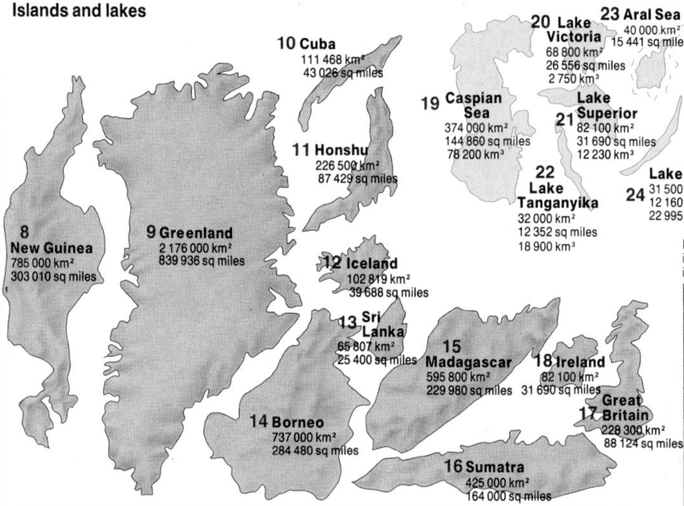

Mount Everest, was surveyed in 1852 and was first climbed in 1953.

From source to sea

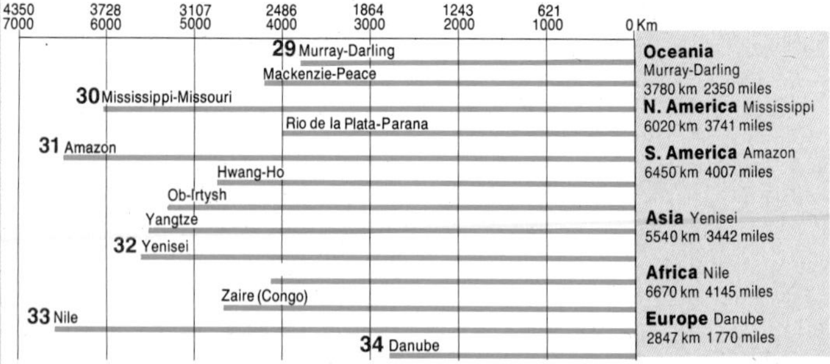

The deep seas

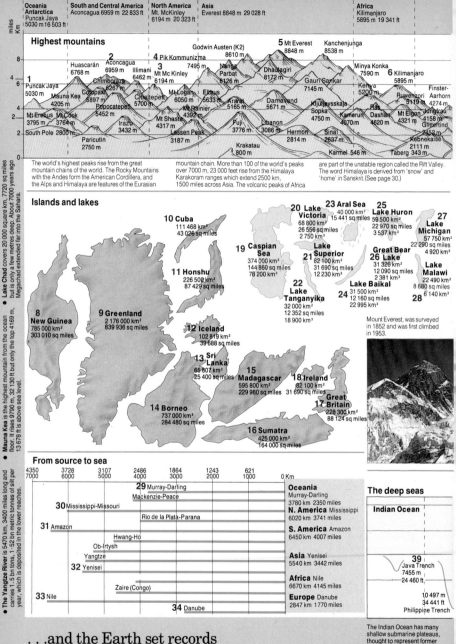

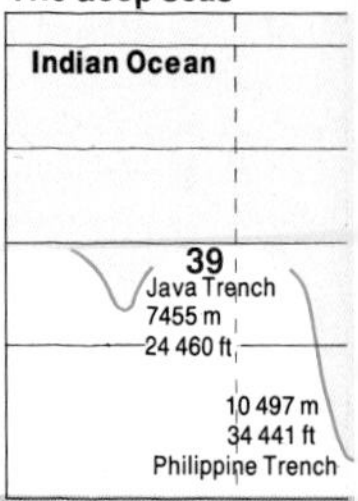

The Indian Ocean has many shallow submarine plateaus, thought to represent former microcontinents.

- **Lake Chad** covers 20 000 square km, 7720 sq miles but is only a few metres deep. About 7000 years ago Megachad extended far into the Sahara.
- **Mauna Kea** is the highest mountain from the ocean floor. It rises 9790 m, 32 130 ft but only the top 4169 m, 13 678 ft is above sea level.
- **The Yangtze River** is 5470 km, 3400 miles long and carries 1·5 bn tons, 1·52 bn metric tonnes of silt per year, which is deposited in the lower reaches.

. . .and the Earth set records

mountains, lakes, rivers

Europe
Mont Blanc
4810 m 15 782 ft

		1 sq mile	= 2·589 km²
1 kilometre	= 0·621 miles	1 km²	= 0·386 sq miles
1 mile	= 1·609 kilometres		
1 metre	= 3·282 feet	1 acre	= 4047 m²

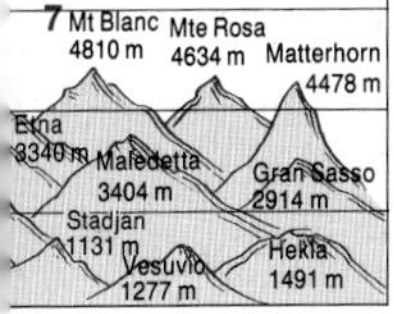

Definitive values

As long ago as 1774, Dr Nevil Maskelyne made the first attempt to weigh planet Earth. He realized that, if the mass of a mountain could be worked out and if a pendulum of known length was held at a known distance from the mountain, its deflection from the vertical would enable the Earth to be weighed. The mountain he chose was Schiehallion in Perthshire, Scotland because of its near conical shape and ease of measurement.

It was in October 1987 that satellite measurements settled the argument over the world's highest mountain. Everest once again reigns supreme over K2 by as much as 238 m, 780 ft. And, in 1960 the United States bathyscape, Trieste descended 10 917 m, 35817 ft into the inky black waters of the Mariana Trench in the Pacific Ocean before hitting the bottom, although a previous Soviet multi-beam sonar estimate had given a depth of 11 033 m, 36 200 ft for this the deepest part of the world's oceans.

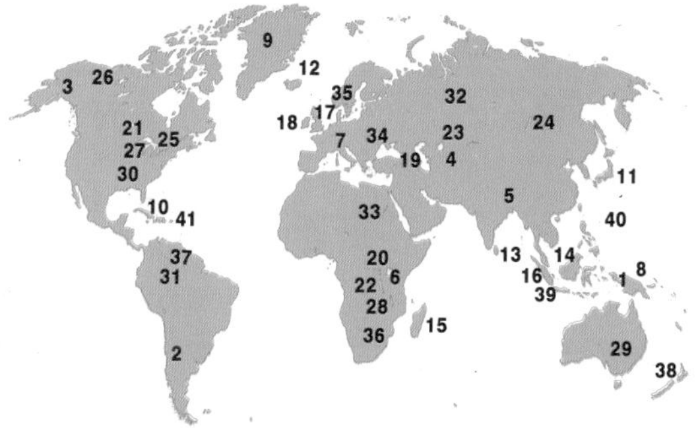

Oceania	N. & S. America	Africa	Europe
Sutherland 580 m 1904 ft	Angel 979 m 3212 ft	Tugela 948 m 3110 ft	Mardalsfossen 657 m 2154 ft

Highest waterfalls

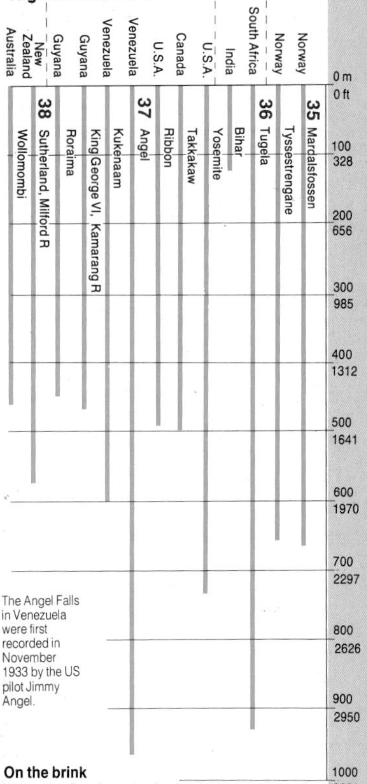

The Angel Falls in Venezuela were first recorded in November 1933 by the US pilot Jimmy Angel.

On the brink

When a river flows from a hard rock to a softer one the water's force carves out an everdeepening channel, while turbulence erodes the banks. Thus at the foot of many falls is a circular pool surrounded by steep banks. A powerful waterfall may wear away the rock so that a cave develops behind it.

Mid ocean ridges indicate the presence of constructive plate margins where material is forced to the surface. This creates pressure at the outer edges. The deepest trenches lie along the margins where ocean crust is subducted beneath an adjacent plate producing great pressure and heat. This is apparent in the evolution of island arcs or lines of volcanoes, such as the Aleutian Islands and the South American volcanoes.

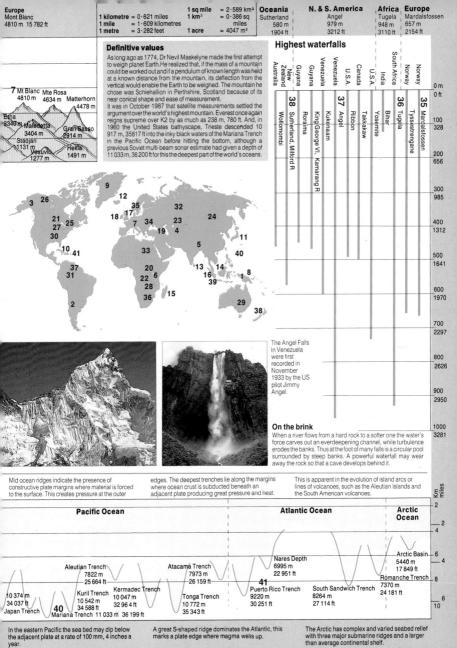

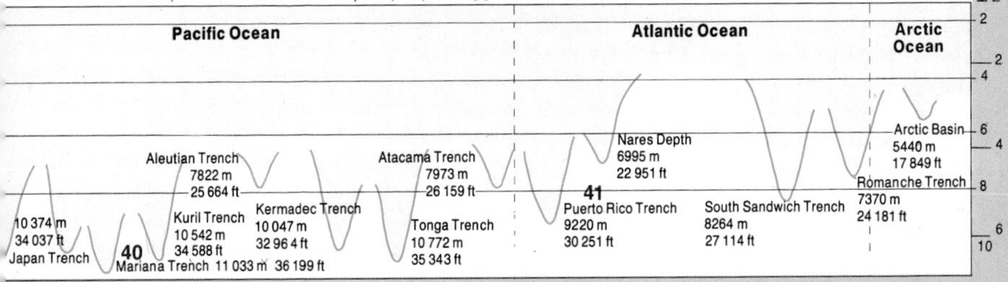

In the eastern Pacific the sea bed may dip below the adjacent plate at a rate of 100 mm, 4 inches a year.

A great S-shaped ridge dominates the Atlantic, this marks a plate edge where magma wells up.

The Arctic has complex and varied seabed relief with three major submarine ridges and a larger than average continental shelf.

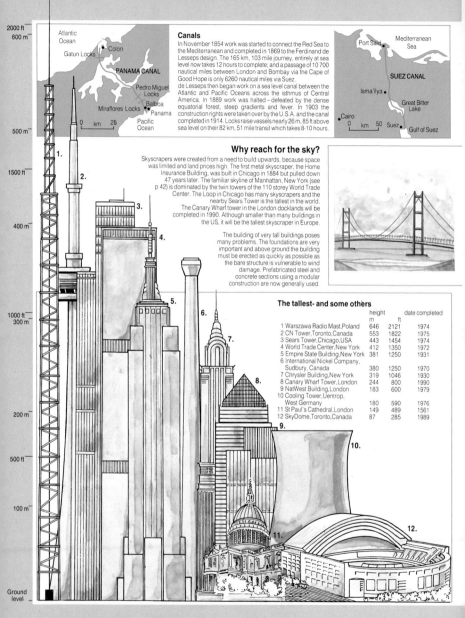

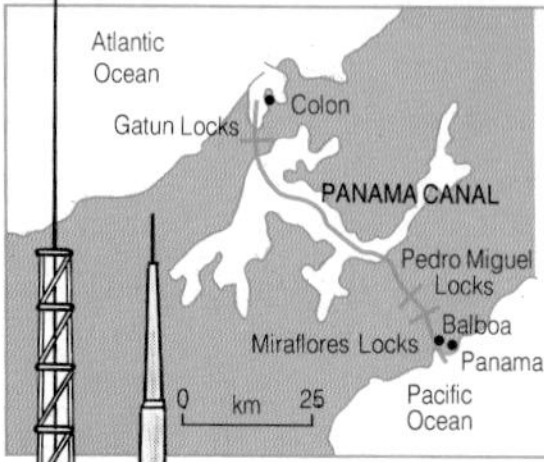

Canals

In November 1854 work was started to connect the Red Sea to the Mediterranean and completed in 1869 to the Ferdinand de Lesseps design. The 165 km, 103 mile journey, entirely at sea level now takes 12 hours to complete; and a passage of 10 700 nautical miles between London and Bombay via the Cape of Good Hope is only 6260 nautical miles via Suez.

de Lesseps then began work on a sea level canal between the Atlantic and Pacific Oceans across the isthmus of Central America. In 1889 work was halted – defeated by the dense equatorial forest, steep gradients and fever. In 1903 the construction rights were taken over by the U.S.A. and the canal completed in 1914. Locks raise vessels nearly 26 m, 85 ft above sea level on their 82 km, 51 mile transit which takes 8-10 hours.

Why reach for the sky?

Skyscrapers were created from a need to build upwards, because space was limited and land prices high. The first metal skyscraper, the Home Insurance Building, was built in Chicago in 1884 but pulled down 47 years later. The familiar skyline of Manhattan, New York (see p 42) is dominated by the twin towers of the 110 storey World Trade Center. The Loop in Chicago has many skyscrapers and the nearby Sears Tower is the tallest in the world. The Canary Wharf tower in the London docklands will be completed in 1990. Although smaller than many buildings in the US, it will be the tallest skyscraper in Europe.

The building of very tall buildings poses many problems. The foundations are very important and above ground the building must be erected as quickly as possible as the bare structure is vulnerable to wind damage. Prefabricated steel and concrete sections using a modular construction are now generally used.

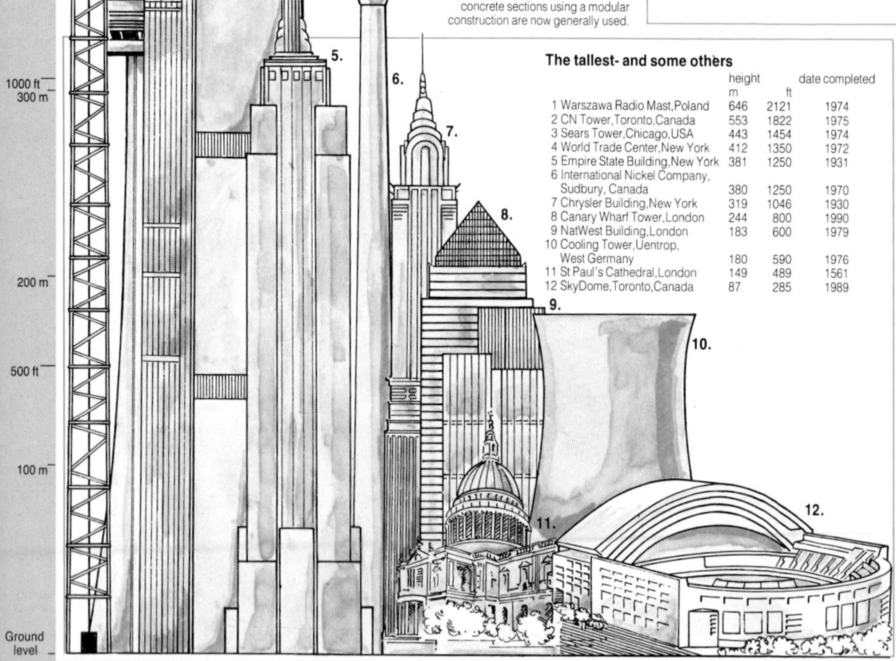

The tallest- and some others

	height m	ft	date completed
1 Warszawa Radio Mast, Poland	646	2121	1974
2 CN Tower, Toronto, Canada	553	1822	1975
3 Sears Tower, Chicago, USA	443	1454	1974
4 World Trade Center, New York	412	1350	1972
5 Empire State Building, New York	381	1250	1931
6 International Nickel Company, Sudbury, Canada	380	1250	1970
7 Chrysler Building, New York	319	1046	1930
8 Canary Wharf Tower, London	244	800	1990
9 NatWest Building, London	183	600	1979
10 Cooling Tower, Uentrop, West Germany	180	590	1976
11 St Paul's Cathedral, London	149	489	1561
12 SkyDome, Toronto, Canada	87	285	1989

Dams and man-made lakes

The Grand Coulee Dam, on the Columbia river in the USA is the largest concrete structure in the world. It is 167m, 550 ft high and contains 8 092 000m³, 10 585 000 cu. yards of concrete. The highest confirmed dam is the Nurek dam in USSR, it is 300 m, 984 ft high. Lake Volta in Ghana is the largest man-made lake, with an area of 8482 km², 3275 sq. miles.

Bridges

The Humber road bridge, opened in 1981, has the world's longest single span. At 1410 m, 4626 ft, it is 112 m longer than that of the Verrazano Narrows Bridge in New York. The Transbay Bridge crossing San Fransciso Bay is 12 km, 7·5 miles long.
One of the oldest suspension bridges still in use was built in 1826 by Thomas Telford. It crosses the Menai strait from Wales to Anglesey and was built with clearance of 33m, 100 ft. This was to allow rigged sailing ships to pass beneath.

Peering into the earth

The deepest working mine is a gold mine at Carletonville, South Africa. This mine is 3777 m, 2·34 miles deep. Bingham Canyon mine in Utah, USA is the deepest open pit mine with a depth of 800 m, 2625 ft. The deepest drilling into the earth's crust is a geological exploration in the Kola Peninsula in USSR. It was begun in 1970, and by 1987 had reached a depth of 13 km, 18 miles. The eventual target depth is 15 km. The temperature at 11 km, 6·4 miles was 200°C, 392°F.

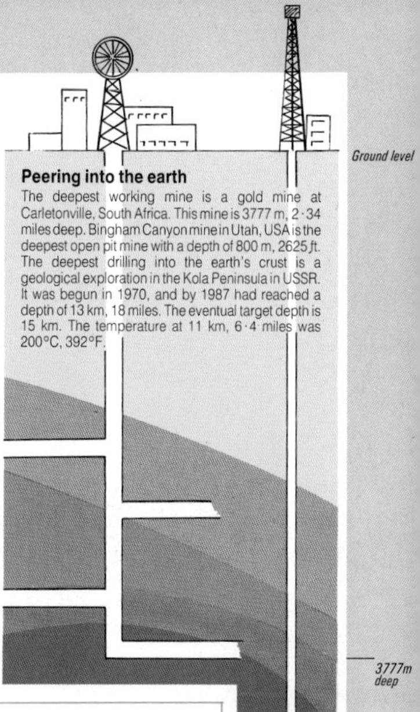

Tunnels

The Euro Tunnel linking Folkestone, England with Coquelles, France will be completed in 1993. The 50 km, 31 miles link beneath the Channel will consist of 3 tunnels, of which two will take trains carrying vehicles, freight and passengers, the central tunnel is for services, ventilation and emergencies.
Tunnelling involves 11 boring machines slicing through 7·5 million cubic metres of chalk marl.

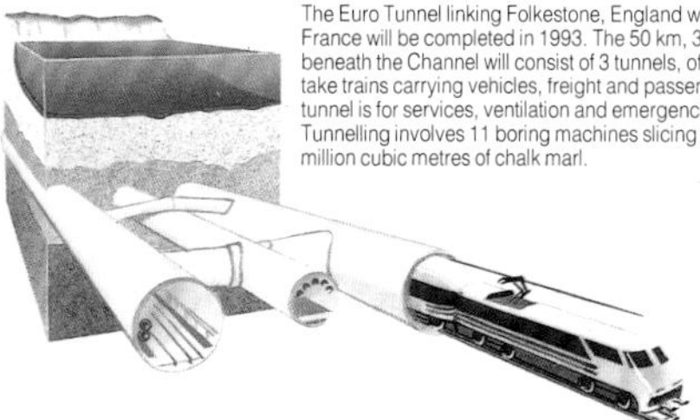

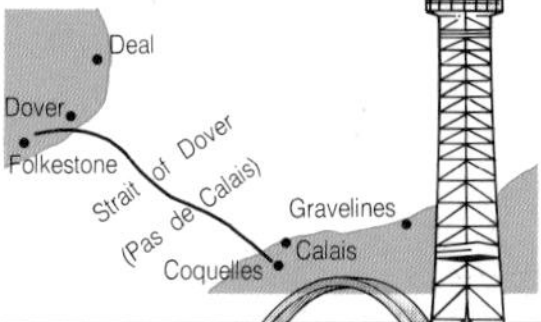

Artwork source: EuroTunnel

The Eiffel Tower and other monuments

'100 ans' was proudly emblazoned on the Eiffel Tower in 1989. 300 skyjacks needed 2·5 million rivets to build it in 1889. It was nearly pulled down in 1909, but its value for communications was realised and in 1916 it was the terminal for the first radio telephone across the Atlantic.

	height m	height ft	date completed
1 Trajan's Column, Italy	33	108	AD 113
2 Arc de Triomphe, France	50	164	1836
3 Totem Pole, Alert Bay, Canada	53	174	
4 Nelson's Column, UK	56	185	1867
5 Motherland Sculpture, USSR	82	26	
6 Statue of Liberty, USA	93	305	1886
7 Pyramid at Giza, Egypt	148	480	c2580 BC
8 Washington Monument, USA	169	555	1884
9 San Jacinto Column, USA	173	570	1939
10 Gateway to the West, USA	192	630	1965
11 Eiffel Tower, France	300	984	1889

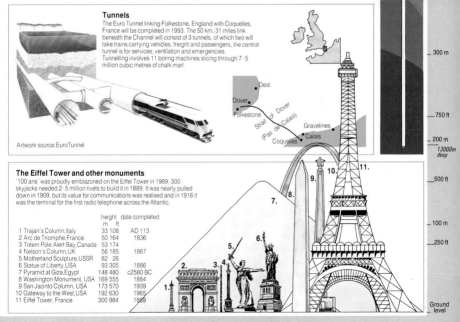

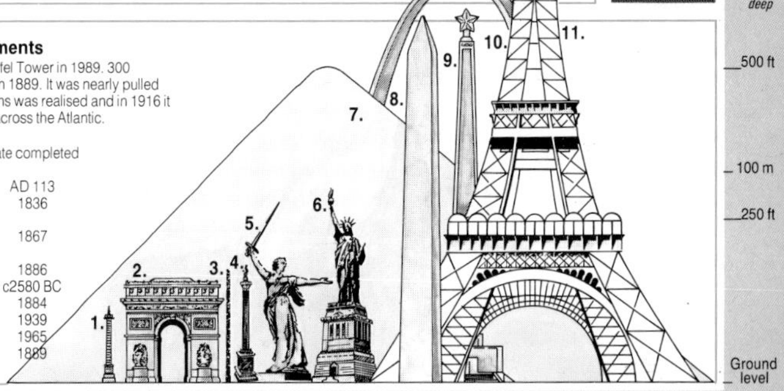

The Great Wall of China is the longest structure ever built, stretching 6400 km, 4000 miles across the northern plain. It can be spotted from the moon by the naked eye.

More than seven wonders

The Seven – or perhaps eight Wonders of the Ancient World were considered the best, the most important or the most spectacular among many other buildings. Some had a religious significance and related to this a quest for immortality. Others reflected more earthly concerns, particularly the safety of shipping for states dominated by trade, or the possession of power and wealth displayed by fine buildings.

In later centuries religion has been the driving force in the creation of many monuments. The wonderful temples of the Orient, the prehistoric works and the Christian cathedrals of Europe, the amazing constructions of the ancient civilizations of Central and South America; all these bear witness to the efforts devoted to honouring the gods. The biggest structure of all, the Great Wall of China, was however built for the mundane purpose of self protection.

What are the wonders of today's world? Possibilities might include shopping complexes, international offices, bridges and dams, sports arenas or even museums. Perhaps in the last of these we show concern for the things of the past which can give a materialistic society an awareness of different values.

The Walls of Babylon
– sometimes the second wonder

These walls and heavily guarded gates were built by King Nebuchadnezzar in 600 B.C., to protect the palaces, temples and quays of his city. They could have been 100 m (335 ft) high. Babylon which stood beside the River Euphrates in what is now Iraq, had been an important city from 2000 B.C. and was wealthier and more magnificent than any other city in Egypt or Asia. The walls of Babylon are sometimes referred to as the second wonder of the ancient world instead of the Hanging Gardens.

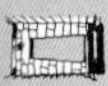

Machu Picchu. This 'lost city' of Peru was built by the Incas fleeing from the Spanish army in the 16th century. The spectacular fortress retreat was discovered by Bingham in the 1920s.

The Colossus of Rhodes
– the fifth wonder

The bronze figure of the sungod Helios was 32 m (105 ft) high, eighteen times as big as a man. It stood beside, probably not astride, the harbour of the Greek city of Rhodes, as is shown in this romanticized print. It was constructed in 340 B.C. and fell in an earthquake in 225 B.C.

The Statue of Zeus – the sixth wonder

Zeus also known as Jupiter was the most important god. This statue of him seated on a richly carved chair, was the centrepiece of the temple in Olympia, home of the Olympic games. It was 12 m (40 ft) high and reached the roof of the temple. The Greek sculptor Pheidas who carved the figure in 480 B.C. embellished it with gold robes, flesh of ivory and precious stones for eyes.

Xian Warriors. The buried terracotta army guarded the grave mound of the first Emperor of China. The distinctive facial features of the 7000 soldiers reflects the ethnic diversity of the Chinese people. This treasure was uncovered in 1974.

The Hanging Gardens of Babylon
– the second wonder

These gardens built by Nebuchadnezzar in 600 B.C. to please one of his queens, were made up of terraces at different levels to form a pyramid. They appeared to be unsupported and hanging in the air. The terraces were perhaps 75 m (300 ft) high with storage tanks at the highest level to water the trees and flowers brought from all the then known world.

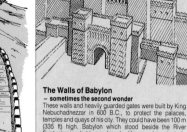

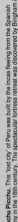

Angkor Wat. Temple-palace of the Khmer kings started in 1200 A.D. Tiered galleries rise to a central tower; the whole a treasure store of statues and bas reliefs.

Stonehenge. Begun in 1800 B.C. and completed by 1400 B.C. The huge stones were arranged in complex patterns based on the movements of the sun.

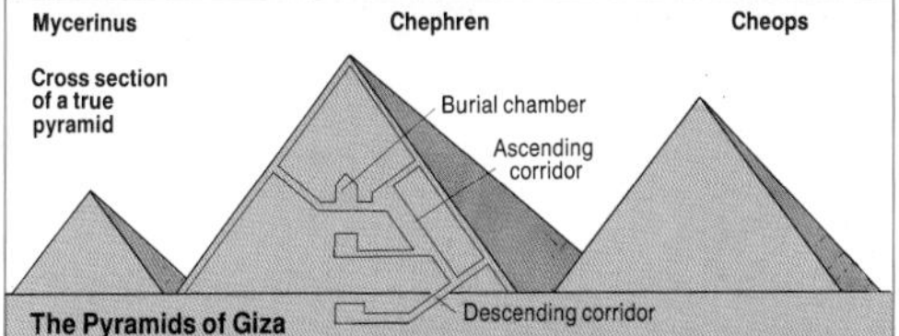

The Pyramids of Giza
– the first wonder

Although there were many pyramids in ancient Egypt, the group of three built at Giza near Cairo were considered the most important of the Wonders of the Ancient World. The Great Pyramid was built for the Egyptian King Kufu about 2580 B.C., but when the Greeks listed the Wonders of the World they turned his name to the Greek form Cheops, a name by which the pyramid is still sometimes known.

The pyramids were constructed over 4 500 years ago as royal burial places, and contained food, drink and personal possessions to ensure comfort in the afterlife.

The sides of the base of the Great Pyramid are 230 m (750 ft) long, and on the area of the base 10 jumbo-jets could be ranged. The pyramid originally reached a height of 155 m (510 ft), though 9 m (30 ft) has now been eroded from the top. The pyramid was originally faced with smooth stones so that the sides were flat. These were later removed to be used for other buildings. The summit of the pyramid was plated in gold when it was built, and it would have glinted in the sun.

The rock core was covered by 230 000 000 blocks of stone each weighing about 2.75 tonnes (3 US tons). This is nearly twice the weight of a modern car. Some of the stones were signed in red ochre by the workmen who quarried and built the monument. The blocks were brought to the site by barge. A quay was built and then a ramp constructed out of resistant granite so that the huge stones could be dragged up the causeway on wooden rollers. The stone was very carefully and accurately cut with copper tools.

The section shows the main burial chamber and the descending corridor by which the workmen left after sealing the ascending corridor with a stone plug.

The Temple of Artemis
– the fourth wonder

The site of this temple was the city of Ephesus near the present town of Sulcuk on the southwest coast of Turkey. Dedicated to Artemis the goddess of fertility and the hunt, the temple was a place of pilgrimage. One solitary pillar remains of the building which once had 127 Ionic columns each 18 m (60 ft). It covered an area four times that of the Parthenon in Athens, and fine sculpture decorated both the inside and outside of the temple. With the spread of Christianity in the 4th century, the temple was abandoned and the marble reused.

Petra in present day Jordan, was the capital of the Nabataean Arabs who settled along the trading routes in the 3rd century B.C. The temples cut in red-brown rock were approached through a narrow ravine only 2 m, 6 ft wide.

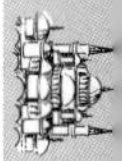

Hagia Sophia. In 532 A.D. the Christian cathedral was built in Constantinople, now Istanbul, but was used as a mosque when the city was captured by the Turks in 1453. The coloured mosaics and the huge dome are magnificent.

The Tomb of Mausolus
– the third wonder

Mausolus was the ruler of Caria, a coastal province in Asia Minor, the modern day Turkey. At his death at Helicarnassus in 353 B.C. his widow Artemisia engaged the finest artists to build a magnificent tomb in memory of her husband. The word mausoleum originates from this gesture. The tomb was about 40 m (130 ft) high and decorated with magnificent friezes on all four sides. The base of the building was in the Carian style, the temple in the Greek style, while the pyramid-shaped roof is Egyptian in character. It survived until the fifteenth century when it collapsed in an earthquake.

The Lighthouse at Pharos
– the seventh wonder

This lighthouse, the first ever built, was constructed in about 250 B.C. on the island of Pharos near the important port of Alexandria in Egypt. This city, on the Mediterranean coast near the Nile delta was one of the leading commercial and academic centres of the period. With a height of 180 m (600 ft) and surmounted by a fire and mirrors, made of bronze sheets that reflected the firelight, the lighthouse would have been invaluable in guiding ships into the harbour. The fire-tower, as it was originally called, was built with 3 towers one above the other. The lower one housed the lighthouse staff as well as soldiers. The second storey had no rooms but the spiral pathway continued up to the top tower where the fire was kept burning brightly. Animals had to be led up the pathway loaded with fuel for the fire and then back down again.

The Seven Wonders of the ancient Greeks have practically all disappeared. However other civilizations and cultures have produced many more than seven wonders; a few are shown here.

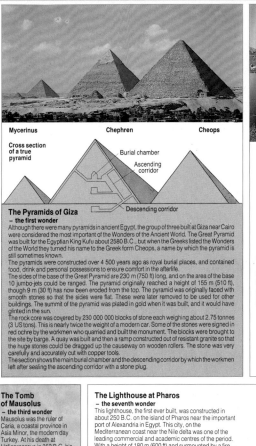

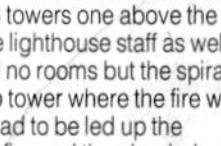

Colosseum. This outdoor theatre in Rome was dedicated in 80 A.D. and for some 300 years was the scene of fights between gladiators or men and animals. There was seating for 45 000 round the four storied arena.

UNESCO

The World Heritage

The World Heritage Convention was adopted by UNESCO in 1972. It is a unique international attempt to preserve the world's most important monuments and natural sites. At present 108 countries belong to the Convention, and so far 315 places worldwide have been chosen for special consideration. The aim is to select those sites that are of such interest and exceptional value that it is the responsibility of the people of the whole world to protect them. These sites are placed on the World Cultural and Natural Heritage List.

Two types of site are recognised, firstly, natural areas and secondly buildings, or groups of buildings; though in many cases there is a strong interaction between the natural and the man-made. Often the most important monuments are complemented and enhanced by the beauty of their natural surroundings. The natural sites possess special geological or biological features, and include habitats of large mammals (Serengeti National Park – see page 37), wetlands with their distinctive animals and plants, and reserves where a new type of environmental management is being practised. These special areas are called Biosphere Reserves and the aim is to reconcile the needs of people while protecting the natural habitats. On the Heritage List there are 5 from the United States, including Yellowstone and the Great Smoky Mountains National Parks.

There are over 200 cultural sites. They are chosen for several reasons, but all are outstanding examples of their kind. Some properties bear witness to vanished cultures, some exemplify architectural excellence and some have special religious or spiritual significance. They include sites such as Mount Taishan in China which combine an exceptional association of natural and cultural features. When a site is chosen by the Committee and put on the List, technical expertise may be available to determine the causes of deterioration, or to plan conservation or restoration measures. Training in conservation may be given to local people. In appropriate cases equipment or materials may be supplied for protecting or restoring a monument. If a site is in particular danger, it may be put on a special list which makes provision for emergencies. The Convention has created a World Heritage Fund which receives obligatory and voluntary contributions from the countries in the Convention. Although much more money is needed the Fund has already supported a number of important conservation schemes in Latin America, Africa, the Caribbean and Europe.

One of the most important functions of the Convention is to heighten public awareness, both nationally and internationally, in order to preserve the past for the future.

The Convention logo symbolizes the interdependence of cultural and natural sites. The square represents a man-made feature and is closely integrated with the circle which represents nature. The whole emblem symbolizes protection.

How the World Heritage Convention works

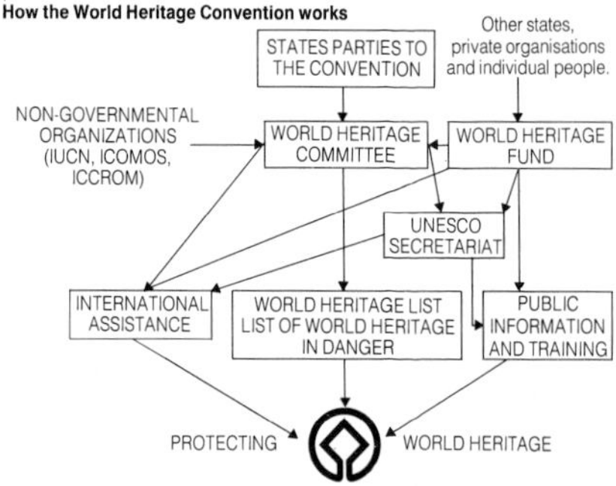

85.Giant's Causeway,Northern Ireland

90.Ironbridge Gorge,England

92.City of Bath,England

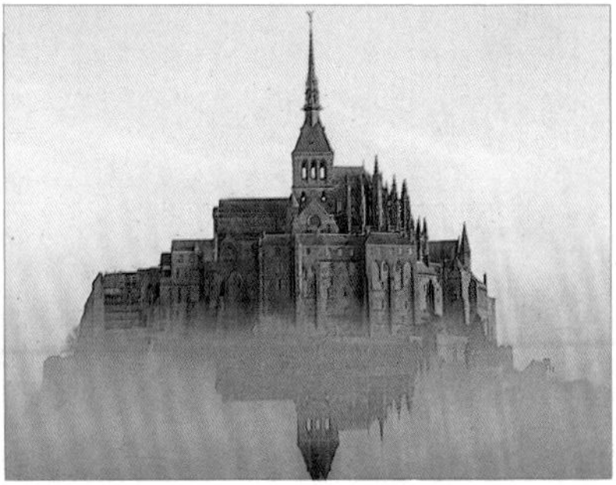

101.Mont St.Michel,France

67 Rock drawings of Alta (Norway)
68 Bryggen area in Bergen (Norway)
69 Roros (Mining town) (Norway)
70 Urnes Stave Church (Norway)
71 Lübeck (Hanseatic city) (Germany)
72 Aachen Cathedral (Germany)
73 Speyer Cathedral (Germany)
74 Pilgrimage Church of Wies (Germany)
75 Würzburg Residence (Germany)
76 Brühl (Augustusburg and Falkenlust Castles (Germany)
77 Trier (Historic monuments) (Germany)
78 St Mary's Cathedral and St Michael's Church at Hildesheim (Germany)
79 Bialowieza National Park (Poland)
80 Auschwitz Concentration Camp (Poland)
81 Cracow Historic Centre (Poland)
82 Warsaw Historic Centre (Poland)
83 Wieliczka Salt Mine (Poland)
84 St Kilda Island (UK)
85 The Giant's Causeway (UK)
86 Hadrian's Wall (UK)
87 Durham (Castle and cathedral) UK)
88 Studley Royal Park and the Ruins of Fountains Abbey (UK)
89 Castles and Town Walls of King Edward in Gwynedd (UK)
90 Ironbridge Gorge (UK)
91 Blenheim Palace (UK)
92 City of Bath (UK)
93 Westminster (Palace and Abbey) and Saint Margaret's Church (UK)
94 The Tower of London (UK)
95 Canterbury Cathedral, St.Augustine's Abbey and St.Martin's Church (UK)
96 Stonehenge, Avebury and related megalithic sites (UK)
98 Girolata and Porto Gulfs and Scandola Reserve (France)
99 Amiens Cathedral (France)
100 Versailles (Palace and park) (France)
101 Mont St Michel and its bay (France)
102 Chartres Cathedral (France)
103 Fountainebleau (France)
104 Places Stanislas, Carrière and Alliance in Nancy (France)
105 Chambord (Chateau) (France)
106 Vézelay (Basilica and hill) (France)
107 Cistercian Abbey of Fontenay (France)
108 Arc-et-Senans (Royal salt works) (France)
109 Strasbourg – Grand Ile (France)
110 St Savin-sur-Gartempe (France)
111 Decorated caves of the Vezère (France)
112 Orange (Roman theatre and triumphal arch) (France)
113 Pont du Gard (France)
114 Arles (Roman and Romanesque monuments) (France)
116 Convent of Christ in Tomar (Portugal)
117 Monastery of Batalha (Portugal)
118 Monastery of the Hieronymites and Tower of Belem (Portugal)
119 Evora (Historic centre) (Portugal)
120 Santiago de Compostela (Spain)
121 Altamira Cave (Spain)
122 Churches of the Kingdom of the Asturias (Spain)
123 Parque and Palacio Güell and Casa Milá in Barcelona (Spain)
124 Burgos Cathedral (Spain)
125 Salamanca (Old city) (Spain)
126 Segovia and its aqueduct (Spain)
127 Old town of Avila with its extra-muros Churches (Spain)
128 The Escurial (Monastery) (Spain)
129 Teruel (Mudejar architecture) (Spain)
130 Toledo (Historic town) (Spain)
131 Caceres (Old town) (Spain)
132 The Mosque of Cordoba (Spain)
133 The Alhambra and the Generalife in Granada (Spain)
134 Sevilla (Cathedral, Alcazar and Archivo de Indias) (Spain)
136 Berne (Old city) (Switzerland)
137 Convent of St Gall (Switzerland)
138 Benedictine Convent of Müstair (Switzerland)
139 Rock drawings of Valcamonica (Italy)
140 Venice and its lagoon (Italy)
141 Santa Maria delle Grazie with 'The Last Supper', of Leonardo da Vinci (Italy)
142 Florence (Historic centre) (Italy)
143 Pisa (Piazza del Duomo) (Italy)
144 Rome (Historic centre) (Italy)
145 The Vatican City (Holy City, Vatican)
146 Durmitor National Park (Yugoslavia)
147 Skocjan Caves (Yugoslavia)
148 Plitvice Lakes National Park (Yugoslavia)
149 Dubrovnik (Old city) (Yugoslavia)
150 Studenica Monastery (Yugoslavia)
151 Stari Ras and Sopocani (Yugoslavia)
152 Split (Historic centre with Diocletian Palace) (Yugoslavia)
153 Kotor and its gulf (Yugoslavia)
154 Ohrid and its lake (Yugoslavia)
155 Budapest (Banks of the Danube with the district of Buda Castle) (Hungary)
156 Hollōko (Traditional village) (Hungary)
157 Delphi (Archaeological site) (Greece)
158 Epidaurus (Archaeological site) (Greece)
159 Mount Athos (Greece)
160 Thessaloniki (Paleochristian and Byzantine monuments) (Greece)
161 The Acropolis, Athens (Greece)
162 Rhodes (Medieval city) (Greece)
163 Meteora (Greece)
164 Temple of Apollo Epicurius at Bassae (Greece)
165 Srebarna Nature Reserve (Bulgaria)
166 Thracian tomb of Svechtari (Bulgaria)
167 Ivanovo Rock-hewn Churches (Bulgaria)
168 Madara Rider (Bulgaria)
169 Nessebar (Old city) (Bulgaria)
170 Boyana Church (Bulgaria)
171 Rila Monastery (Bulgaria)
172 Kazanlak Thracian Tomb (Bulgaria)
173 Pirin National Park (Bulgaria)
174 Historic areas of Istanbul (Turkey)
175 Hierapolis-Pamukkale (Turkey)
176 Hattusha (Hittite city) (Turkey)
177 Göreme National Park and the rock sites of Cappadocia (Turkey)
178 Great Mosque and hospital of Divrigi (Turkey)
179 Nemrut Dag (Turkey)
180 Xanthos-Letoon (Turkey)
181 Paphos (Archaeological site) (Cyprus)
182 Painted churches in the Troodos region (Cyprus)
183 Valetta (Old city) (Malta)
184 Hal Saflieni Hypogeum (Malta)
185 Ggantija Temples (Malta)
186 Medina of Fez (Morocco)
189 Djemila (Roman ruins) (Algeria)
190 Al Qal'a of Beni Hammad (Algeria)
191 Timgad (Roman ruins) (Algeria)
192 Tipasa (Archaeological site (Algeria)
195 Ichkeul National Park (Tunisia)
196 Amphitheatre of El Djem (Tunisia)
197 Kerkwan (Punic town necropolis) (Tunisia)
198 Carthage (Archaeological site) (Tunisia)
199 Medina of Tunis (Tunisia)
200 Medina of Sousse (Tunisia)
201 Kairouan (Tunisia)
251 Aleppo (Old city) (Syria)
256 Hatra (Iraq)

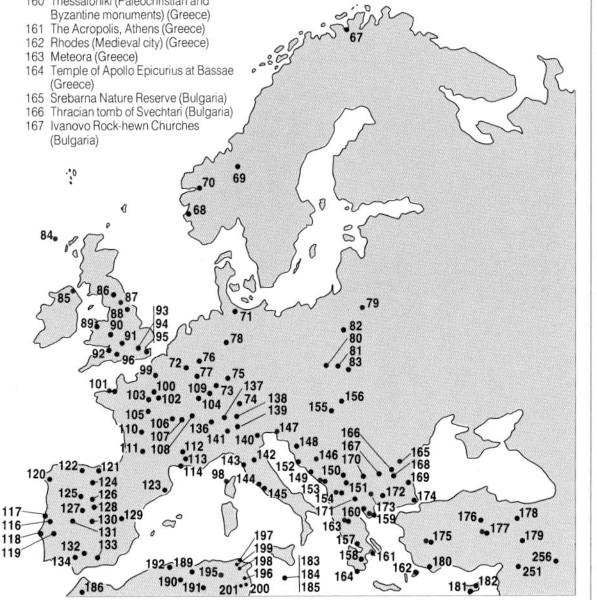

IUCN:- International Union for Conservation of Nature and Natural Resources
ICOMOS:- International Council for Monuments and Sites
ICCROM:- International Centre for Conservation in Rome

1 Nahanni National Park (Can)
2 Wood Buffalo National Park (Can)
3 Canadian Rocky Mountains Parks (Can)
4 Head-Smashed-in Bison Jump (Can)
5 Anthony Island (Can)
6 L'Anse aux Meadows Historic Park (Can)
7 Gros Morne National Park (Can)
8 Dinosaur Provincial Park (Can)
9 Quebec (Historic Area) (Can)
10 Kluane/Wrangell-St Elias Parks (Can/USA)
11 Olympic National Park (USA)
12 Redwood National Park (USA)
13 Yosemite National Park (USA)
14 Yellowstone National Park (USA)
15 Grand Canyon National Park (USA)
16 Mesa Verde National Park (USA)
17 Chaco National Historical Park (USA)
18 Cahokia Mounds Site (USA)
19 Mammoth Cave National Park (USA)
20 The Statue of Liberty (USA)
21 Independence Hall (USA)
22 Monticello and the University of Virginia in Charlottesville (USA)
23 Great Smoky Mountains Nat. Park (USA)
24 Everglades National Park (USA)
25 Puerto Rico (Historic site) (USA)
26 Hawaii Volcanoes National Park (USA)
27 Old Havana and Fortifications (Cuba)
28 Trinidad, Valley de los Ingenios (Cuba)
29 Citadel, Sans-Souci and Ramiers Historic Park (Haiti)
30 Mexico (Centre and Xochimilco) (Mex)
31 Teotihuacan (Pre-Hispanic city) (Mex)
32 Puebla (Historic centre) (Mex)
33 Oaxaca (Historic Zone) and Monte Alban (Archaeological site) (Mex)
34 Palenque (Pre-Hispanic city) (Mex)
35 Sian Ka'an (Biosphere Reserve) (Mex)
36 Chichen-Itza (Pre-Hispanic city) (Mex)
37 Guanajuato (Historic town/adjacent mines) (Mex)
38 Antigua Guatemala (Guatemala)
39 Quirigua (Archaeological Park and ruins) (Guatemala)
40 Tikal National Park (Guatemala)
41 Maya Ruins of Copan (Honduras)
42 Río Platano Reserve (Honduras)
43 Talamanca Range and La Amistad Reserve (Costa Rica)
44 Darien National Park (Panama)
45 Portobello and San Lorenzo Fortifications (Panama)
46 Cartagena (Colombia)
47 Galapagos Islands Nat. Park (Ecuador)
48 Sangay National Park (Ecuador)
49 Quito (Old city) (Ecuador)
50 Chan Chan (Archaeological area) (Peru)
51 Huascaran National Park (Peru)
52 Chavin (Archaeological site) (Peru)
53 Convent Ensemble of San Francisco de Lima (Peru)
54 Machu Picchu (Historic sanctuary) (Peru)
55 Manu National Park (Peru)
56 Cuzco (Old city) (Peru)
57 Potosi (Mining town) (Bolivia)
58 Olinda (Historic centre) (Brazil)
59 Brasilia (Brazil)
60 Centre of Salvador de Bahia (Brazil)
61 Ouro Preto (Historic town) (Brazil)
62 Sanctuary of Bom Jesus do Congonhas (Brazil)
63 Iguacu National Park (Brazil)
64 Jesuit Missions of the Guaranis (Argentina/Brazil)
65 Los Glaciares National Park (Argentina)
66 Iguazu National Park (Argentina)
97 Henderson Island (UK)
115 Angra do Heroismo (Azores, Portugal)
135 Garajonay National Park (Canary Islands) (Spain)
187 The Medina of Marrakesh (Morocco)
188 Aït-Ben-Haddou (Morocco)
193 M'Zab Valley (Algeria)
194 Tassili n'Ajjer (Algeria)
202 Cyrene (Archaeological site) (Libya)
203 Leptis Magna (Libya)
204 Sabratha (Archaeological site) (Libya)
205 Ghadames (Old town) (Libya)
206 Rock-art sites of Tadrart Acacus (Libya)
207 Abu Mena (Christian ruins) (Egypt)
208 Ancient Thebes necropolis (Egypt)
209 Islamic Cairo (Egypt)
210 Memphis necropolis/Pyramid fields (Egypt)
211 Abu Simbel to Philae (Egypt)
212 Old towns of Djenné (Mali)
213 Timbuktu (Mali)
214 Djoudj Bird Sanctuary (Senegal)
215 Niokolo-Koba National Park (Senegal)
216 Gorée Island (Senegal)
217 Mount Nimba (Guinea/Ivory Coast)
218 Comoe National Park (Ivory Coast)
219 Tai National Park (Ivory Coast)
220 Ashante Traditional Buildings (Ghana)

259. Old city at San'a, Yemen

238. Ngorongoro Plain, Tanzania

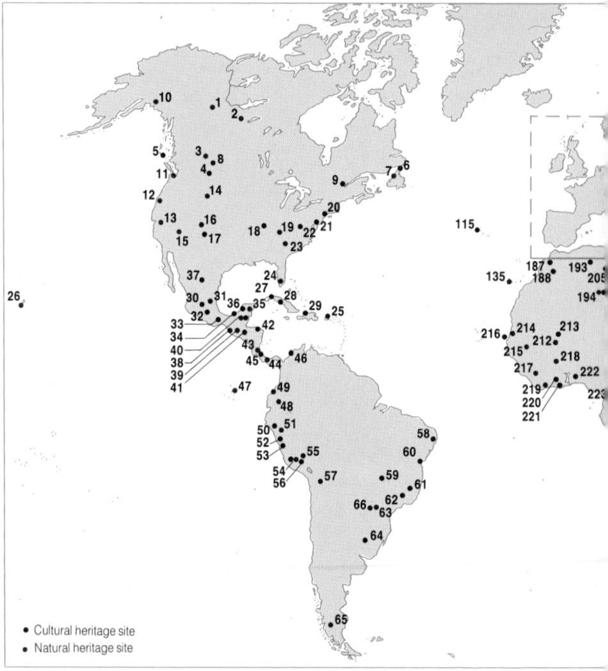

. . .and we protected our heritage

– UNESCO World Heritage Convention THE WORLD

40.Tikal National Park,Guatemala

294.Sigiriya(Ancient city),Sri Lanka

221 Forts and castles of Ghana (Ghana)
222 Royal palaces of Abomey (Benin)
223 Dja Faunal Reserve (Cameroon)
224 Parc national du Manovo-Gounda St.Floris (Central African Republic)
225 Simen National Park (Ethiopia)
226 Aksum (Archaeological site) (Ethiopia)
227 Fasil Ghebbi/Gondar monuments (Ethiopia)
228 Lalibela Rock-hewn Churches (Ethiopia)
229 Tiya (Carved steles) (Ethiopia)
230 Awash Lower Valley (Ethiopia)
231 Omo Lower Valley (Ethiopia)
232 Garamba National Park (Zaire)
233 Kahuzi-Biega National Park (Zaire)
234 Virunga National Park (Zaire)
235 Salonga National Park (Zaire)
236 Serengeti National Park (Tanzania)
237 Kilimanjaro National Park (Tanzania)
238 Ngorongoro Area (Tanzania)
239 Selous Game Reserve (Tanzania)
240 Kilwa Kisiwani and Songo Mnara Ruins (Tanzania)
241 Lake Malawi National Park (Malawi)
242 Mana Pools, Sapi and Chewore Reserves (Zimbabwe)
243 Khami Ruins (Zimbabwe)
244 Great Zimbabwe Ruins (Zimbabwe)
245 Aldabra Atoll (Seychelles)
246 Vallée de Mai Reserve (Seychelles)
247 Anjar (Archaeological site) (Lebanon)
248 Baalbek (Lebanon)
249 Byblos (Lebanon)
250 Tyr (Archaeological site) (Lebanon)
252 Bosra (Ancient city) (Syria)
253 Damascus (Old city) (Syria)
254 Palmyra (Archaeological site) (Syria)
255 Jerusalem (Old city and its walls)
257 Quseir Amra (Jordan)
258 Petra (Jordan)
259 Sana'a (Old city) (Yemen)
260 Shibam Old Walled City ((PDY)
261 Bahla Fort (Oman)
262 Bat,Al-Khutm and Al-Ayn (Oman)
263 Esfahan (Meidam Emam) (Iran)
264 Persepolis (Iran)
265 Tchoga Zambil (Iran)
266 Lahore (Fort and Gardens) (Pakistan)
267 Thatta (Historical monuments(Pakistan)
268 Mohenjo Daro (Archaeological site) (Pakistan)
269 Takht-i-Bahi Buddhist Ruins (Pakistan)
270 Taxila (Archaeological remains) (Pakistan)
271 Sagarmatha National Park (Nepal)
272 Kathmandu Valley (Nepal)
273 Chitwan National Park (Nepal)
274 Keoladeo National Park (India)
275 Agra Fort (India)
276 Taj Mahal (India)
277 Fatehpur Sikri (Mongol city) (India)
278 Khajuraho monuments (India)
279 Ellora Caves (India)
280 Ajanta Caves (India)
281 Elephanta Caves (India)
282 Goa (Churches and convents) (India)
283 Monuments at Pattadakal (India)
284 Hampi monuments (India)
285 Brihadisvara Temple, Thanjavur (India)
286 Mahabalipuram (Monuments) (India)
287 The Sun Temple at Konarak (India)
288 Sundarbans National Park (India)
289 Kaziranga National Park (India)
290 Manas Wildlife Sanctuary (India)
291 Nanda Devi National Park (India)
292 Buddhist Vihara (Bangladesh)
293 Bagerhat (Bangladesh)
294 Sigiriya (Ancient city) (Sri Lanka)
295 Polonnaruwa (Ancient city) (Sri Lanka)
296 Anuradhapura (Sacred city) (Sri Lanka)
297 Sinharaja Forest Reserve (Sri Lanka)
298 Sacred City of Kandy (Sri Lanka)
299 Galle (Town and fortifications) (Sri Lanka)
300 Imperial Palace of the Ming and Qing Dynasties (China)
301 Peking Man Site at Zhoukoudian (China)
302 Mount Taishan (China)
303 The Great Wall (China)
304 The Mausoleum of the First Qin Emperor (China)
305 Mogao Caves (People's Republic of China)
306 Kakadu National Park (Australia)
307 Wet Tropics of Queensland (Australia)
308 Great Barrier Reef (Australia)
309 East Coast Temperate and Sub-tropical Forests (Australia)
310 Lord Howe Islands (Australia)
311 Uluru National Park (Australia)
312 Willandra Lakes Region (Australia)
313 Western Tasmania National Parks (Australia)
314 Westland and Mount Cook National Parks (New Zealand)
315 Fiordland National Park (New Zealand)

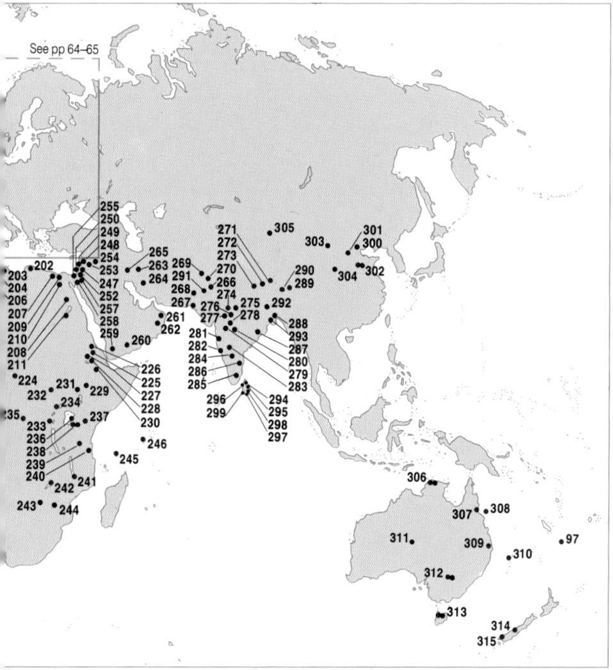

Physical exercise is important for health, and in the prevention of many illnesses. Energy used is given in kilocalories/minute.

Activity	kcal/min
• Walking	4
Climbing stairs	8
Cross-country walking	7
• Mountain climbing	10
Bicycling	8
Dancing	6
• Jogging	8
Sprinting	14
Water skiing	6
• Swimming	8
Downhill skiing	9
Cross-country skiing	12
• Skating	7
Tennis (singles)	7
Tennis (doubles)	6

Areas where malaria could occur in the past

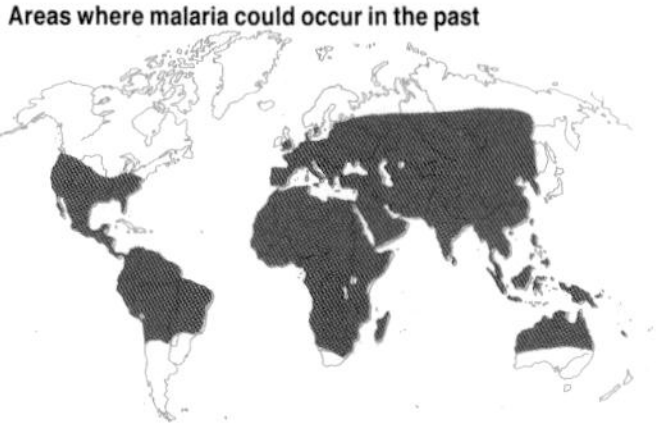

Malaria transmission occurs 1980

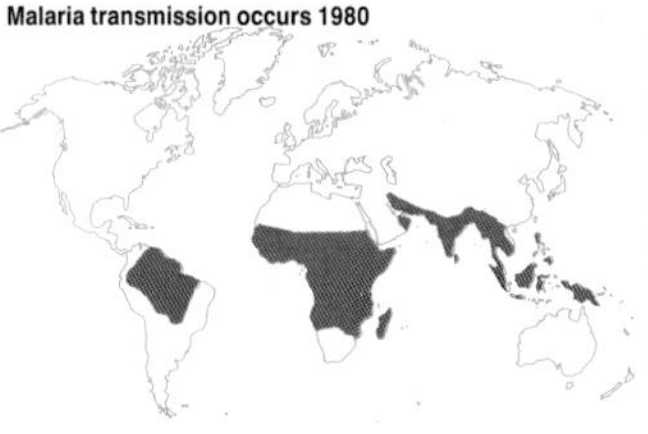

Malaria

Malaria is widespread but particularly prevalent in the Tropics. It is carried by the Anopheles mosquito. An intensive DDT spraying programme from 1955, has meant that some areas have become virtually malaria free. DDT has become ineffective against the mosquito and increased resistance of the disease to drug treatment has resulted in a resurgence.

1967:Smallpox endemic or imported

1976:Smallpox endemic or imported

Smallpox

A disease of great antiquity, smallpox was at its most devastating during the 17th and 18th centuries. Jenner discovered a vaccine in the late 18th century. In 1966 the Intensive Smallpox Eradication programme was launched. The disease was eradicated in 1978, and there is debate whether the laboratory virus should be destroyed.

Epidemic waves

Diseases are either infectious or noninfectious. The latter, group includes heart disease, cancer and many occupational and nutritional diseases, as well as disorders that a baby is born with.

Infectious diseases are the commonest and cover many illnesses which are caused by bacteria and viruses, gonorrhea, measles, influenza, the common cold and AIDS are just a few.

Important changes have taken place in the pattern and incidence of disease. In developed countries typhoid and tuberculosis caused many deaths in the early 19th century, but heart disease and other noninfectious illnesses have now replaced them.

Historically, diseases have been spread by trade and by war; and where soldiers, merchants or pioneers have moved among people who had never had contact with European diseases, the result has often been catastrophic. The original inhabitants of North and South America, as well as the native people of Oceania, were decimated by such infections as measles.

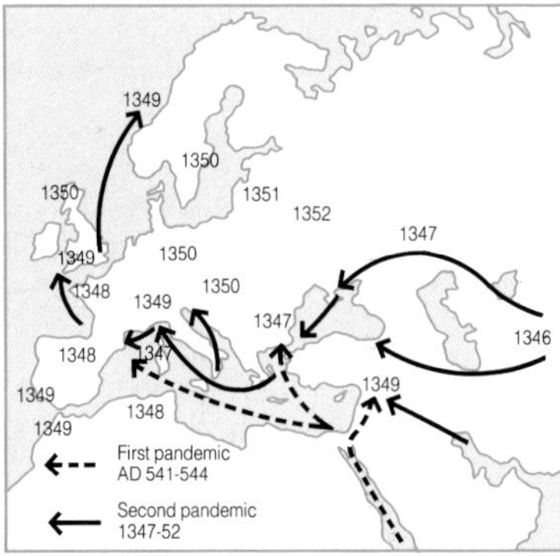

Plague -the first and second pandemic

The deadly plague is passed to man by fleas from rats. In the 6th century the first pandemic swept Europe and populations were decimated. Again several centuries later plague devastated Europe and many millions died particularly in the cities. In the 19th century the disease again spread across the world. The use of sulpha drugs and antibiotics has now brought the disease under control.

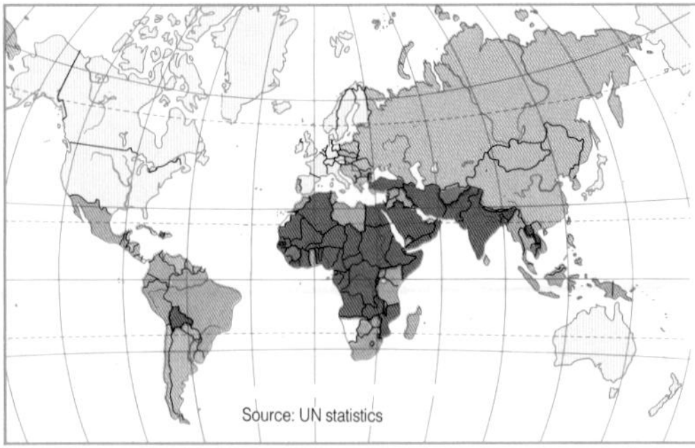

Infant mortality

The survival rate of young babies is one of the best indicators of the general health of a country. Over 4 million children under 5 die each year, and many deaths are related to intestinal infections and illnesses associated with malnutrition, unsafe water and poor sanitation.

Infectious diseases can often be prevented; and a WHO campaign is under way to immunize all children against 6 common diseases by 1990 – in 1977 only 5% were so treated. Many deaths could also be prevented by cheap ORT. Better health education and fewer infant deaths should lead to smaller families. In developed countries a small but significant number of young babies die from sudden 'cot death'. A combination of circumstances, not yet fully understood, appears to cause these tragedies.

Deaths per 1000 in first year of life

- 0-15
- 15-30
- 30-50
- 50-100
- 100-150
- >150
- No data available

. . .and plagues were sent upon us

development of world major diseases

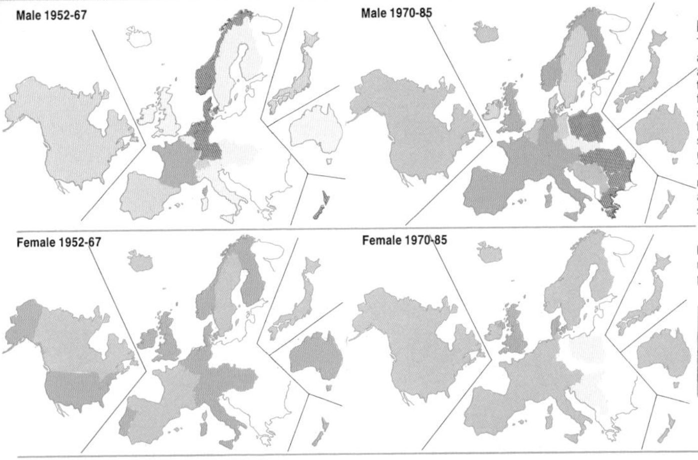

Heart Disease

This is primarily a disease of affluent countries where it accounts for more than half the deaths. The death rates for men have declined significantly since the 1960s. For women there has been a steady decrease in the death rate since 1950.
The incidence of the disease may be decreased by attention to diet, exercise, stress and smoking.

Percentage change in death rate from heart disease for people aged 30-69

>30%
15-30% } Increase
0-15%
0-15%
15-30% } Decrease
>30%

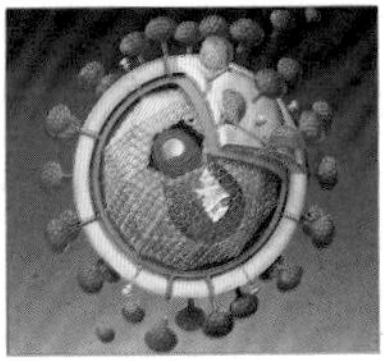

HIV

The agent of AIDS is the Human Immunodeficiency Virus. It has chemical outer layers with protein and a fatty layer with proteins and sugars embedded. Together with the genetic material at the centre are molecules of an enzyme called *reverse transcriptase* which causes the virus to multiply.

AIDS – history repeats itself

The earliest evidence of AIDS comes from blood collected in Central Africa in 1959. However, since 1981 when AIDS was formally identified it has spread rapidly through the world, and has become the modern pandemic. The virus is spread chiefly through sexual contact, intravenous drug use and through transfusion of infected blood. WHO estimates that there are 500 000 cases and suggests that 5 to 10 million are HIV infected. Numbers may rise to 20 million over the next decade. Most will not develop symptoms of the disease for several years, but it is believed that virtually all HIV infected people will eventually develop AIDS.
Earlier pandemics sweeping the world affected the entire population, but AIDS strikes at young and middle aged adults. The potential loss of the economically most important section of society raises serious concern for social, economic and political stability in some countries.
It is relatively easy to avoid contracting AIDS and as there is as yet no known cure, prevention is best.

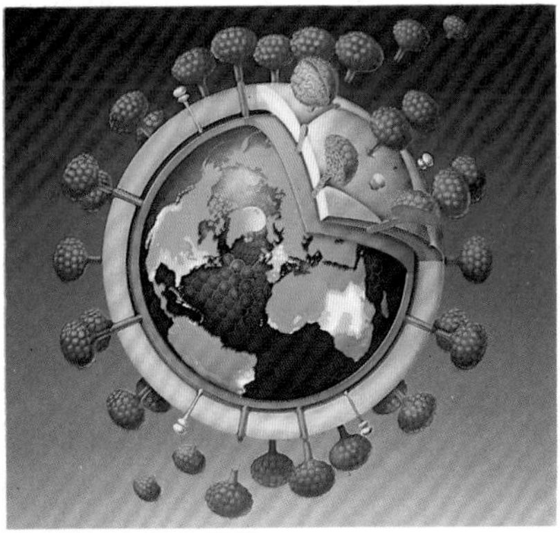

The basic structure of the Human Immunodeficiency Virus (HIV), the causative agent of AIDS. A map of the world showing regions with a high incidence of AIDS in yellow is superimposed over the protein core shell

Causes of death among young children

Estimated total of 5 million deaths of children under 5 years old

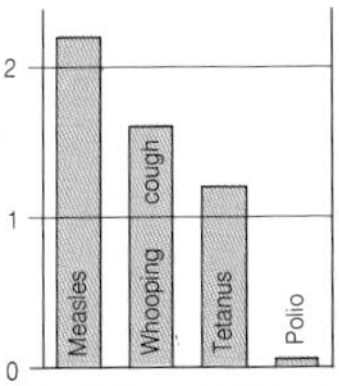

The AIDS Pandemic

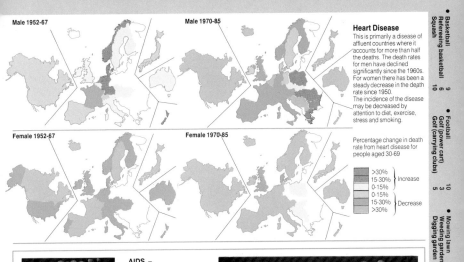

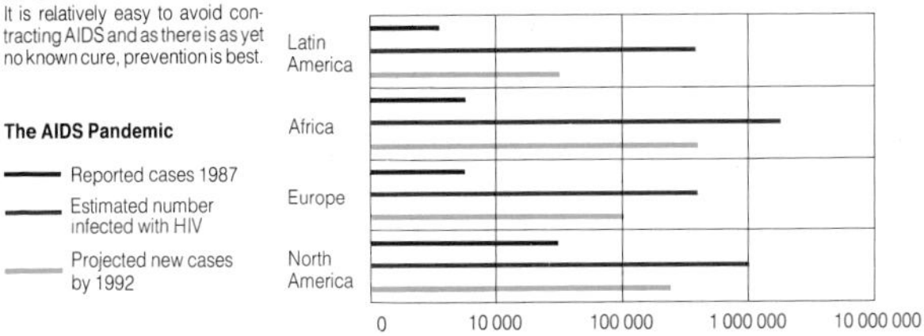

- Basketball 9
  Refereeing basketball 6
  Squash 10
- Football 10
  Golf (power cart) 3
  Golf (carrying clubs) 5
- Mowing lawn 4
  Weeding garden 5
  Digging garden 6
- Shovelling snow 7
  Wood chopping 8
  Carpentry and painting 4
- Fishing 3
  Hunting 4
  Horse riding 6
- Car washing 5
  Snowshoeing 13
  Table tennis 4

Expanded Programme of Immunization 1974. Immunization for all children against tetanus, measles, polio, whooping cough, diphtheria and tuberculosis.

ORT. Oral Rehydration Therapy is a cheap and simple procedure for restoring body fluids lost through intestinal infections.

Pandemic. A disease which is prevalent over a whole continent or over the whole world. An epidemic is limited to a smaller area.

1902 First hearing aid produced, but it was too large to be worn
1923 Banting and Macleod receive Nobel prize for work on insulin
1924 Electrocephalograph

1927 'Iron lung'
1929 Heart-lung machine
1930s Electron microscope

1887 Electrocardiograph
1891 Incubators first used in France
1901 Discovery of blood groups

1600 Clinical thermometer
1851 Hypodermic syringe introduced into France
1869 Modern clinical thermometer containing mercury

Some more of the major medical inventions and discoveries.

Pasteur at work

Antibiotics.

In 1928, a discovery was made which revolutionised the treatment of bacterial disease. Alexander Fleming (1881-1955) found that a mould had grown on a culture in his laboratory, which had killed the bacteria, but left the healthy cells untouched. Folk medicine had used mouldy bread and cobwebs to treat festering wounds but this had never been accepted by conventional medicine. It took over ten years before Fleming's discovery, penicillium, was made stable enough to be used as a drug, but it was available for widespread use during the Second World War.

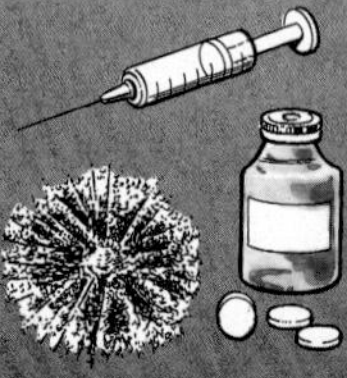

Vaccines

The principles of vaccination had been understood at least partially in China many centuries before Edward Jenner (1749-1823) experimented with a smallpox vaccine. The Chinese discovered that, if they blew the powdered scab of a smallpox victim up the nose of someone who hadn't had the disease, then they would acquire some immunity. Jenner found that cowpox victims were unlikely to catch smallpox, so he deliberately injected a boy with cowpox. A few months later, the boy was found to be immune to smallpox. Other vaccines were later developed which wiped out many highly infectious and debilitating diseases, such as tuberculosis and polio.

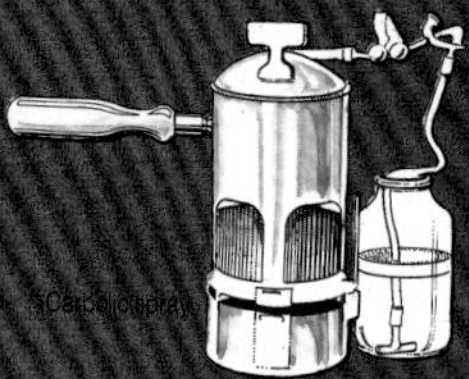

Carbolic spray

Antiseptics

Ignaz Semmelweis, working in the mid-nineteenth century, was one of the first doctors to connect disease and hygiene. When he used chlorinated water in his maternity wards he noticed a marked fall in the incidence of puerperal fever. Joseph Lister (1827-1912) pioneered the introduction of antiseptics into the hospital routine, although it took many years before his ideas gained general acceptance. He noticed that simple fractures healed easily, while compound fractures which broke the skin, were more likely to fester. He began to use carbolic sprays in the operating theatre to sterilise everything that came into contact with the wound and the wounds healed perfectly.

The 'Father of Medicine'.

Hippocrates (born on the Greek Island of Kos, 460 B.C.) helped medical practise to break away from its ties with magic and religion. He believed that illness was a natural biological event, not based on the whims of gods or spirits, and that the body had a natural capacity to heal itself. He stressed the importance of observation and recording symptoms and medical histories. But Hippocrates' scientific approach was impeded by another Greek, who lived in Rome. Galen (c.131-200 A.D.) propounded many theories which were mainly inaccurate, but they were adopted for well over a thousand years after his death. One of those erroneous theories was that the body was composed of four humours: blood, which was hot; phlegm, which was cold; yellow bile, which was dry; and black bile, which was moist. The task of the doctor was to keep a balance between the humours. The extraction of 'vicious humours' by dry and wet cupping was widely practised by the Greeks and Romans.

Wet and dry cupping

Christianity and medicine

After Galen's death and during the Middle Ages European medicine entered a period of decline. This period covers the rise to supremacy of the Roman Catholic Church, centred on Rome, and the development of the papacy as an international religious and political power. The Catholic Church provided hospices to care for the sick, but they strongly opposed Greek medicine, teaching that Christianity offered spiritual and physical healing. Arab doctors such as Avicenna, Rhazes and Maimonides, kept the teachings of Galen and Hippocrates alive. Avicenna's Canon of Medicine based on Roman and Arabic medicine and his own medical knowledge became a popular text throughout the Middle East and Europe. Maimonides was a Jewish philosopher and physician to the Egyptian court in Cairo. However, although these three made original contributions to medical treatment, and backed by their knowledge of chemistry they enlarged the range of effective medicines they lacked a basis of accurate anatomical knowledge.

Medical gods.

A mixture of magical thinking and careful medical observation was evident in Ancient Egypt, where priests practised as doctors, using texts ascribed to the medical god Thoth. Many 'cures', such as the use of dung, or fat from black snakes to repel evil spirits seem remote from modern practise. But a link with the past can be seen in the symbol R on prescriptions which represent the Egyptian god Horus, whose blindness was cured by Thoth; and the British Medical Association and the American Medical Association have a staff with a snake coiled round it as their official symbol, a design dating back to Aesculapius, the mythological god of medicine, who carried a staff entwined with a single snake, representing life-giving powers.

Medicine men: magicians, priests and doctors.

Today, we can see the great advances which have been made in medical science. Many diseases have been brought under control or eliminated, and man's ingenuity has created machines which increase our understanding of the complexities of the human body. But the body is not a mindless machine, capable of being tuned to perfection. Sugar pills can cure, if they are given under appropriate psychological conditions; priests and witchdoctors have healed by prayer and spellcasting. It is clear that, although our understanding of the human body has advanced considerably, there are still many mysteries concerning the role of the mind in regulating our health.

Beaked man

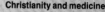

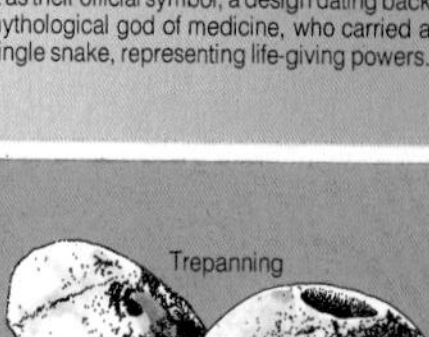

Trepanning

1935 First wearable hearing aid
1935 Sulpha drugs discovered by Domagk
1942 Use of ultrasonics pioneered in the United States
1944 Kidney machine
1952 First contraceptive pill introduced
1954 First organ, a kidney, successfully transplanted

Inside the body

Early doctors could trace the patterns of different illnesses only by observing and recording the experiences of their patients. The invention of the stethoscope in 1816 made it possible for doctors to listen to a patient's heart and lungs. The Anatomy Act of 1832 made more bodies available to British doctors for the study of anatomy. With the discovery of X-rays in 1895, photographs could be taken of the inside of the body. The first anaesthetics (ether, 1842 and chloroform, 1847) gave surgeons time to perform more careful, precise surgery.

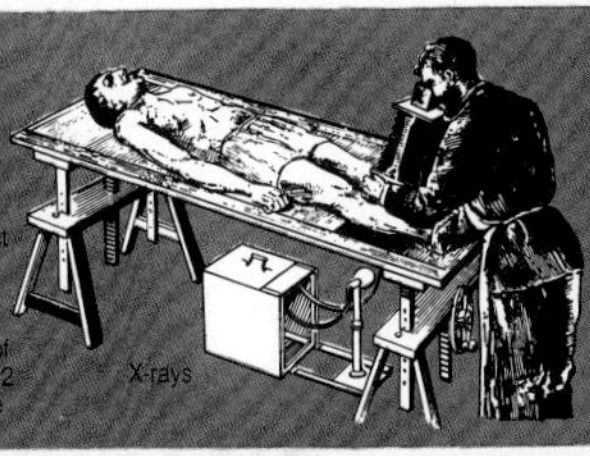
X-rays

Medicine today

Science has enabled us to discover much about the causes of certain illnesses and technological advances such as the computerized tomograph scan have enabled us to make precise diagnoses of many medical conditions. The study of genetics has produced a revolution in preventative medicine. Heart disease has become a major threat to health especially in the developed countries.

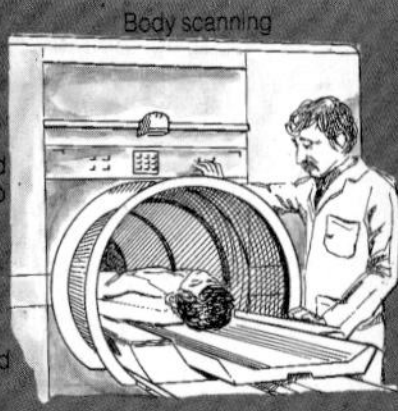
Body scanning

Amputation

Microbiology.

In 1683, a home-made microscope was used to discover the existance of tiny crawling creatures in a drop of canal water. It was nearly two hundred years later that Louis Pasteur discovered that these creatures, which he called 'germs', were agents of fermentation, and also of disease. He found out that germs could be killed by heat and established the process called pasteurization, by which our milk is made safe to drink. Until the discovery of germs, no-one had realised that dirt could be responsible for causing illness. Hospitals were dirty and surgery had always been a very brutal process, used only as a last resort; because of the lack of hygiene, many wounds turned gangrenous or septic and the chances of surviving an operation were very small.

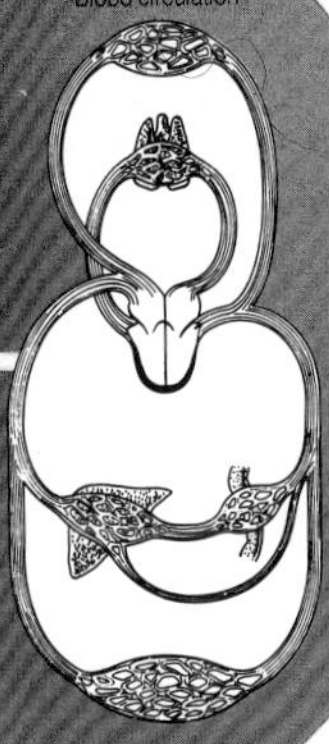
Blood circulation

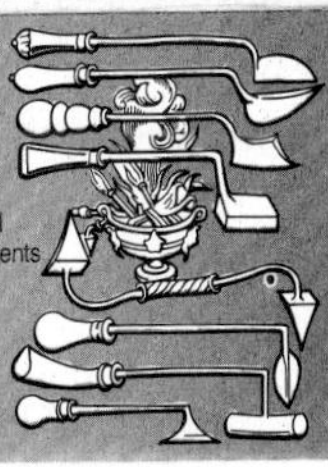
Surgical instruments

New discoveries.

For many centuries, the Church opposed the dissection of bodies so inaccurate theories could not be tested out. But in the mid-sixteenth century, Andreas Vesalius published a series of books on anatomy, the Fabrica, based on his dissection of executed criminals. Accurate surgery was now a possibility, and the new understanding helped later physicians, such as William Harvey (1578-1657) who discovered that the blood was pumped around the body by the heart. Previously, Galen's theory had been accepted, that the blood originated in the heart, and somehow ebbed and flowed through veins and arteries. The microscope, invented in 1590, also helped doctors to question the old ideas which had been accepted for many centuries.

1955 Salk developed successful polio vaccine
1961 Airbed – for the treatment of burns
1964 Bloch and Lynen receive Nobel prize for work on cholesterol and fatty acids

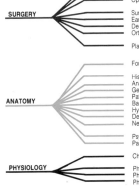

Releasing the spirits.

Excavations show that primitive man performed surgery on the skull (cranial trepanning) probably to release evil spirits from diseased people. Skulls have been found which show that the wound healed over and the person survived for some time after the operation, so considerable skill must have been used.

SURGERY

- Gynaecology – study, diagnosis & treatment of diseases in women
- Obstetrics – care of women before and after child birth
- Opthalmology – study, diagnosis & treatment of diseases of the eye
- Surgery – treatment by operation of disorders and injuries
- Ear, nose and throat – specialized study of related organs
- Dentistry – care of teeth, gums and mouth
- Orthopaedics – treatment of deformities by disease or injury to bones and joints
- Plastic surgery – repair or reconstruction of malformed or missing parts of the body

ANATOMY

- Forensic medicine – application of medical knowledge to legal matters
- Histology – study of tissues
- Anatomy – study of structure of living organisms
- Genetics – study of heredity and variation
- Pathology – study of disease
- Bacteriology – study of microscopic organisms causing disease
- Hygiene – principles of health and cleanliness
- Dermatology – study of skin
- Neurology – study of structure, function and diseases of nervous system
- Psychiatry – study and treatment of mental disorders
- Paediatrics – problems and illnesses of children from birth to adolescence

PHYSIOLOGY

- Chemotherapy – treatment of disease, particularly cancer, by drugs
- Physiology – study of function of living organisms
- Pharmacology – study of action of drugs
- Pharmacy – science of preparation and supply of medicinal drugs

1967 Barnard performs first heart transplant. Patient lives for a few weeks
1968 Second transplant patient lives for 19 months
1970s Investigations of viruses as possible causes of cancer

• **Minamata Bay, Japan 1950s.** Chemical pollution. Many people died or suffered nervous disorders from eating fish poisoned by methyl mercury which had been discharged into the sea.

Seveso, Italy 1976. Chemical cloud. Atmospheric pollution caused by the accidental release of a cloud of poisonous dioxin. An area of 18 km², 7 sq miles was contaminated and more than 900 people evacuated.

• **Torrey Canyon, 1967.** Oil spill. Tanker ran aground near the Scilly Islands off UK. 120 000 tons of oil spilled and British and French coasts heavily polluted.

• **Amoco Cadiz, 1978.** Oil spill. Tanker aground off Brittany, NW France. Entire cargo of 220 000 tons of oil lost. Coastal pollution much worse than in the 1967 disaster.

• **Exxon Valdez, 1989.** Oil spill. Worst North American spill with disastrous effects on fishing, seabirds and American confidence.

Sadly, we have to paint a new world map

Deforestation, desertification, and sea-water pollution

Great harm is done to the environment by human activity. Vast areas of forest are cleared; exploitation of marginal land causes deserts to spread and the spillage and dumping of oil and other toxic compounds are major problems.

These problems along with issues such as 'the greenhouse effect' and the thinning of the ozone layer (see page 107) are of international concern. The survival of plants, animals, and humans is at stake.

Sea pollution has become a serious problem of world-wide proportions. Sewage outlets, radioactive waste and industrial effluents as well as the run-off from agricultural processes all contribute to marine pollution.

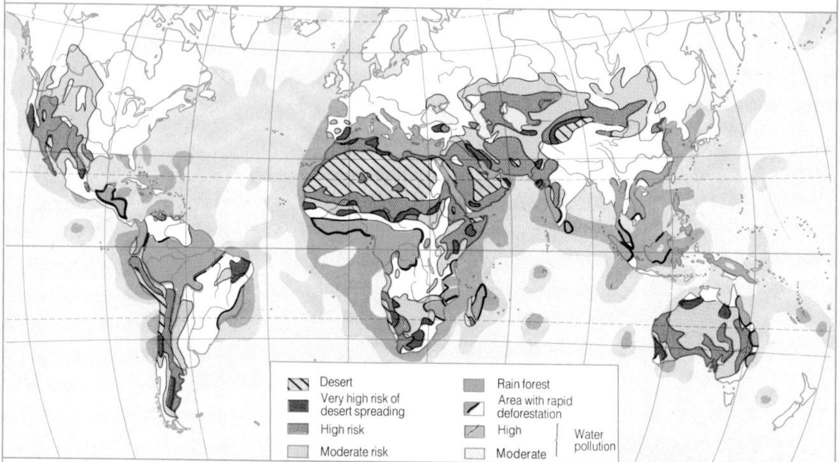

Oil pollution. In March 1989 the Exxon Valdez grounded in the Prince William Sound off Alaska. There was enormous damage to wildlife from the oil.

Rain forest destruction. In Indonesia forests are being cleared, using heavy machinery. The effect on the indigenous people is disastrous.

The Trans-Amazon Highway is 5600 km, 3500 mile long and was designed to open up the Brazilian interior for ranching.

Deforestation

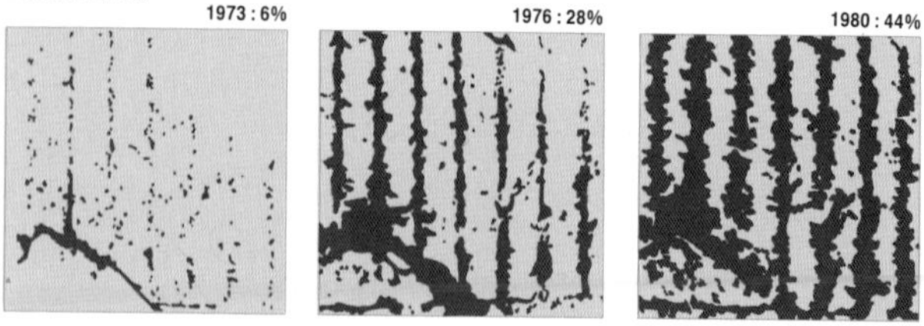

The diagrams (left) show the increasingly rapid deforestation of an area of Amanzonas. The brown is cleared land. Trees are cleared along ever expanding strips.

One species a day. About 40% of all tropical rain forests have already been cleared and each year about 75 000 km², 30 000 square miles of forest are destroyed. Tropical forests are so rich in plant and animal species that numerous species have not yet been identified. At present one species a day is lost, and as the destruction goes on more and more species are becoming extinct. The build-up in the atmosphere of carbon dioxide, the most important of the 'greenhouse gases' is accentuated by forest burning. Although Brazil has said that it will take action to protect the forest, it maintains its right to exploit its own resources.

- **Bhopal, India 1984.** This city was overwhelmed by the poison gas MIC. Up to 10 000 people were killed and many more disabled by lung conditions and many other disorders.

Sandoz, Switzerland 1986. Chemical spill. A warehouse fire caused the explosion of drums of dangerous chemicals. Toxic pesticides, fungicides and dyes poured into the Rhine making it dead for 200 km.

Desertification

Too intensive grazing, over cultivation and burning may contribute to the spread of deserts. As the population grows so marginal land is farmed. This land may have inadequate or unreliable rainfall. When the surface vegetation is destroyed by cultivation or by grazing the soil quickly becomes degraded. Soil erosion is a constant threat. Before long the land has become useless and is transformed into barren desert.
Dry tropical areas are most at risk.

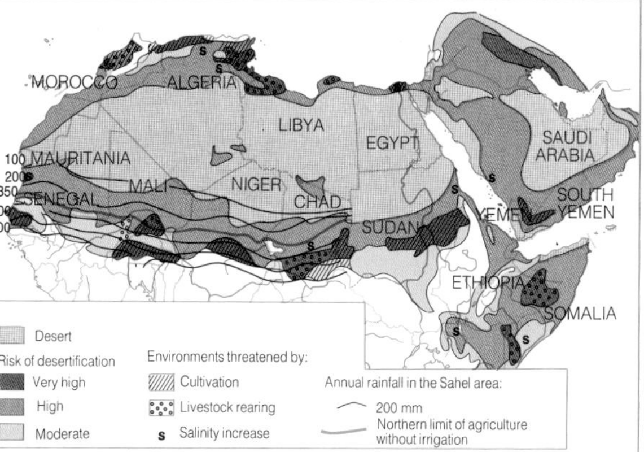

Atmospheric pollution
Sulphur covered Europe

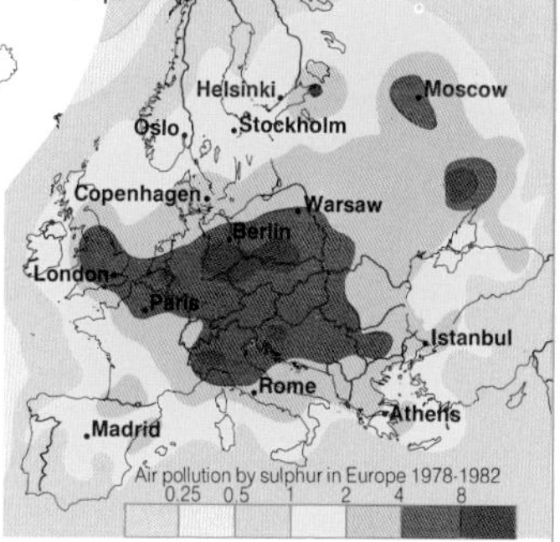

A threat which knows no boundaries. Acid rain is one of the most damaging forms of pollution. Power stations and smelters using fossil fuels (see p.81) emit chemicals which return to earth as poisonous acids. Britain has been blamed for damage to the Black Forest. Eastern Europe, where coal-fired power stations are used is also affected by acid rain. It is estimated that 3.2 million tonnes of acid-releasing pollutants blow into eastern Canada from the US each year, and as many as 80% of the lakes in eastern Canada are acid or becoming so.

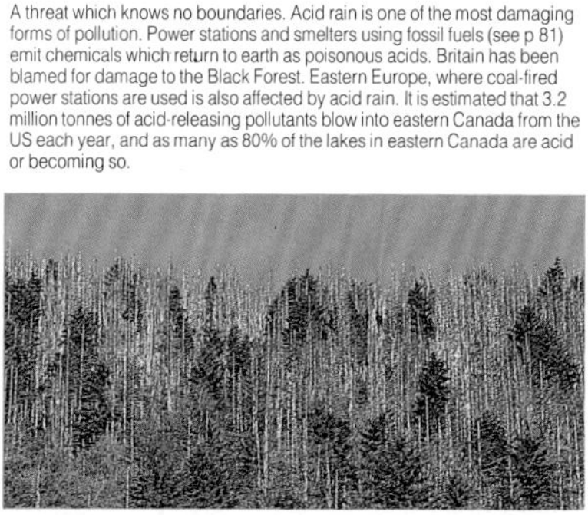

Chernobyl accident
April 26 1986

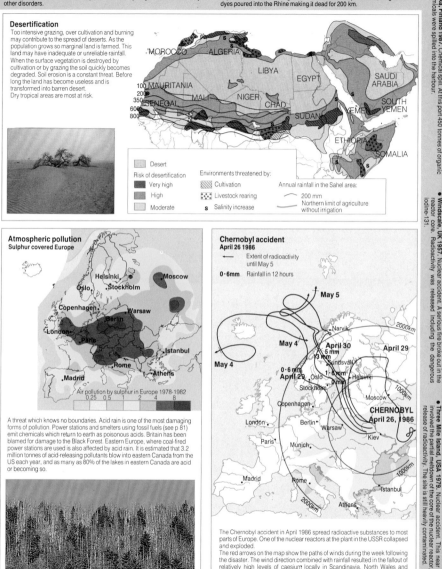

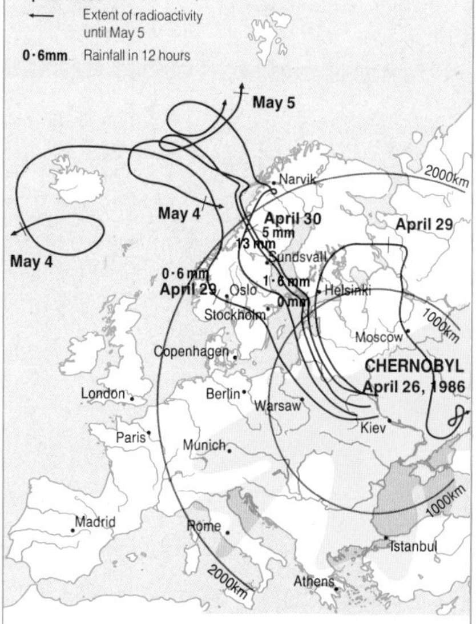

The Chernobyl accident in April 1986 spread radioactive substances to most parts of Europe. One of the nuclear reactors at the plant in the USSR collapsed and exploded.
The red arrows on the map show the paths of winds during the week following the disaster. The wind direction combined with rainfall resulted in the fallout of relatively high levels of caesium locally in Scandinavia. North Wales and Scotland were similarly affected. The yellow area on the map shows the parts of Europe affected when the wind direction changed.

Acid rain. Power stations using fossil fuels emit sulphur and nitrogen oxides, which return to earth as rain containing sulphuric and nitric acids.

Greenhouse effect. The global warming caused by the 'blanket' influence of gases. Predictions vary widely but the results are likely to be far-reaching.

Greenhouse gases. Carbon dioxide is the most important gas. It is released into the atmosphere by the burning of fossil fuels.

- **Kotka, Finland 1987.** Chemical spill. At this port 450 tonnes of organic chemicals were spilled into the harbour.
- **Windscale, UK 1957.** Nuclear accident. A serious fire broke out in the reactor core. Radioactivity was released including the dangerous iodine-131.
- **Three Mile Island, USA 1979.** Nuclear accident. This near disaster involved the partial meltdown of the core of the nuclear reactor, and the release of radioactivity. The site is still heavily contaminated.

- **Lindane.** An insecticide used for public health purposes. Toxic to fish and birds. In 1989 a ship containing lindane sank off France.
- **PCDFs.** Polychlorinated dibenzo-furans are responsible for the worst effects of PCBs. They result from PCBs having been heated.
- **Mercury.** Widely used as a seed dressing to prevent mould. Bioconcentration in food chains. Can cause death in humans.

- **DDT.** A once widely used pesticide. Concentration has caused a failure to reproduce in birds such as the osprey.
- **TBT.** Tributyl tin is used as a fungicide in wood preservation and in anti-fouling paints for boats. Toxic in low doses to oysters, whelks and salmon.
- **CFCs.** Chlorofluorocarbons are chlorine based chemicals used in aerosol sprays and refrigeration equipment.
- **Nitrite.** Babies who absorb excessive nitrate from drinking water can suffer from nitrite poisoning which damages blood haemoglobin.

Dumping waste – but where?

Both developed and developing countries are generating ever-increasing amounts of household and industrial waste, some of it toxic. One problem is that the world is running out of places to dispose of the rubbish. The danger to human health is rising, and, more insidiously, there is the threat to plant and animal life, a danger about which we are becoming increasingly aware.

There are a number of different types of waste, each posing particular problems of disposal. Household waste, from rotten apples to paper and cans is very bulky, but not intrinsically harmful. Waste including solid material and liquid discharges from chemical and petro-chemical plants is often toxic and can be highly dangerous to people and to the environment. The disposal of nuclear waste poses particular problems as some high level waste can remain radioactive for thousands of years.

In industrialized countries there may be considerable local resistance of the NIMBY type to new disposal sites. A solution is sometimes seen in dumping at sea or taking toxic waste to another country. The revenue is welcomed by the recipient country but, arguably, there should be a ban on exporting hazardous waste to countries without the resources to dispose of it safely.

In some developing countries there are few regulations covering the output of toxic waste. The provision of plant or personnel to deal with hazardous material has low priority both for foreign investors and for local companies.

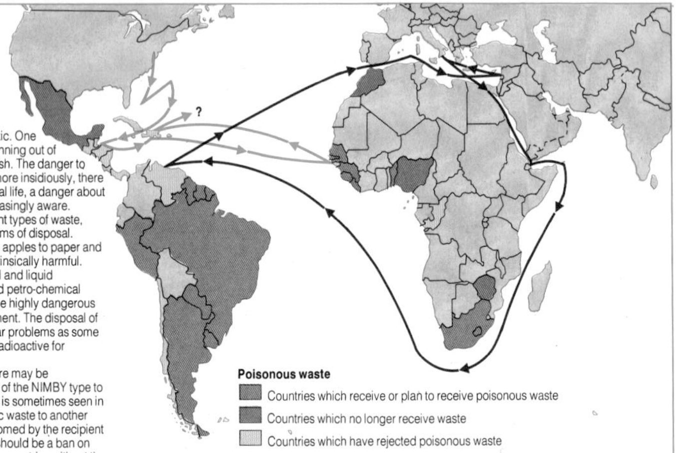

The wandering waste ships

A cargo of toxic waste containing PCBs left Italy in early 1987 bound for Djibouti. Unloading it there proved impossible and the ship sailed on to Venezuela. During the trip the crew suffered skin and respiratory problems from the fumes. The containers were unloaded and left for three months during which time some corroded and leaked. The cargo was then reloaded, and the vessel returned to Italy in 1988.

In 1986 a ship loaded with 13,500 tonnes of incinerator waste set out from Philadelphia, USA. The Bahamas refused to accept the cargo of contaminant laden ash and so the Khian Sea continued onto Bermuda, Honduras, Dominican Republic; crossing the Atlantic to Guinea Bissau and back to Haiti. They all refused and probably the cargo was dumped in the Caribbean Sea.

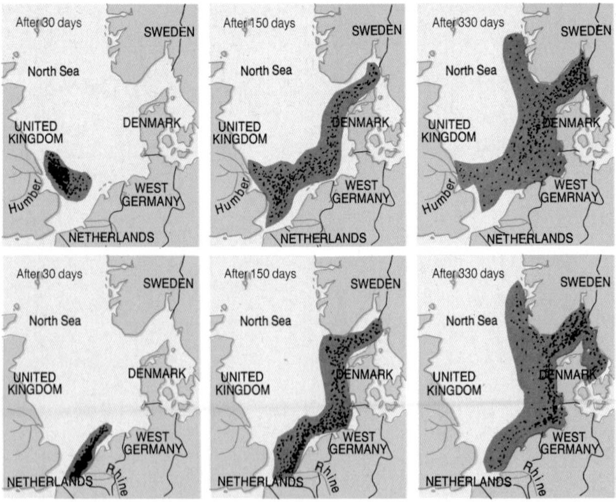

Neighbourly relations

Scientists from the Institute for Marine Research in Hamburg report that, regardless of where in the North Sea polluting waste is discharged, it is carried by currents to Skagerrak north of Skagen. Computer modelling, using data relating to winds, tides and air pressure, can predict the direction of currents and the probable distribution of pollution.

The original aim of the research was to investigate the movement of waste discharged by the Rhine and the Humber. One month after discharge chemical pollutants could still be found near the mouths of the rivers. After 150 days much of it had spread along the Dutch and West German coasts and finally after 330 days it finished up in Skagerrak.

In addition to discharges through the Humber and the Rhine, substantial amounts of toxic chemicals and heavy metals are regularly dumped into the North Sea by Britain and other European countries. This, too, contributes to pollution in Skagerrak. It had been thought, erroneously, that pollution in the North Sea was evenly distributed and toxicity could not reach dangerous levels.

- **Aldicarb.** A highly toxic insecticide which may cause miscarriages and could interfere with the immune system in humans.
- **Asbestos.** Widely used in the building industry but the fibres cause lung disease and lung cancer.
- **Cadmium.** Widely used in the metal and plastic industries. Can occur in high levels in sewage sludge.

Looking for a home, draped with a protest banner, an unwanted garbage barge floats off New York.

Dumping nuclear waste
Low-level waste can be stored in metal containers filled with concrete and the material buried in a pit in a geologically suitable area. Drums are also dumped at sea. The Pacific region is regarded as a favourable area because of its vast extent. Future methods of disposal may include placing containers into shafts drilled into sediments on the deep sea floor. As yet no ideal method of disposal or destruction exists.

Household garbage
Much of the 200 million tons of household waste generated each year in the United States is buried or burnt. The picture (above) shows a garbage barge off New York. The barge is draped with a banner suggesting a more acceptable alternative to dumping the garbage.

and industrial waste
Industrial wastes which are carelessly dumped can be particularly dangerous. Corrosion can cause poisons to leak from containers, and so pollute groundwater and contaminate farmland.

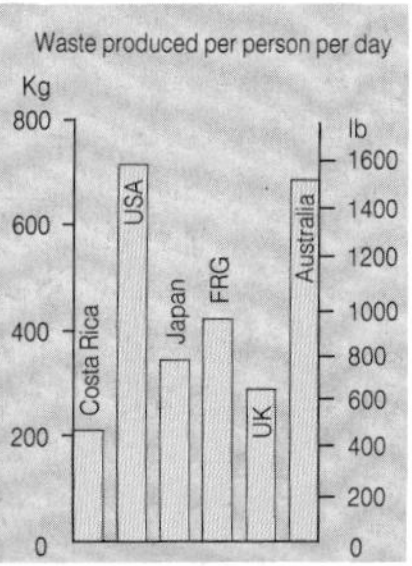

Waste disposal
Landfill is one of the most widely used methods of waste disposal. However, as sites fill up new ones become scarcer and more expensive.

Incineration of waste requires heavy investment. Burning can produce polluting gases but the ash must still be disposed of. Energy is generated.

Nuclear waste may also be disposed of, at least temporarily, by storing it in caves in remote mountain areas where it will not contaminate the ground, or pollute water or air.

To live on garbage
To many of the poorest in the Third world garbage is their lifeline. From the malodorous, disease-ridden and rat-infested piles, children and adults collect metal, glass or household objects; anything in fact than can be used or sold. The hungry will also pick out anything that appears edible. To the affluent society it is horrifying that rubbish heaps must support victims of war and poverty.

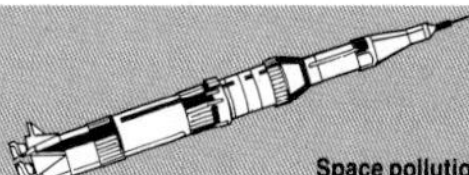

Space pollution
Space debris is building up so fast that it may be unsafe for astronauts to walk in space eventually. At present 6000 objects are being tracked in space, but only 5% are operating satellites. Of the rest, some are satellites that have ceased working but most are parts of spacecraft jettisoned at launch, or fragments from exploded rockets. There are also fears that space may become polluted by gases released from rockets. There is concern that the Earth will be surrounded by a permanent ring of debris.

Regular garbage collection
Most householders in developed countries take regular rubbish collection for granted. Disposal is a major difficulty but a partial solution is waste recycling.
In 1988 in Japan 50% of waste paper and 66% of cans were recycled. Much of the remaining rubbish was turned into fertilizers and recycled metals. Such success requires heavy investment combined with the willingness of the public to follow stringent regulations in sorting different types of waste.

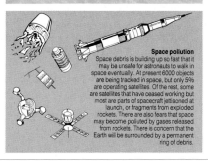

- **Dioxins.** 75 related compounds of which tetrachlorodibenzopara-dioxin (TCDD) is the best known. A byproduct of other processes particularly burning. Probably carcinogenic.
- **PCBs.** Polychlorinated biphenyls are used in a wide range of products, particularly heavy electrical equipment. Stable until heat produces very poisonous compounds.
- **Heavy metals.** Include mercury, cadmium and lead. The elements bioconcentrate and particularly affect shellfish and some plants.

NIMBY. Not In My Back Yard. A very common reaction from the public when it is proposed to site waste disposal facilities nearby.

Bioconcentration. Poisons concentrate within the tissues of living organisms. Further up the food chain there is a cumulative effect.

Carcinogenic chemicals and heavy metals are those which appear to cause cancer.

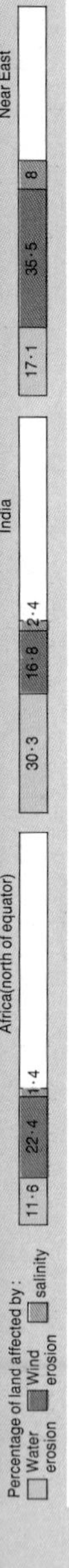

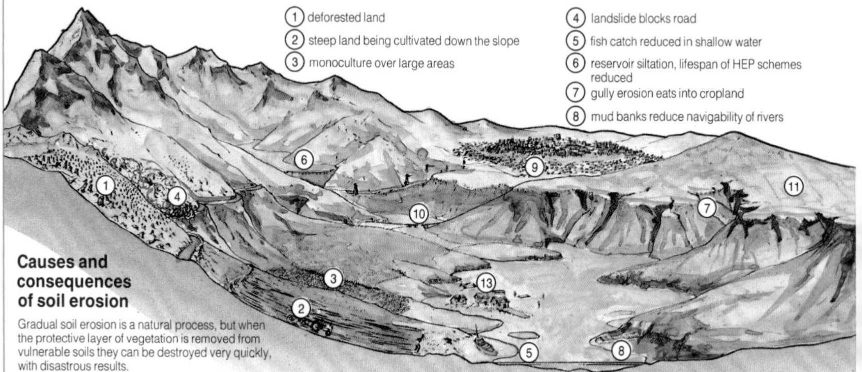

Causes and consequences of soil erosion

Gradual soil erosion is a natural process, but when the protective layer of vegetation is removed from vulnerable soils they can be destroyed very quickly, with disastrous results.

The problem is most severe in arid and semi-arid areas where overgrazing or inappropriate farming practices leave the soil exposed and unprotected by binding plant roots. Frequently the situation arises when too many animals are kept in an area which becomes overgrazed. If the vegetation is not allowed to regenerate, desert conditions soon appear and the soil is easily lost.

Farming practices which include deep ploughing and planting down, rather than along the contour can also leave the soil in a vulnerable condition. If steep forested land is cleared for agriculture or used for fuelwood, surface runoff can easily erode gullies. Wind can remove huge quantities of dry unprotected soil and the particles in the air then act like a sand blaster to erode the surface even more.

Water can be an even more destructive force. Sheet erosion occurs when the whole surface of a field is gradually eroded away in repeated storms. More noticeable on steep land is the creation of gullies which gradually enlarge, working their way back into the hillside. Streambank erosion by rivers can create shallow slow-moving water-courses and a considerable loss of farming land.

The consequences of erosion affect not only the rural population, but also the urban people living in the lower parts of the river basin. The loss of farming land can also have far reaching consequences for the national economy and planning.

The first sign of trouble may be a drop in crop yields, and as gullies develop grazing replaces cultivation. If the land is overgrazed it soon becomes useless. Less food leads to malnutrition, if not starvation, and the poor leave for the city. As shanty town dwellers,

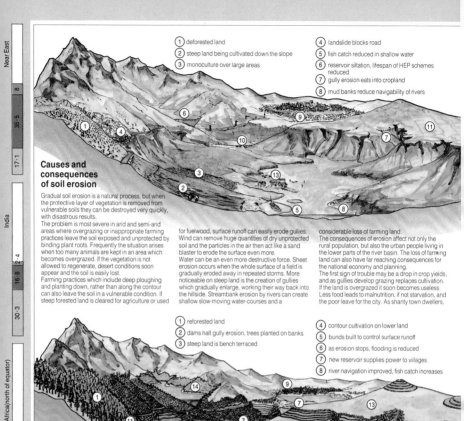

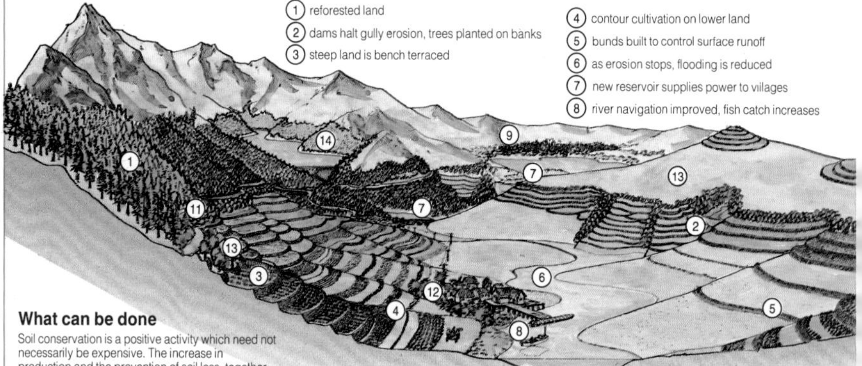

What can be done

Soil conservation is a positive activity which need not necessarily be expensive. The increase in production and the prevention of soil loss, together with the improvement of the quality of the soil, are results that can be seen in the relatively short term.

There are two main groups of techniques for controlling soil erosion; the biologically based and the physically based.

The most important decision from the biological point of view is to choose the correct crop for the site. Due consideration must be given to the steepness of the ground and the soil and climatic conditions; it is important to choose crops which have roots which bind the soil, and enough growth to ensure adequate ground cover to prevent erosion. Of course the food requirements of the people and the needs of a cash economy must also be considered.

While there is a wide choice of crops suitable for growing on flat land, it may be advisable to grow tree crops such as olives or citrus fruits on steeper land. On very steep slopes it may be possible to replant with a quick growing species such as the ipit-ipit tree.

Good management techniques such as ploughing along the contour can reduce erosion. Intercropping is the practice of planting crops with differing growth periods in strips. This too retards and helps to avoid the disease problems associated with monoculture.

In some low lying areas a cheap and effective way of preventing erosion is to plant mattoral, a dense shrubby vegetation. When the plants are established they protect the soil and encourage other plants by providing organic material.

The second group of techniques involves physical protection and there is a battery of measures that have proved effective. The cheapest form of terracing involves the construction of 'eyebrow'

9. urban slums grow as peasants move to the city
10. bridge destroyed by floods
11. crops grown on large unprotected fields
12. badly managed pasture suffers from wind erosion
13. frequently flooded village becomes deserted

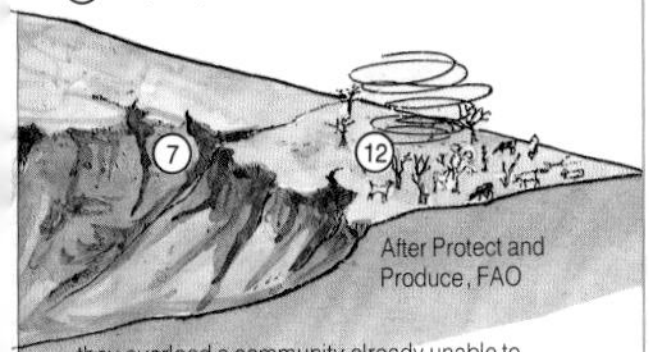

they overload a community already unable to provide jobs, or indeed, health, housing and other facilities. Such conditions can lead to crime and social unrest.
Further problems arise as the soil washed away fills the rivers or reservoirs. The shallower silty water may cause a loss in fishing stocks, reduce navigation and increase the risk of flooding. The silting of the reservoir (see page 84) may result in hydro-electric and irrigation schemes failing to come up to the planners' expectations.
The effect on the national economy of food shortages, and an inability to grow cash crops may delay development plans, and plunge the country even further into debt in the international market.

9. urban services improve as population stablises
10. shelter belts reduce erosion, pasture improved
11. better roads with fewer floods and landslides
12. rural services improve as villages expand
13. crops rotated in strips along contours
14. forested slopes reduce reservoir silting

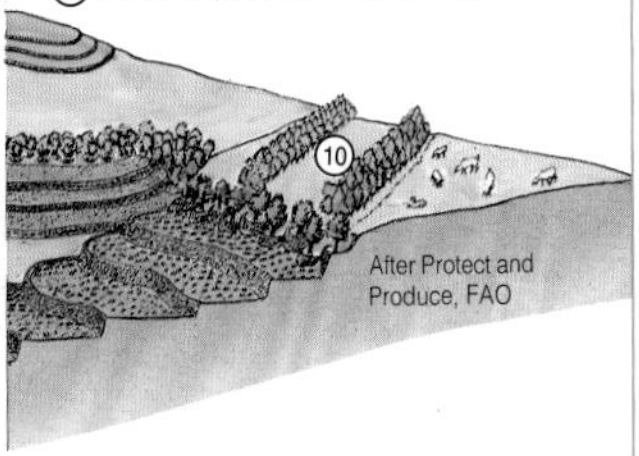

terraces to protect single trees. More substantial terracing is very effective, and there are a number of ridge and ditch techniques which may be appropriate. However maintenance is essential as blocked ditches or broken terraces can cause huge loss of land.
Wind erosion in dry flat areas can be counteracted by the planting of wind breaks, possibly in a chequerboard configuration. This slows down soil loss and may encourage plant growth.

Monumental decay

Even 'pure' rain is acidic and does, over the years cause stonework to dissolve. However the problem is exacerbated by acid deposition from sulphur dioxide released from power stations. Chemical salts such as gypsum can trigger other forms of decay through crystallization, and changes due to humidity and temperature variations. Gypsum combining with soot leads to 'black crusts' which must be cleaned off.
Water and sand blasting, using silica powder have been the usual method of cleaning stonework. New microblasting techniques using a jet of aluminium powder blown by air through a ·65mm nozzle are safer, highly accurate and very costly.
After treatment it is necessary to stablise the stonework to prevent further decay, and repel water while still allowing the stone to 'breathe'. Tests on silicones and other compounds are in progress, but caution is needed as it has been found that some chemicals accelerate decay.
Where stonework has become so worn away that the sculpture or design is unrecognisable, highly skilled masons may carve replacement stones. Old stonework, like old woodwork is in constant need of maintenance.

Scrub down for Big Ben
During the 1980's parts of the Houses of Parliament in London needed cleaning. The Clock tower, popularly known as Big Ben was washed and the four dials, each 7m, 23 ft wide were refurbished.

Treat textiles tenderly

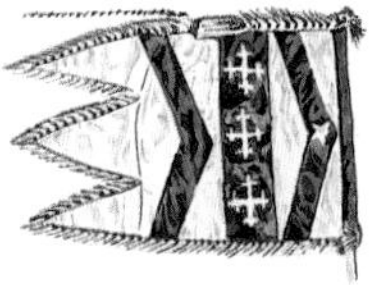

Preservation is the prime consideration where the care of textiles is concerned. Old textiles are often delicate and conservation involves keeping them clean and if possible away from light. Restoration is not recommended but fabrics can be strengthened with interlinings and backings attached using a quilting technique.
Many textiles have distintegrated and been thrown away and so particular gratitude must go to collectors such as Hazelius. At the end of the 19th century this Swedish collector sought out and preserved many tapestries, hangings and costumes from remote regions of Sweden.

Metals need protection

Precious metals such as silver and gold are affected by hydrogen sulphide in the atmosphere. Minute amounts of the chemical in the air react with the metals to form a sulphide compound which is deposited as a dark coating. Cleaning can remove the deposit from vases or other metal articles but repeated restoration removes surface layers and destroys fine detail. Precious metal objects should be protected from atmospheric pollution where possible.

Repairing books

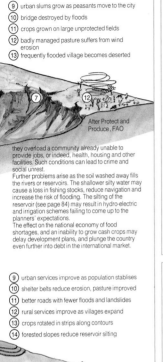

When books were a rarity they were kept in the libraries of monasteries and cared for. However age and accident mean that some may need conservation. Paper may be treated to remove stains or prevent mould spreading, and tears can be invisibly mended. Leather bindings can be treated or, if the conservator wishes, a bookbinder can replace a spine, restick pages or completely renew the binding.

In 1981 sandstone samples were placed in 5 European locations. Two years later pollution levels were measured. The figures are in mg.

Location	Sulphur (S)	Nitrogen (N)
Bondeska Palatset, Stockholm	154·6	4·2
St Pauls Cathedral, London	248·9	12·4
Basilica of San Marco, Venice	154·6	12·5
Pantheon, Rome	156·6	3·7
Acropolis, Athens	102·8	10·7

Gypsum. If limestone is attacked by sulphuric acid, the soluble compound calcium sulphate, commonly known as gypsum is formed.

Calcium chloride and calcium nitrate are other compounds formed when weak acids react with limestone. These compounds are easily washed away and the stone crumbles.

- **Black Rhinoceros.** *Diceros bicornis.* Numbers have dropped from 65 000 in 1970 to fewer than 3000 due to demand for ornamental daggers.
- **Big Leaf Palm.** *Marojejya darianii.* Only 50 plants survive in Madagascar. Half the habitat was recently cleared for rice growing.
- **Philippines Eagle.** *Pithecophaga jeffreyi.* One of the largest eagles has become almost extinct because of logging. Nests on platforms of ferns.

1 Dahl's toad-headed turtle, *Phrynops dahil,* (Bolivia). Extremely rare, through loss of habitat.

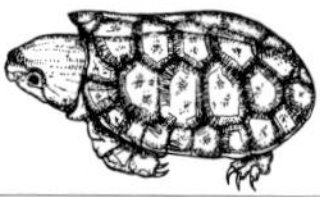

2 Kouprey, *Bos sauveli,* (Indo-China). Small population sighted in Thailand.

3 Ocelot, *Felis pardalis,* (Southern Texas to Northern Argentina). Rare, but numbers are unknown. Threatened by loss of habitat and hunted for its pelt.

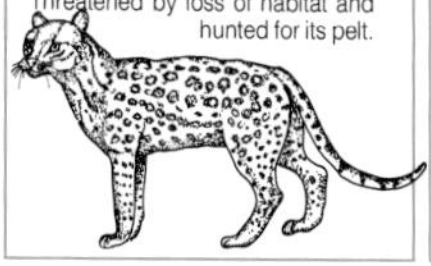

4 Chinese alligator, *Alligator sinensis,* (China). Critically endangered through wilful killing and loss of habitat.

9 Short tailed chinchilla, *Chinchilla brevicaudata,* (Andes). Rare. Hunted for its fur.

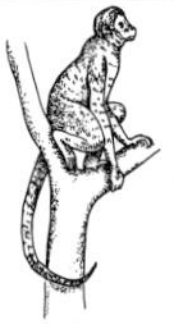

10 Woolly spider monkey, *Brachyteles arachnoides,* (Brazil). A few hundred survive in 1987 – from c. 3,000 in 1972.

11 Giant armadillo, *Priodontes giganteus,* (eastern South America).

12 Spectacled bear, *Tremarctos ornatus,* (South America).

Life in a zoo – no alternative

If there had been conservationists in the time of the dinosaurs, would they have seen them as a threatened species and tried to save them? Species of life have come and gone since life began on Earth, so should we be concerned that this continues to happen? The difference now is that the human race is interfering with this process. As animals disappeared through evolutionary processes, others appeared, but now, we are threatening and destroying species far faster than they are being replaced. We cannot argue that we are only threatening animal life in order to feed the growing human population, because human life is too intricately linked with the natural world.

The World Conservation Strategy aims to maintain viable stocks of all animals and plant species; to maintain pure air and water and fertile soil; and to preserve stocks of wild and domestic animals and plants which can be harvested, cropped and enjoyed indefinitely.

In small, simple societies, man has managed to live side by side with animals without posing a threat to their existence. The real problems began with industrialisation and population pressure on the land. In North America, the Indians had little effect on the wildlife, but the arrival of the Europeans saw widespread destruction.

Vast areas of forest have been destroyed, swamps have been drained and dustbowls created, all at unknown cost to the wildlife, and ultimately to all life. The Europeans who arrived in Australia caused great damage when they introduced familiar European animals like the rabbit, which then had to be controlled by foxes, weasels and stoats. Between them, these animals wiped out at least 9 Australian marsupial forms. In Asia, the rhinoceros, the orang-utan, the pygmy hog, the tiger and the elephant are just some of the animals which have become scarce because of over-hunting and the loss of habitat. To balance this, we must look at the success stories. The koala in Australia, hunted for its fur, and the American alligator, hunted for its hide, have both been saved by state legislation restricting the degree of exploitation. In India, the habitat of threatened creatures, particularly the tiger, has been protected. Other animals are being saved from extinction by captive breeding. The Arabian Oryx has been reintroduced into the wild after successful breeding in zoos, and so saved from extinction.

21 Pygmy hog, *Sus salvanus,* (Himalayan foothills). Under 100 surviving?

22 Amsterdam albatross, *Diomedea amsterdamensis,* (Amsterdam Islands in the Southern Indian Ocean). Only 5/6 pairs nesting.

23 Golden lion tamarin. *Leontopithews rosalia,* (Brazil). Under 100 surviving.

24 Puerto Rican amazon, *Amazona vittata.* About 30 in the wild.

27 Galapagos hawk, *Buteo galapagoensis.* Total population of about 130 pairs of groups.

28 Volcano rabbit, *Romerolagus diazi,* (Mexico).

29 Californian condor, *Gymnogyps californianus,* (California). The last wild bird was captured in 1987 to join 26 others for captive breeding.

30 Przewalski's horse, *Equus p. przewalskii.* There are 660 in 70 zoological collections. Plans are being made to reintroduce them into the wild, where they are probably extinct.

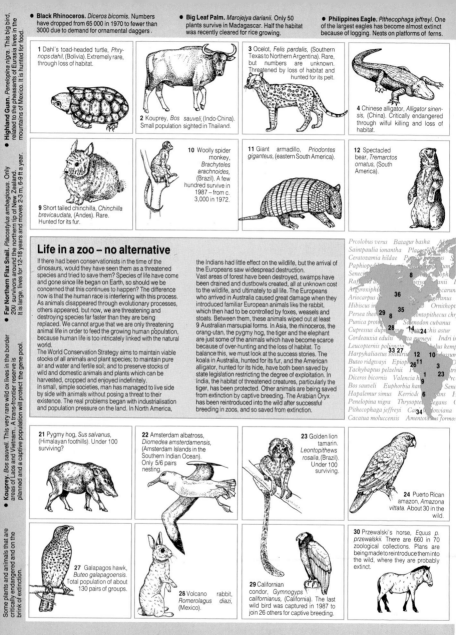

Some plants and animals that are critically endangered and on the brink of extinction.

- **Kouprey.** *Bos sauveli.* This very rare wild ox lives in the border areas of Laos and Vietnam.. A trans-frontier reserve is planned and a captive population will protect the gene pool.
- **Far Northern Flax Snail.** *Placostylus ambagiosus.* Only 200 survivors around the northern tip of New Zealand. It is large, lives for 12-18 years and moves 2-3 m, 6-9 ft a year.
- **Highland Guan.** *Penelopina nigra.* This big bird, related to the pheasants of Eurasia lives in the mountains of Mexico. It is hunted for food.

. . .and we threatened life

animals on the verge of extinction

- **Greater Bamboo Lemur.** *Hapalemur simus.* Until 1964 when one was found in a market this Madagascar lemur was thought extinct.
- **Salmon-crested cockatoo.** *Cacatua moluccensis.* A popular cage bird. Over 7000 birds were exported a year until recently.
- **Chiapas Slipper Orchid.** *Phragmipedium exstaminodium.* Extremely rare Mexican orchid threatened by collectors and cultivators.

5 Snow Leopard, *Panthera unica,* (area between China and Nepal). Threatened by loss of habitat. There are now around 5,000 surviving.

6 Marine otter, *Enhydra lutris,* (Peru and Chilean coastal areas). Scattered remnant populations.

7 Polar bear, *Ursus maritimus,* (Arctic). Population is threatened by over-hunting and attack on the environment during the search for oil. c. 20,000 surviving.

8 Whooping crane, *Grus americana,* (Northwest Territories, Canada, winters in Texas). About 200 in 1987.

13 Baiji dolphin, *Lipotes vexillifer,* (China). The world's most threatened whale species.

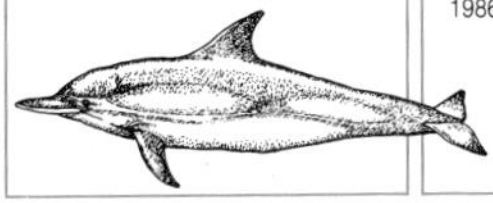

14 Ivory billed woodpecker, *Campephilus principalis,* (Cuba). Presumed extinct, but sighted in 1986.

15 Jentinck's duiker, *Cephalophus jentinki,* (Liberia/Ivory Coast).

16 Pygmy hippopotamus, (Liberia/Ivory Coast).

17 Gorilla, *Gorilla gorilla,* (Equatorial West Africa and East Central Africa). Threatened by forest clearance.

18 Black rhinoceros, *Diceros bicornis,* (Sub-Saharan Africa). Fewer than 3,000 surviving.

19 Wild yak, *Bos grunniens,* (China, India). Numbers in the low hundreds?

20 Asiatic lion, *Panthera leo persica,* (Gujarat State, India). 180 are protected in the Gir Forest Sanctuary.

25 Hawaiian duck. *Anas wyvilliana,* surviving on a few of the Hawaiian Islands.

26 Peruvian penguin, *Spheniscus humbaldti,* (Peru). Seriously declining numbers.

31 Madagascar teal, *Anas bernieri,* (Madagascar).

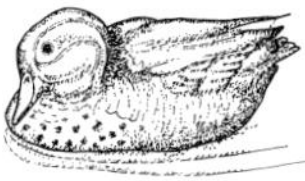

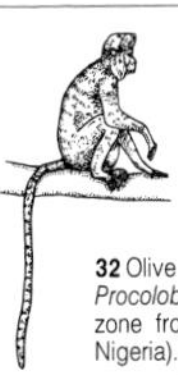

32 Olive colobus monkey, *Procolobus verus,* (forest zone from Sierra Leone to Nigeria).

Survivors corner

34 Southern fur seals, *Arctocephalus.* Laws restricting sealing have led to a recovery in numbers.

33 Giant tortoise, *Geochelone elephantopus.* The 1000th captive bred giant tortoise was released into the wild in 1988.

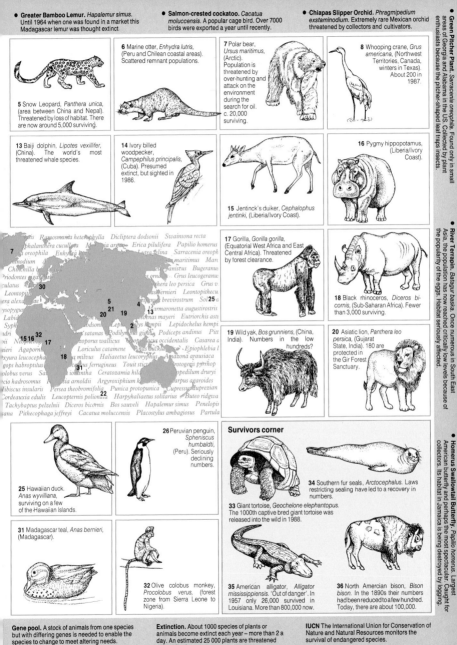

35 American alligator, *Alligator mississippiensis.* 'Out of danger'. In 1957 only 26,000 survived in Louisiana. More than 800,000 now.

36 North American bison, *Bison bison.* In the 1890s their numbers had been reduced to a few hundred. Today, there are about 100,000.

- **Green Pitcher Plant.** *Sarracenia oreophila.* Found only in small areas of Georgia and Alabama in the US. Collected by plant enthusiasts because the pitcher-shaped leaf traps insects.
- **River Terrapin.** *Batagur baska.* Once numerous in South East Asia, the population has now reached critically low levels because of the popularity of the eggs. Habitat seriously affected.
- **Homerus Swallowtail Butterfly.** *Papilio homerus.* Largest American butterfly and perhaps the most spectacular. Caught for collectors. Its habitat in Jamaica is being destroyed by logging.

Gene pool. A stock of animals from one species but with differing genes is needed to enable the species to change to meet altering needs.

Extinction. About 1000 species of plants or animals become extinct each year – more than 2 a day. An estimated 25 000 plants are threatened with extinction.

IUCN The International Union for Conservation of Nature and Natural Resources monitors the survival of endangered species.

Energy consumption in petajoules.	oil	coal	hydro
Middle East	4535	92	109
Western Europe	24 582	10 468	4 233
Latin America	9019	925	3442
North America	34 078	19 725	6 582
Africa	3421	2772	674
Asia and Australasia	17 137	10 957	2 378
USSR	18 632	15 751	2 198

1 million metric tons of oil are equal to 41·87 petajoules.

Energy from six sources

Hydro-electric power

Hydropower generated from falling water accounts for about 25% of the world's electricity. The Guri Dam in Venezuela, completed in 1986 is the world's largest HEP project.

Thermal power

Oil, coal and natural gas are by far the largest sources of energy. Since 1979 oil consumption has fallen but it is still the most important source. Coal is used in power stations for generating electricity. Unlike other sources coal is widely distributed and is relatively easy and cheap to produce. The USA and the USSR are the biggest producers of natural gas.

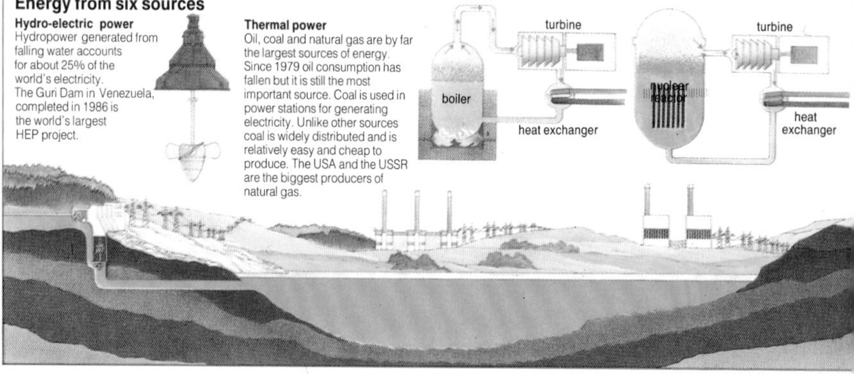

We need light and warmth

The use of energy, often in the form of electricity for providing homes with heat, light and power for domestic appliances is obvious. However, far greater amounts are needed for industry, for transport and for agriculture, for instance in the manufacture of fertilisers. World demand for energy continues to grow but more slowly than a decade ago.

Energy may be derived from renewable or non-renewable sources. There is a finite supply of fossil fuels but there are still large proven resources and probably sources undiscovered. Fuelwood is the prime energy source for almost half the world's people. It is gathered daily for cooking and domestic heating, but in many areas supplies are depleted causing hardship and increasing damage to the environment.

There are problems associated with the use of fossil fuels. These include the deposition of acidic substances, atmospheric pollution, ozone layer depletion and climate change. However renewable energy sources are not likely to become an important alternative in the near future. The prospect of nuclear power, once regarded as the solution to the energy problem, is not now universally welcome. Recent accidents have dented public confidence although there are many safety precautions that can be incorporated into nuclear power projects.

The challenge for the 90's is to find more efficient ways of using energy and to promote public awareness of the need to conserve resources. The possibility of using superconductors for ultra efficient electricity transmission is a distinct possibility. Perhaps one day nuclear fusion will become a viable commercial energy source.

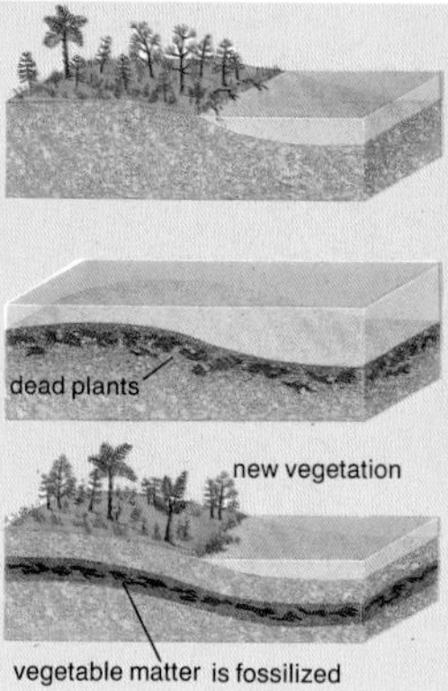

From trees to coal seam

During the Carboniferous period, 300 million years ago, swamp forests covered large areas. (see left) When water levels rose the trees died. As oxygen could not reach the vegetation it did not decay and instead formed a layer of organic material. Over time the band would be transformed into a coal seam by temperature and pressure. New forests would grow and eventually another layer of coal would form. Areas such as the Amazon basin might one day be coalfields.

From plankton to oil drop

In shallow lakes and lagoons many tiny organisms lived. When they died they sank to the bottom forming a layer of mud rich in organic material. Dinosaurs probably went the same way but they are not a major contributor to the process. Lack of oxygen in the deep water inhibited decay of the hydrocarbons. Pressure and heat converted the material into the larger molecules of petroleum, thus forming oil-bearing sandstone or shale.

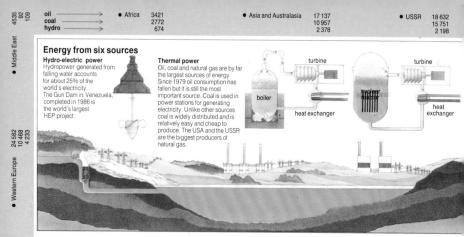

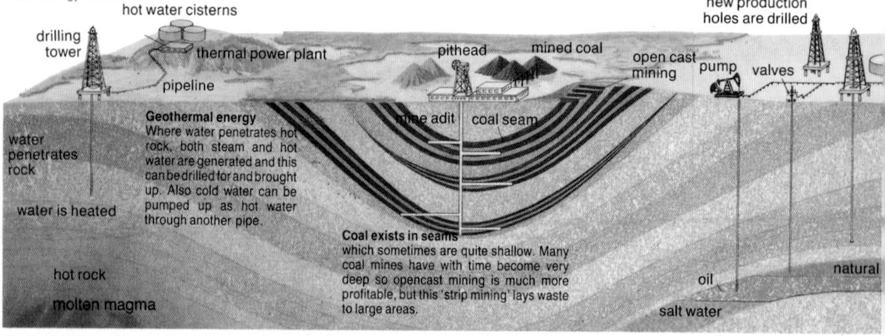

Geothermal energy
Where water penetrates hot rock, both steam and hot water are generated and this can be drilled for and brought up. Also cold water can be pumped up as hot water through another pipe.

Coal exists in seams which sometimes are quite shallow. Many coal mines have with time become very deep so opencast mining is much more profitable, but this 'strip mining' lays waste to large areas.

Ratio of reserves to 1986 production in years		North America	Latin America	Western Europe	Middle East	Africa	Asia and Australasia
	oil	9	37·7	12·2	85·5	29·3	14·3
	coal	313	241	212	X	362	211
	natural gas	15·5	71·6	35·9	100	100	54·4

Nuclear power

Of the world's energy 2% is supplied from nuclear power stations. Since the Chernobyl disaster (see page 73) the consumption of nuclear power continues to grow, but at a lower rate. At present 27 countries have 398 plants. There is public disquiet about the safety of nuclear energy, but also the realisation that power stations using fossil fuels, rather than nuclear energy, can damage the earth's protective ozone layer.

Wind, solar and geothermal power

Ecologically these are the most acceptable forms of power generation.
California utilises wind power and wind farms are planned for Europe, India and China. Solar power is widely used for heating buildings and water. Geothermal power is used for heating but only a fraction of the resource is commercially exploitable.

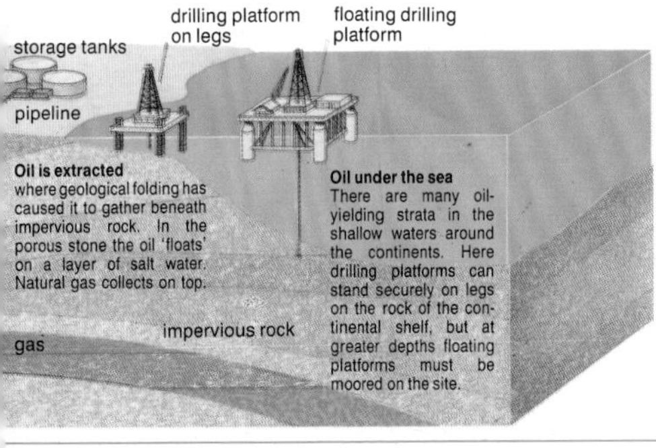

Oil is extracted where geological folding has caused it to gather beneath impervious rock. In the porous stone the oil 'floats' on a layer of salt water. Natural gas collects on top.

Oil under the sea
There are many oil-yielding strata in the shallow waters around the continents. Here drilling platforms can stand securely on legs on the rock of the continental shelf, but at greater depths floating platforms must be moored on the site.

Energy from the sea

Wave power may be exploited by using the energy as sea water moves from a chamber on one side of a barge to one on the other. Another method employs floats. Their movement can be used to generate power.

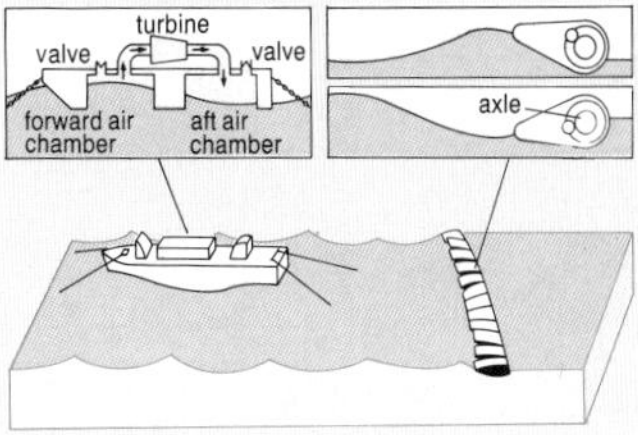

Solar energy is a technique which could be particularly useful in tropical waters. Ocean Thermal Energy Conversion (OTEC), shown right and below, uses warm surface water to turn a suitable liquid to gas. The gas drives a generator and the liquid is condensed by deep cooler water. It is brought to the surface by a siphon. This heat exchange system can be used to generate electricity which is then taken ashore by cable.

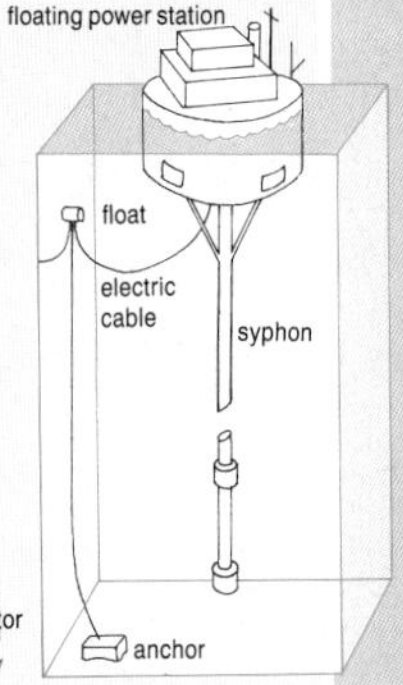

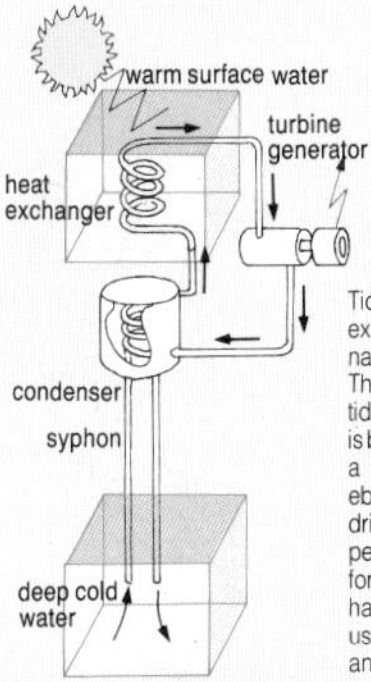

Tidal energy has been exploited where there is a narrow entrance to a bay. This can create a big inter tidal range. A power station is built with the bay forming a reservoir. As the tide ebbs and flows, the power drives reversible propellors in a turbine. This form of power generation has been successfully used in France, Canada and China.

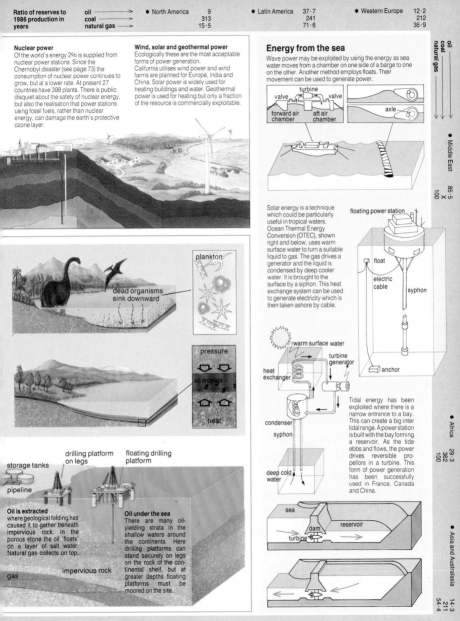

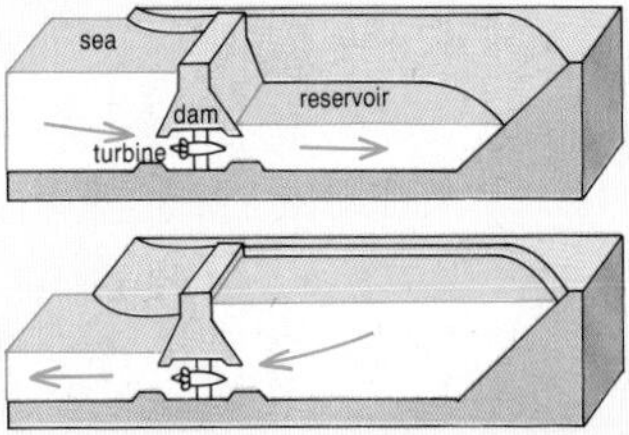

Fossil fuels. Solar energy stored millions of years ago, now used as coal and oil.

Renewable sources. Energy derived from wind, waves, falling water and the sun.

Non-renewable sources. Fossil fuels will eventually all be used up.

Biomass Plant or animal material used as fuel – principally wood.

Mohs scale of hardness. The intervals do not represent equal differences in hardness.

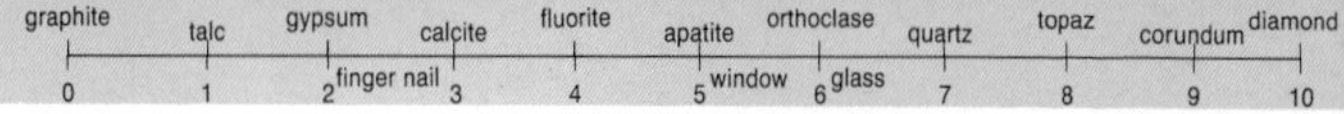

Percentage of annual world production by the 3 leading producing countries

• **Copper**	Chile	17·0
	USA	15·5
	Canada	9·4
• **Lead**	USSR	13·6
	Australia	13·4
	USA	10·9
• **Zinc**	Canada	18·9
	USSR	11·8
	Australia	9·7
• **Gold**	South Africa	47·9
	USSR	18·6
	Canada	5·6
• **Silver**	Mexico	15·5
	Peru	13·8
	Canada	5·6
• **Diamonds**	Zaire	29·0
	Botswana	20·2
	USSR	18·8

Zinc and lead mining

The photograph shows a lead mine, now closed, in the Lake District, England. Adits were dug into the hillsides and ore often taken on a mineral rail line for crushing.

There are many old mines in Europe as the Romans used a great deal of lead for water pipes; the latin name for lead is plumbum. The richest source of lead is the ore galena, but most ores contain other metals such as zinc and silver as well as lead. The ore is first crushed and concentrated, before being taken to the refinery.

Zinc blende is zinc's main ore. More than 6 million tonnes of zinc are produced each year.

Copper mining

Copper is often mined by the open pit method, which is used at the mine illustrated at Chuquicamata in Chile (see page 89). Draglines with buckets attached scrape the loose gravel which is carried by truck to the smelter. If the rock is hard, explosives will be used and the boulders moved with power shovels. Often huge pits are dug in this way and a series of connected steplike benches form a road round the sides of the mine.

Silver mining

The richest silver ore is argentite which contains 2 parts silver to 1 part sulphur, but the metal is generally found in combination with other minerals. Bingham Canyon (see page 61) has ore containing copper with molybdenum, gold and silver.

Silver-bearing ores may be close to the surface or deep underground as in the photograph from a mine in Mexico. In Level and Shaft mining a vertical shaft is constructed and a system of horizontal passages dug. It is important that only the veins containing the mineral bearing ores are removed and the surrounding rock left intact. After they are mined the ores are concentrated. Smelting removes many of the impurities and the metal is then refined. Some of these processes produce polluting sulphur dioxide.

Gold mining

The photograph from northern Brazil shows how shallow ores are carried to the top of a ditch by hand. Such work is highly dangerous.

The richest goldfields are in South Africa where some mines are over 3 km, 2 miles deep. For 30g, 1 oz of gold 3 or 4 tonnes of rock have to be brought to the surface. Placer deposits may be mined using a dredge with buckets which scoop up a mix of sand and gold which must then be separated.

Valuable rocks

All the rocks in the Earth's surface contain metal, but the concentration of them is usually too low for it to be worth exploiting them. Almost 75% by mass of the Earth's crust consists of the elements oxygen and silicon: metals account for only 16%. Metals vary greatly in their abundance; aluminium, iron and magnesium make up about 94% of all metals. Where geological processes have concentrated the metals to many times their natural strength mineral-rich ores are found. Because there are differences of density, solubility and melting point between different elements, a variety of different ores may be found together. Minerals are rarely found in a pure form but often combined with sulphur to form sulphides.

Valuable minerals are not uniformly distributed over the Earth. Regions of folded mountains and igneous activity are often rich in minerals.

Many valuable rocks are found where molten material from deep in the crust has cooled. As the mass hardens some minerals settle as a layer at the bottom. The nickel deposits in Sudbury, Canada were formed this way. When the molten material is nearly hard, metal rich fluids seep into cracks and crystallize slowly to form veins rich in metal yielding ores.

Erosion wears away all rocks and the minerals they contain. Some dissolved metals reached the sea millions of years ago, and were deposited as mineral sulphides, the copper fields in Zambia (see page 89), and the lead and zinc in Australia were formed this way.

Minerals eroded from rocks will be sorted by size and density as they are carried by a river, and will be deposited where the current slackens. This was the gold that brought so many to California in Gold Rush days.

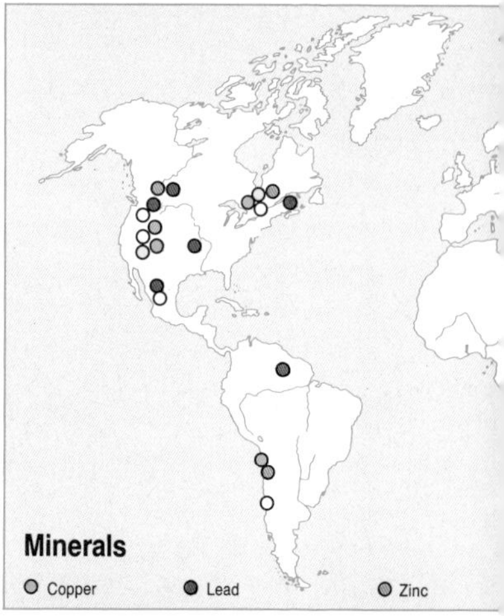

Gold and silver production

There are many stages in the transformation from an uninteresting-looking ore containing gold to a sparkling gold ring or a solid bar of gold.

Separating gold from ore may be done in several ways but first the ore must be crushed and milled to a fine powder and water added. In the flotation process, chemicals are added so that when air is blown into the mixture the gold particles rise to the surface on air bubbles. The froth is then skimmed off and the metal further refined. Another process involves dissolving the gold with cyanide. The metal is then separated out using zinc, and the gold precipitate is smelted to remove impurities. The gold can then be cast into bars of crude bullion.

Silver ores are usually concentrated, smelted, and then further refined to remove associated metals.

Beauty and strength

Gemstones have been valued for their beauty and rarity since ancient times. The Egyptians fashioned jewellery using turquoise and amethyst. They imported lapis lazuli from Afghanistan and the mines are producing fine gems 6000 years later. They are the oldest mines still working.
Most gemstones are minerals formed within the earth. Each mineral has a unique chemical composition and crystal structure, so that the way light is affected is the same for a particular gem. Colour results from chemical impurities. Sapphires and rubies are both varieties of corundum, and their colour may range from the palest cream through reds to blue and green depending on the amount of iron, chromium, or titanium present.
Some gems such as emerald, a variety of the mineral beryl, are brittle and cleave easily, while others such as agate are much tougher. The hardness of minerals is also very variable and is measured on the Mohs scale. The weight of a gem is measured in carats, 1 carat equals ·2g of diamond. The finest gold is 24 carats.
Sparkle in gems results from the interaction of light with the mineral. The beauty of a jewel is revealed by skilful cutting and polishing.

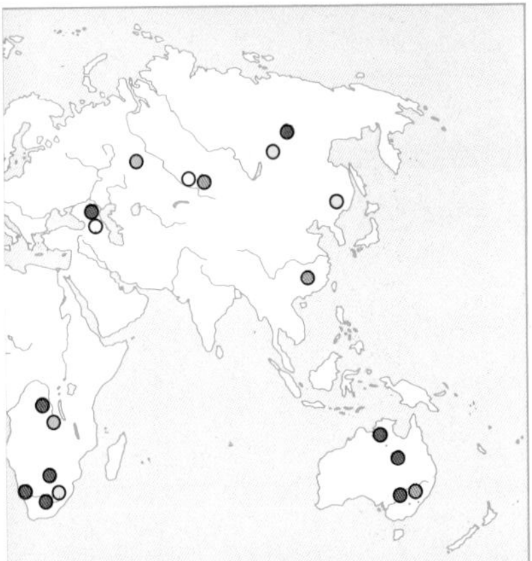

Diamonds in Namibia

The diamond bearing rock in Namibia, in southwest Africa, is overlain by a thick overburden of sand. This layer is removed by huge bucket scoops and the sand is deposited to provide a protective wall. The floor of the vast open pit is an ancient rock beach terrace. The workers search the rock, using brushes and sharp eyes. Vast amounts of material must be removed, but most of the diamond is of gem quality.

Sieved diamonds

If a river contains diamonds which have been eroded out of the rock, they will tend to collect, where the current slackens. The gravels at the bends of a slow moving river are sieved by hand and the diamonds picked out. Many attempts may be needed before a single diamond is found. The photograph shows an overseer watching workers in Venezuela. Similar techniques are carried out to recover diamonds in other parts of South America as well as in Africa.

Kimberlite pipe

In South Africa diamonds are found in 'carrot-shaped' pipes containing kimberlite rock. These pipes may vary from a few metres to 1·5 km, 1 mile wide, and may reach a depth of 3 km, 2 miles. The pipes are mined from open pits at the top as shown in the photograph. Below 300 m, 1000 ft, a variety of underground mining techniques are used to extract the diamond bearing ore. For each 100 000 000 grams of rock removed there will be 1 gram of diamond recovered. The ore is crushed and the heavy minerals separated out. They are passed over a grease belt and water washes away all the material except the diamond.
Over 80% of diamonds mined are used industrially, mainly for cutting and grinding tools. The largest diamond yet found is the Cullinan diamond (3106 carats).

Cutting and polishing

A skilful lapidary can exploit the shape of a rough stone, and, using its cleavage planes and flaws, produce a gem of exquisite beauty.
When light enters a diamond it is split into various wavelengths. Each of these are refracted or bent differently and are seen as different colours; the red light is the least refracted and the violet the most. This causes the fiery sparkle of diamonds. Some gemstones such as tourmaline show pleochroism and the rays of light emerge as several different colours or shades of the same colour.
Most gems are cut with a pattern of flat surfaces called facets which act as mirrors. The facetting must be planned to take full advantage of the physical and optical qualities of the stone while aiming for the largest possible gem. Size as well as clever cutting affect value.
Stones may first be cut using a diamond crystal, and then the facets ground and polished on a revolving wheel or scaife. Diamond is ground and polished using ground diamond; this can be done as diamond is less hard in certain directions in the crystal.
The brilliant and the step cuts are popular today. The ideal shape for a brilliant diamond was set in 1919 by Marcel Tolkowsy, who worked out the best ratios between the facets, and the ideal cutting angles.

Styles of cutting are influenced by the optical properties and cleavages of a mineral.

Round brilliant

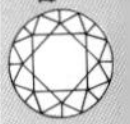

Oval brilliant

Marquise or navette cut

Pear shaped brilliant

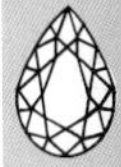

Emerald or modified step cut

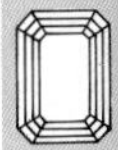

Scissors or cross cut

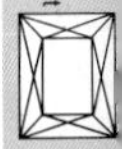

A lapidary is a craftsman who can appreciate the potential of a rough stone and cut and polish it to the best advantage.

Placer deposits occur when minerals which are eroded from the rock and carried by rivers to be deposited with river gravels.

Pleochroism. When light enters a mineral it is split into rays which may be differently absorbed by the stone, and are then seen as different shades.

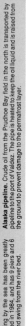

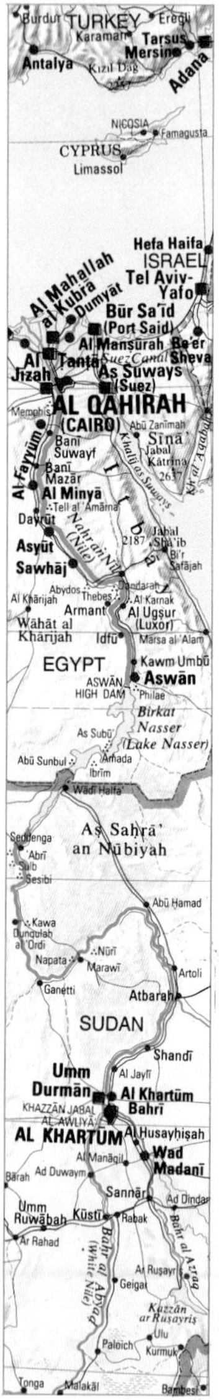

Constructing giant dams

The damming of a river can bring a number of benefits. Control of the river can alleviate flooding and provide stored water for domestic, industrial and agricultural purposes. There is also great potential for the development of hydro-electric power.

But there is often a price to pay in socio-economic terms. Flooding for the reservoir may involve mass relocation. The Volta Dam in Ghana necessitated the displacement of 80 000 people. Siltation, soil waterlogging and salinization are other problems which can be reduced by good soil management.

The Aswan dam

Egypt has reaped many benefits from the Aswan Dam, completed in 1968; the most important is the irrigation of old and reclaimed land. The power generated provides 50% of Egypt's needs.

Many problems have however emerged. Lake Nasser experiences serious siltation while the fine sands and clay no longer fertilize the land. Artificial fertilizers must be imported.

Waterlogging and salinization cause problems and schistosomiasis has spread. Offshore sardine fishing has disappeared.

The Nile is 6670 km, 4135 miles long. The long-term water flow at Aswan is 90 km³ a year.

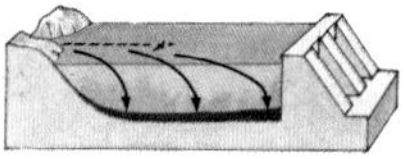

When a river is dammed the speed of the current is reduced and particles of silt and sand sink.

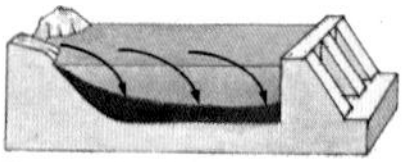

The reservoir bed is covered with a layer of sediment. As the layer becomes thicker the water becomes shallower.

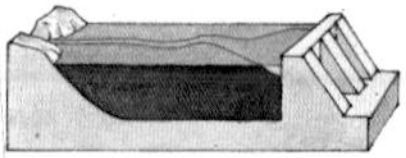

Eventually the reservoir becomes completed silted up, apart from a central channel. The Nile and the Huang He are affected in this way.

Death of the Aral Sea

– a last minute reprieve?

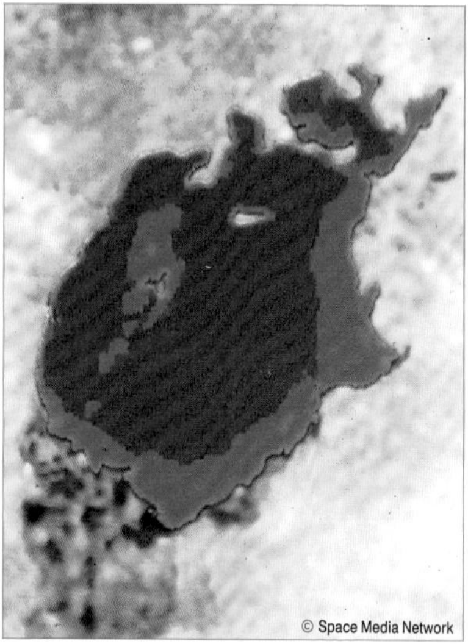

© Space Media Network

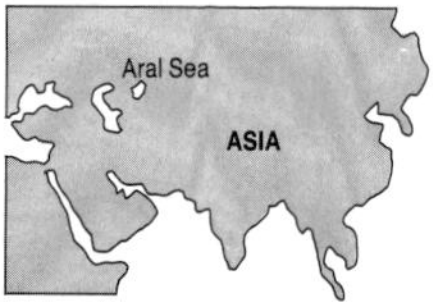

The Aral Sea is under threat. It has shrunk by 40% since 1960 and could become just a few pools of very salty water. Salinity has increased from 1 to 2.7%.

The Sea is in Uzbekhistan which is a dry area. To produce the cotton required by the Soviet Economic Plan needed irrigation. Water from the rivers Amu Darya and Syr Darya which formerly drained into the Aral Sea was used, allowing no fresh water to enter the sea.

The picture above has been made by merging 2 weather satellite pictures, one from 1973 and one from 1988. It shows the shrinkage of the Aral Sea but cannot show that the average depth of the sea, once 16 m, 53 ft is now only 9m, 30 ft. In 1960 the Aral Sea had an area of 68 000 km², 26 250 sq miles. It was the world's fourth largest lake.

The stopping of the freshwater inflow to the lake has had several harmful effects. One has been the loss of plant and animal species that cannot tolerate the increased salinity. The fishing industry is no more. Furthermore, about 70 million tonnes of salt-laden dust fall in the surrounding area causing immense problems for farmers.

Soviet scientists have now highlighted the plight of the Aral Sea and plans are in hand to safeguard its future. In 1987 the inflow to the sea was 10 km³ and this volume should treble by 1990. The aim is to cut the water requirement for irrigation by half. Improved irrigation methods are being introduced to prevent water wastage so that less water is required to achieve the same crop yield.

There is a proposal to divert some water from the Siberian rivers such as the Ob and its tributary the Irtysj to the south. The cost however would be immense and the environmental effect both nationally and internationally can only be guessed at. It is unlikely that this plan will go into operation in the short term, despite enthusiasm for it from the people of Uzbekhistan.

The Great Man-made River

By 1990 the first phase of Libya's Great Man-made River (GMR) should be in operation. It will take water from an underground natural reservoir through a pipeline 1900 km, 1180 miles long to the coast.

Most of Libya's 4 million people live along the coast. Growing numbers and the need to expand agriculture have put a strain on water supply.

Initially water will be taken from the 234 wells in the Sarir and Tazerbo fields. Twin pipes will take the water, by gravity, to a reservoir at Ajdabiya and from there by single pipes to the coastal towns of Sirte and Benghazi. The pipe is planned to carry 2 million m³ of water per day, to provide water for domestic and industrial purposes and to irrigate vast areas of barren land.

Future phases of the GMR will involve taking water from the Fezzan field to Tripoli. There are also plans to extend the pipeline to the Kufra field, and also to link Tobruk to Ajdabiya.

Libyan advisers say that the underground reservoir is being replenished by rainfall and underground seepage, and that the supply should last many decades. Other geologists suggest that the water originated at a time when the area was much wetter, and that it is about 30 000 years old. Tests also show that the rate of replenishment may be very much less than the rate of extraction.

There are suggestions that as the depth to water increases so pumping it up will become very expensive. It is possible that there may also be adverse effects in Egypt and Sudan.

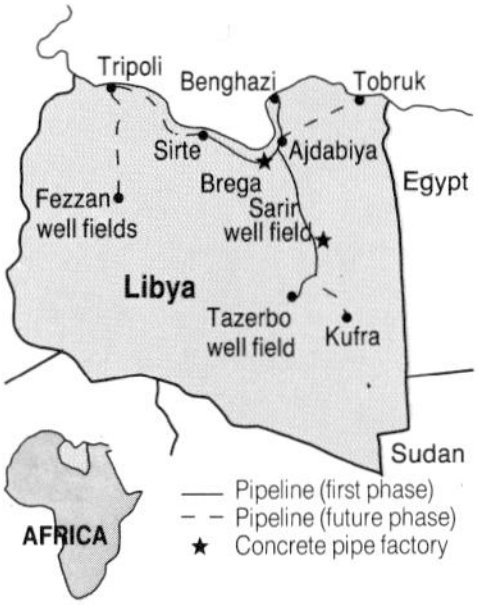

Automatic digging machine which is specially designed to dig a trench of the correct depth, into which the pipes will be laid, and then bury them.

Just some of the thousands of pipe sections waiting in convoy between Benghazi and Sirte. Much of the equipment such as the trailers and cranes for laying the pipes have been designed for this project.

1 pipe laid every 15 minutes
The main pipes are 4 m, 13 ft in diameter and 7.5 m, 25 ft long. Each weighs over 70 tonnes. The pipes are made of steel which is enclosed in prestressed concrete. Steel wire is wrapped round and anti-corrosive mortar provides further protection. Two local plants will make the 250 000 pipes needed. Every 15 minutes a pipe is being buried in the sand.

Building an ice island

In 1986 the American oil company Amoco built an artificial island made entirely from ice. The island, called Mars Ice Island was located 8 km, 5 miles from the Arctic coast of Alaska 160 km, 100 miles west of Prudhoe Bay. It was built to serve as a platform for oil exploration drilling. The process used was similar to that used for making snow on ski slopes. Four pumps spraying seawater at 20 000 litres per minute took 46 days to make the island, which was 16 m, 53 ft thick and 300 m, 1000 ft in diameter.

The cost of the island, the ice road, the airstrip and the construction camp was far less than for a conventional artificial gravel island. Two inexhaustible Arctic resources were used, seawater and cold. The construction method was non-polluting, and the island was completely degradable: in the summer of 1986 it melted and disappeared for ever.

Mars Ice Island was built to serve as a platform for oil exploration.

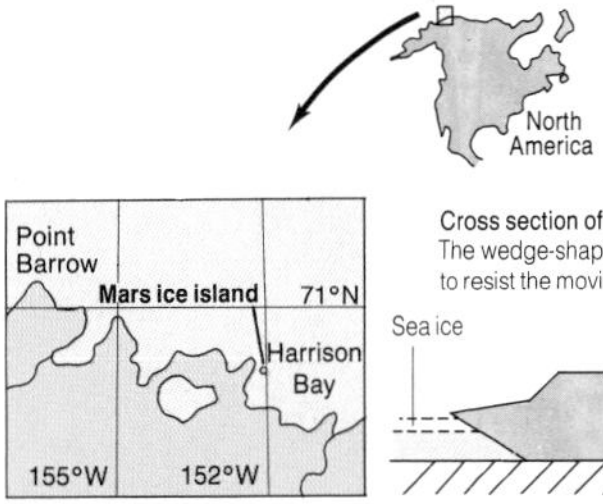

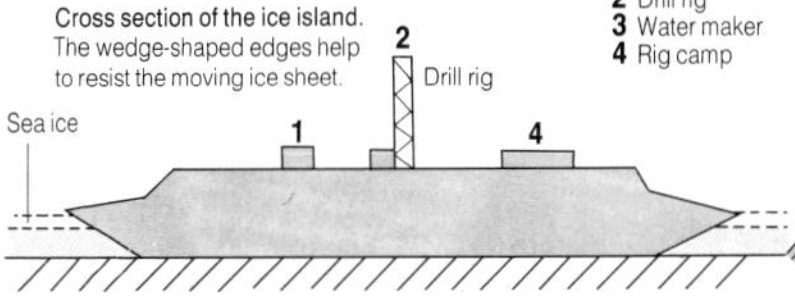

Jonglei Canal. This scheme involves cutting a canal round the swamps of the Sudd region of the southern Sudan, to increase the flow of the Nile. To preserve as far as possible the flooding regime of the surrounding swamp requires careful control of water diversions. The scheme has been interrupted by civil war.

Bondhusbreen, SW Norway. Summer snowmelt from the glacier provides power for HEP station. The water intake, which is beneath the glacier, carries about 70 000 metric tons of coarse sediment a year. The material collects in a huge sedimentation chamber (capacity 10 000 tons) built into the bedrock below the glacier.

Soil waterlogging can occur if excessive irrigation water is applied to land which is inadequately drained.

Salinization. All irrigation water contains dissolved salts and all soils will become saline unless extra water is applied to flush out accumulated salts.

Permafrost layer is the subsurface layer which is always frozen though the surface level may thaw in summer. This is characteristic of tundra regions.

One year of major natural disasters – 1987

• January	Winter damage	UK	34 dead
• February	Typhoon 'Uma'	Vanuatu	50 dead
• February	Floods	USSR	110 dead
• March	Earthquake	New Zealand	Damage
• March	Earthquake	Ecuador	1000 dead
• April	Floods	Canada, USA	Extensive damage
• April	Avalanches	Switzerland	6 dead
• May	Floods	Poland	Extensive damage
• May	Forest fires	China	193 dead
• May	Floods	Indonesia	131 dead
• May	Floods	USSR	Many homeless
• June	Floods	China	200 dead, many homeless
• June	Forest fires	USA	Extensive damage
• July	Floods	Italy	44 dead
• July	3 Typhoons	Korea	333 dead
• July	Tornado	Canada	
• July	Heatwave	Greece	
• July/Aug	Monsoon floods	Bangladesh	

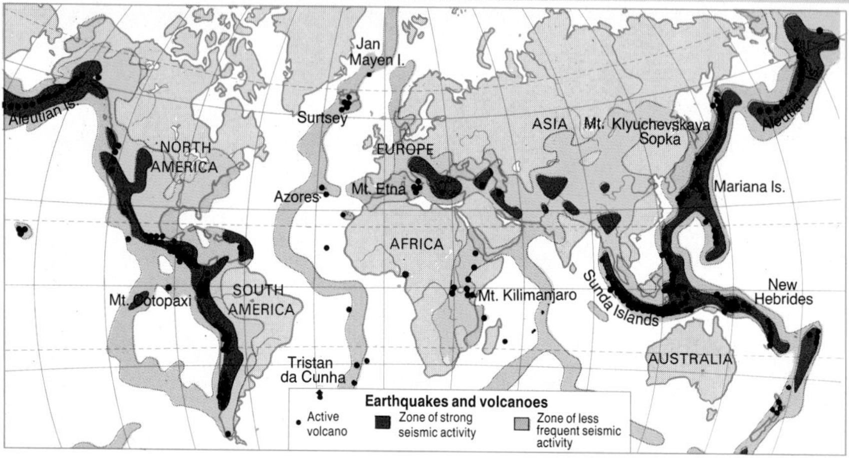

The earth shudders

The Earth is covered by huge tectonic plates which are constantly pushing together, pulling apart and sliding beside each other. The driving forces are convection currents circulating in the liquid mantle beneath the plates. When the plates collide or adjacent ones move alongside each other tensions build up and the earth's energy is released in earthquakes and as volcanic activity. Such areas of instability are to be found in a band through Eurasia and all round the Pacific. Two earthquakes occur each minute, but few, fortunately, are so severe or in such heavily populated places as to be classed as disasters.

Armenia – a geologically unstable region

In December 1988 an earthquake in Armenia claimed 50 000 lives and destroyed whole towns. The cause was instability along the junction between the Arabian plate which is splitting away from Africa and sliding past the Eurasian plate. The force of the earthquake was 6·9 on the Richter scale

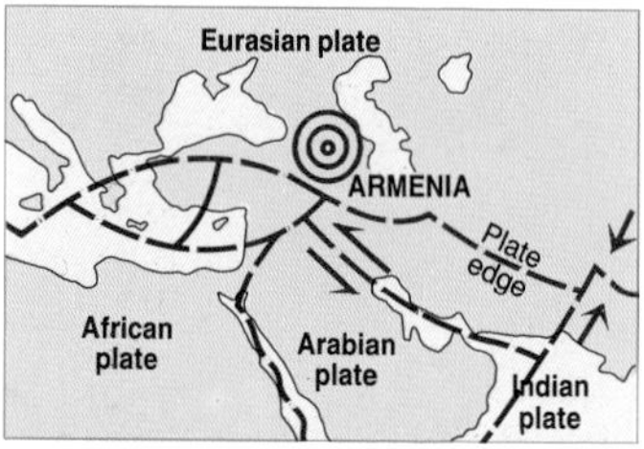

Hurricane in Nicaragua

The US National Weather Service gives girls' names to hurricanes, and it was Hurricane Joan that hit Nicaragua on October 22nd 1988 on her way from the Caribbean to the Pacific where she picked up speed again as Tropical Storm Miriam. The spiralling winds and heavy rain caused extensive damage to several towns, most power lines and much of the coffee crop. But the worst damage was caused by the sea. Low pressure in the eye of a hurricane causes sea level to rise and when the eye reaches shore the storm surge is disastrous. In Nicaragua the damage was extensive but only about 50 people were killed.

Flooding in Bangladesh

The 1988 floods in Bangladesh broke all previous records. In August heavy monsoon rains caused peak flows on the Brahmaputra and Ganges rivers. Deforestation in the upper catchment may cause soil erosion and this could contribute to siltation of the river bed making flooding more likely.

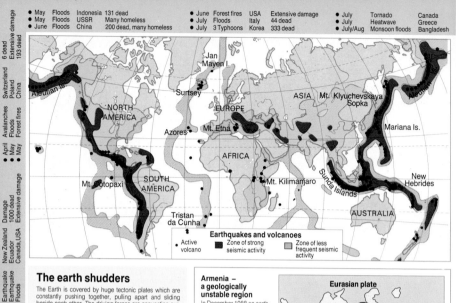

26 dead ● August Floods Switzerland Extensive damage ● October Storm UK Extensive damage
1000 dead ● August Heatwave Greece 1000 dead ● November Typhoon 'Nina' Philippines 650 dead
1600 dead ● September Floods South Africa 487 dead

● December Snowstorm USA Extensive damage
● December Heavy rain Egypt 24 dead

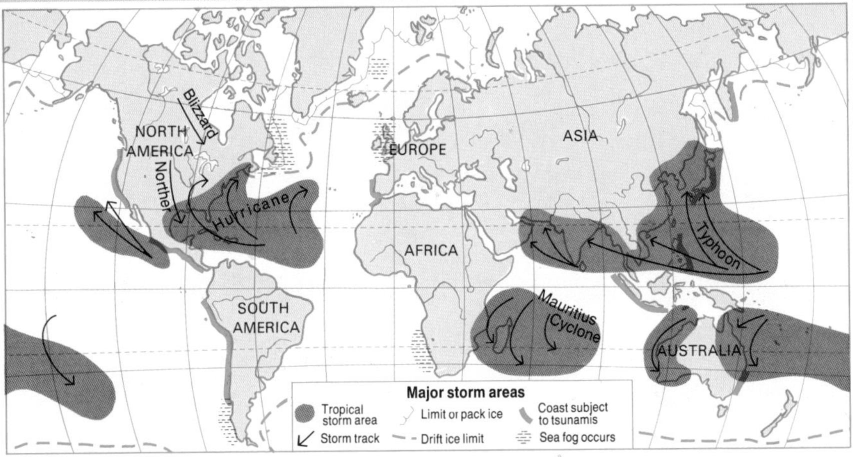

Seasonal winds

The monsoons (see page 14) are large scale sea and land breezes activated by annual rather than daily temperature changes.
The northeast monsoon develops during winter when an area of high pressure builds up over central Asia. Winds blow southwards over Asia towards the sea, but pick up little water over the land. They bring dry weather during the winter months.
The southwest monsoon develops in summer, when a low pressure area over the dry regions in southwest Asia causes moist winds from the Indian Ocean and the South China Sea to blow in over the land with very heavy rain.

The summer monsoon is very unreliable and the timing and amount of rain can vary from year to year. When the rain is late or the amount small crops fail and famine will follow. In other years there will be flooding.

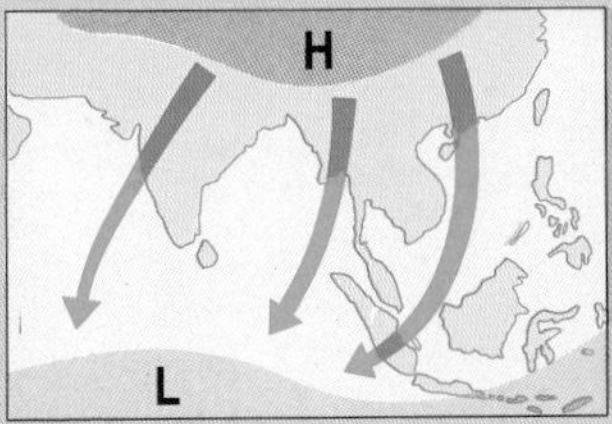

Northeast monsoon

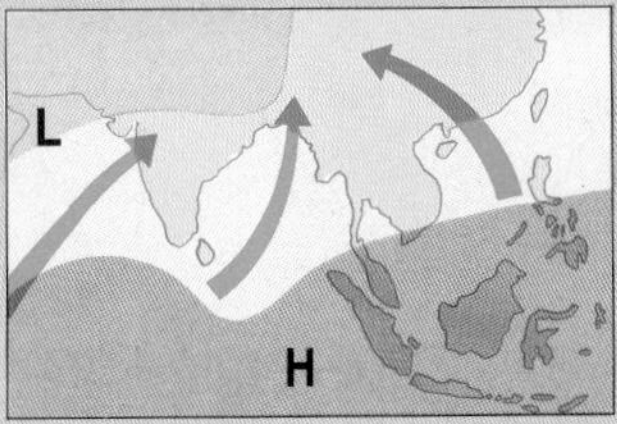

Southwest monsoon

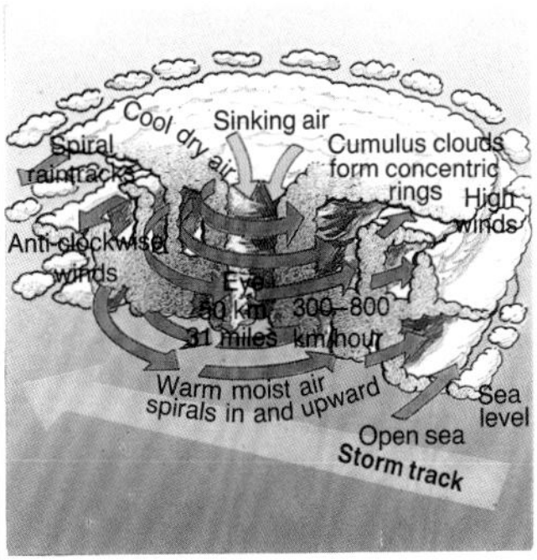

The anatomy of a hurricane

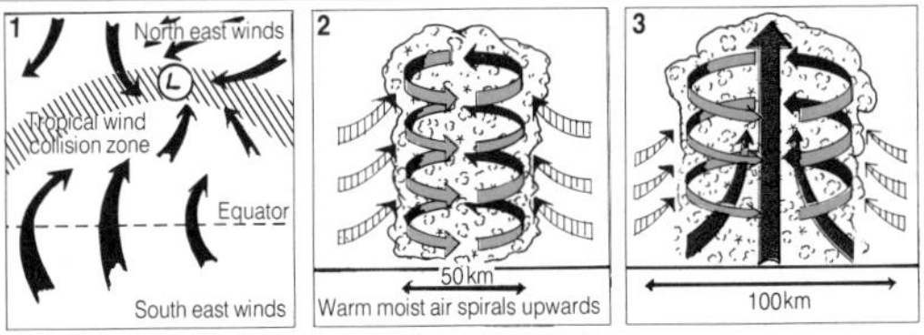

1 In the hot moist air of the tropics there are regions where the trade winds tend to collide. Sometimes the earth's rotation forces these colliding winds into a cyclonic mass of air. **2** The rotation lowers the air pressure at the centre and this causes the hot moist air to rise. **3** When the hot moist air rises condensation will occur. Condensation of water vapour to water droplets releases heat into the surrounding air. This produces even more lifting capacity. The storm starts to develop by itself and makes the column grow more quickly. The growth of the hurricane depends on temperature of the sea surface and the colliding air masses. **4** Over the open sea it may reach a diameter of several hundred kilometres and develop wind speeds of 300 kph, 186 mph. The winds in a hurricane rotate in an anti-clockwise direction.

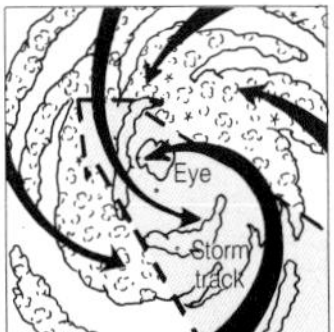

Tropical cyclones	Beaufort scale	12
Severe tropical storm	"	10-11
Moderate tropical storm	"	8-9

Richter scale. Devised in 1935 to measure the strength of earthquakes. The strongest yet recorded registered 8·6 on this scale.

Eye of the storm. The centre of a tropical cyclone is about 25 km, 15 miles wide. It is calm and clear.

Tsunami. High speed ocean waves (150km/hr) of low amptitude (0·5m) which are caused by submarine earthquakes.

Gross National Product per capita in $US (1986)

Algeria	2570	Bolivia	540	El Salvador	820	Ethiopia	120	Philippines	570
Burkina Faso	150	Colombia	1230	Panama	2330	Morocco	590	France	10740
Congo	1040	Bahrain	8530	United States	17500	South Africa	1800	Iceland	13370

Income distributions. The gap between the rich and the poor in a country can be seen by the ratio between the income of the wealthiest 20% and the poorest 20%. Larger numbers show greater inequality.

- Egypt 8.3
- Kenya 23.3
- Brazil 27.7
- China 2.6
- Sweden 5.1
- UK 5.8
- Denmark 7.1
- Peru 32.1
- US 9.5
- Japan 4.1
- Turkey 16.2
- Belgium 4.6
- Canada 9.0
- Mexico 15.4
- India 10.1
- Australia 6.0
- New Zealand 8.8
- Philippines 10.1

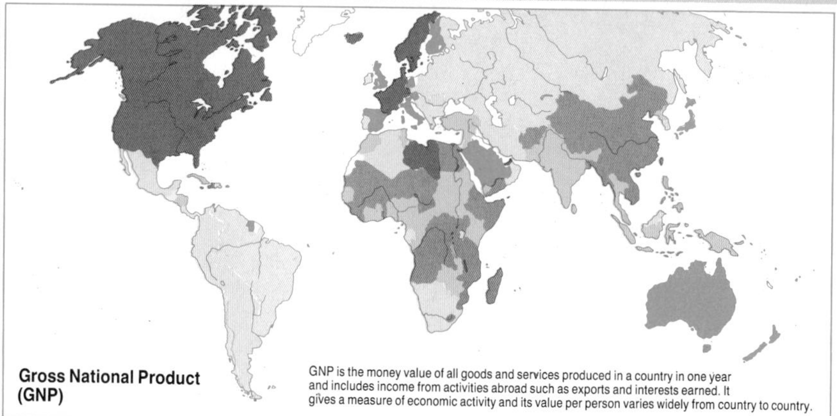

Gross National Product (GNP)

GNP is the money value of all goods and services produced in a country in one year and includes income from activities abroad such as exports and interests earned. It gives a measure of economic activity and its value per person varies widely from country to country.

How they live	GNP in US $ per capita	What they eat
	**> $9000** **Highest income countries** Canada has a GNP of $US 14 120 per capita. Life expectancy at birth is 76 years, and all children attend secondary school. A large proportion of the population works in service industries. Food accounts for 11% of the family budget. The daily calorie supply is 3443 per person (see page 54).	
	**$6000-9000** **High income countries** Saudi Arabia has a GNP of $US 6950 per capita and the standard of living has risen greatly in the last 50 years. Life expectancy at birth is 63 years and infant mortality is relatively high. The daily calorie supply is 3057 per person.	
	**$3000-6000** **Upper middle income countries** Greece has a GNP of $US 3680 per capita and most people live comfortably, but the annual rate of inflation 1980-1986 was 20%. Life expectancy is 76 years and 86% of children attend school. Nearly a third of the family income is spent on food and the daily calorie supply is 3216 per person.	
	**$1500-3000** **Middle income countries** Argentina has a GNP of $US 2350 per capita but has had an average annual rate of inflation 1980–1986 of 326%. Life expectancy is 70 years. 70% of secondary age children attend school. Service industries provide jobs for 44% of the population. More than a third of family income is spent on food. The daily calorie supply is 3368 per person.	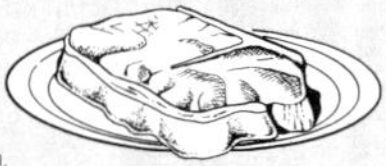
	**$500-1500** **Lower middle income countries** Thailand has a GNP of $US 810 per capita and a very low rate of annual inflation of 3%. Life expectancy is 64 years and infant mortality is high with 43 deaths per 1000 births. 30% of children attend secondary school. Food accounts for 34% of the family income. The daily calorie supply is 2399 per person.	
	**< $500** **Lower income countries** Mozambique has a GNP of $US 210 per capita. Life expectancy is only 48 years and the rate of infant mortality is high with more than 1 in 10 babies dying. Secondary school is attended by 9% of the children, and most of the people live on the land. The daily calorie supply is 1617 per person.	

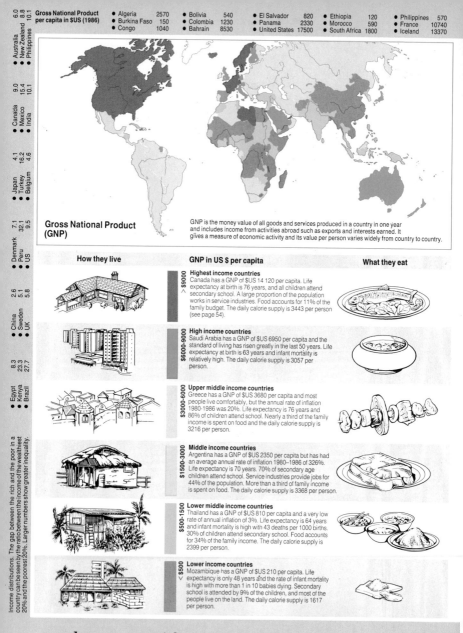

. . .and some were richer and some were poorer
developed versus developing world

Copper production in thousand tonnes

Country	Copper ore	Smelter ore	Refined copper
• West Germany	208	422	
• Japan	980	980	
• Zaire	502	472	218
• Chile	1391	988	786
• Papua New Guinea	171		
• Zambia	536	523	461
• US	1159	1249	988
• Philippines	217	461	
• USSR	1180	1180	1180

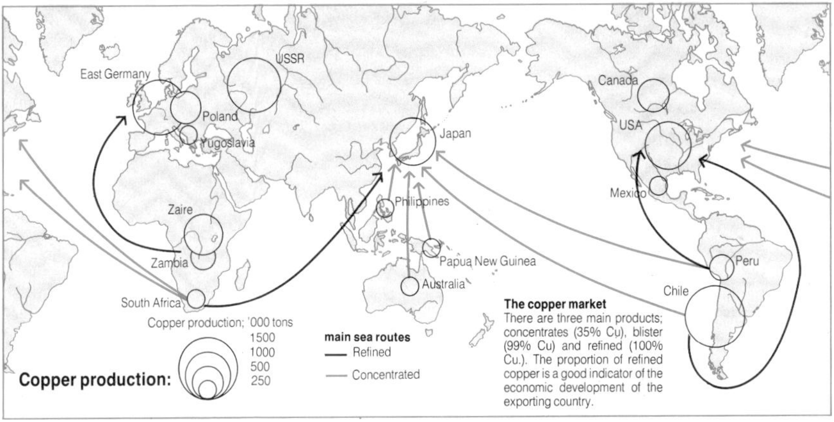

The copper market
There are three main products; concentrates (35% Cu), blister (99% Cu) and refined (100% Cu.). The proportion of refined copper is a good indicator of the economic development of the exporting country.

Copper producers

Since 1974 world copper prices have fallen. Producing countries have invested in smelting equipment to process copper into the more profitable refined product. Low prices, expensive operating costs and pollution have resulted in many plant closures.

Chile – an efficient operation
In 1980 Chile became the world's largest exporter of copper. 80% of production is from 4 mines near the coast. Good management, modern refining facilities and the use of local coal have enabled the industry to survive falling world prices. However, revenue from copper exports has fallen to 60%.

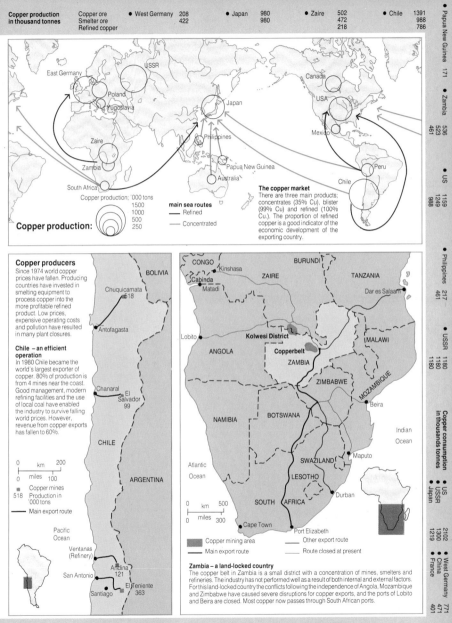

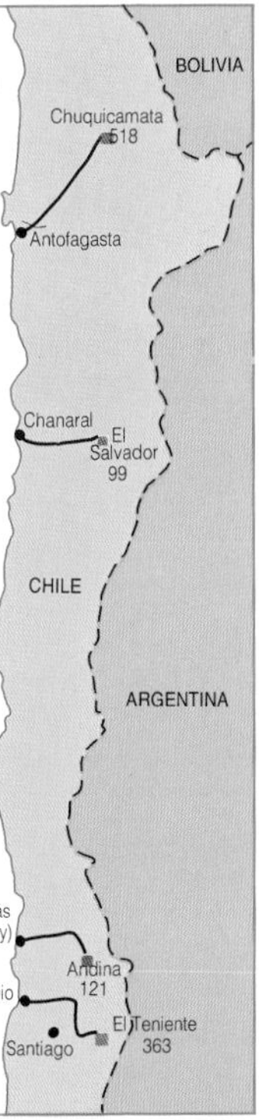

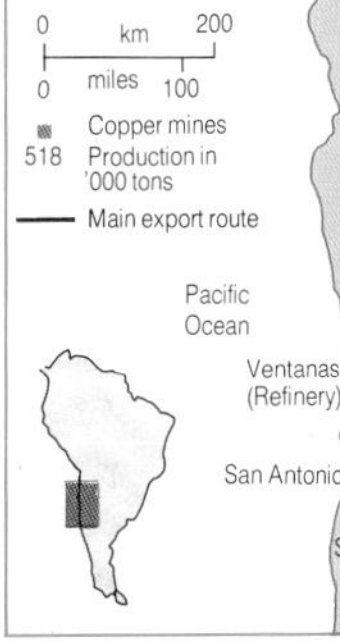

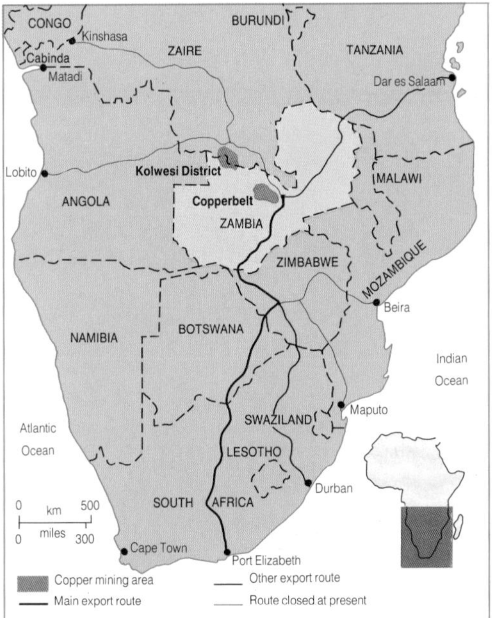

Zambia – a land-locked country
The copper belt in Zambia is a small district with a concentration of mines, smelters and refineries. The industry has not performed well as a result of both internal and external factors. For this land-locked country the conflicts following the independence of Angola, Mozambique and Zimbabwe have caused severe disruptions for copper exports, and the ports of Lobito and Beira are closed. Most copper now passes through South African ports.

Copper consumption in thousands tonnes

• US	2102
• USSR	1300
• Japan	1219
• West Germany	771
• China	471
• France	401

Properties of copper Copper can be easily hammered (malleable), or drawn into wire (ductile). It has good electrical and thermal conductivity, and is widely used in the electrical industry and for pipes in plumbing.

Service industries include retailing, transport and banking. Richer countries have a higher proportion of people in these industries.

The Roman Emperor Domitian (A.D. 81-96) was hated, and after his death his image was destroyed.

This silver penny of Cnut has peck-marks on it, a test for purity done in Scandinavia.

The word 'Dollar' comes from this silver 'thaler' of 1510, short for Joachimsthaler.

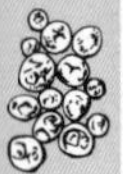
The earliest known hoard of coins was buried during the building of the Temple of Artemis at Ephesus, west Turkey.

These aragonite stones are 50 cms (20") across. They were used by the Yap islanders as money, or as displays of wealth.

The word money is derived from the goddess Juno Monets, whose temple was a mint in ancient Rome.

Banks and banknotes

The earliest known example of a banking transaction is from Mesopotamia, about 1750 B.C., where silver was loaned out for interest. Banks started as sidelines to other businesses; the word comes from the Italian 'banco' (bench), where Venetian merchants sat to do business in the marketplace. The earliest paper money, from 11th century China, replaced copper coins which were always in short supply there. Mulberry bark was used in 14th century China, and silk in Kharzem (USSR) in 1920. The Bank of England was founded in 1684 by London merchants who were dismayed by the actions of Charles II. In 1672 he had forced them to lend him £1,300,000, and no interest was paid for 18 years! In 1816 the Bank adopted 'The Gold Standard', which fixed the price of gold at £3,17s,10½d (£3.89), and gave security to banknotes issued. This lasted until 1914. Since 1945 a different 'Gold Bullion Standard' has been in existence.

Clay tablet from Sippar recording loan of silver made by the Temple of the Sun God and the interest due to them.

Money business

'Shares' in a company started as just that – merchants got together to contribute cash or kind to finance voyages of exploration. By the late 16th century in Europe there were active markets where these shares could be bought, sold or exchanged. A statute of Verona dated 1318 mentions a "forward market"; shares in German mines were quoted at Leipzig fair in the 15th century; and by 1688 Joseph de la Vega of Amsterdam could boast 'speculation and gambling in shares takes place between the new 'coffy' houses and our Stock Exchange which was built in 1631'.
Today, the shares market is enormous – in the UK in 1988 75% of adults had money invested in the London Stock Exchange, most via pension funds or life assurance policies.

A modern bank note

This cheque dated 1663 is the earliest European bank-note, from the Stockholm Banco.

Making money

Coins are made either by stamping a blank metal disc between two dies; or by casting, that is pouring molten metal into a mould. The latter is quicker but less accurate, and was mostly used in the Far East, where gold and silver were not used. Until the 16th century minting (the process of making coins) was done by hand. The picture here shows (from L to R): smelting the metal alloy, hammering it to the correct thickness, coining it using 2 iron dies, and presenting finished coins for testing, or 'assay'.

Wall-painting from Thebes, 14th century B.C., showing gold rings being weighed

The earliest records

As soon as written records start, we have evidence of money being used. The earliest is a Sumerian stone tablet from Mesopotamia, about 2600 B.C.; it details how many baskets of barley land is worth for tax or rent. In Egypt, about 2200 B.C., weighed amounts of silver, gold and copper were used for payments. The Bible records silver weighed in shekels at about 1900 B.C., when Abraham bought a burial plot for Sarah, for 400 shekels – more than 3 kilos, 7 lbs. Markets and trade routes exist from prehistoric times linking Europe, Africa and Asia; money passed freely around. By 1641, Holland had 400 different currencies in circulation, and France over 80.

Silver hoard from El-Amarna, 14th century B.C.

Early money

Most tribal societies were self-sufficient in basic needs, so 'money' was used for special, often ceremonial purposes such as dowries ('bride payment') or compensation. Cattle and pigs are still used in this way in some places. The display of wealth was important – bead, shell or feather 'money' could be used as decoration and worn to show the status of the owner.

Bridal head-dress

The Swedish East-India Company

The activities of the company illustrate the way goods were traded world-wide, with money as an intermediary.
Founded in Gothenburg in 1731, the company had a base in Canton. From 1731-1813, 132 trips were made using 40 ships, of which 8 were lost (see page 100). The journey took approx. 18 months. Iron bars and goods from Sweden were sold in Cadiz or Seville for Spanish silver coin, the currency in use throughout the Far East, and particularly valued by the Chinese. These Spanish piastres ('pieces of eight') or reales were minted in Spain, using silver loot or bullion from mines in Mexico and Brazil. Once in China, luxury goods such as tea, porcelain, and silks (together called 'spices') were bought, and on return to Gothenburg were auctioned. From Sweden, the goods were distributed and re-sold all over northern Europe – Dutch traders were prominent in this part of the trading cycle.

Cowrie shells and ideograms

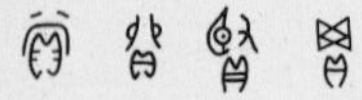

What is money?

Exchange and barter must be as old as human society, though the earliest written record is the 3rd millennium B.C. Money can be defined as any object regularly used to make payments, and in the past shells, stones, beads, tea, salt and cattle have all served. Cowrie shells are one of the earliest widely used forms of currency. The mollusc *Cuprea moneta* lives in the Indian Ocean, but was traded as far apart as West Africa and China at least 4500 years ago. 4000 years ago pictograms of cowries were incorporated into many words to do with payments.

Iron was produced in bars for export from Sweden.

. . . and we paid with shells and plastic cards
currency development

Forgery is as old as money. This silver-plated copper denarius is from a hoard of forgeries found in London, dated early 1st century A.D., just after the Roman invasion.

Greek drachma, 450 B.C., showing a table piled with coins. This is a pun on the name of the city (Trabezus), as 'trapeza' (table) was, and still is, the word for a bank.

Bruges Place de la Bourse. The origin of the word 'borse' comes from this building used as the exchange.

The present and the future

Today every country on earth uses some form of currency, and the majority of monetary transactions are made via computer records rather than paper or metal money. Plastic credit cards make it easy for us to get at our money – or to borrow. The latest developments include 'Direct Debit', where shop tills are linked to bank records and allow money to be taken immediately from an account. Refinements to credit cards will mean each card is itself a mini-computer, capable of carrying out many of the functions we still have to make a visit for. We may enjoy beads or shells on our desert-island holiday, but we will have paid for it with plastic, micro-chip and advanced technology.

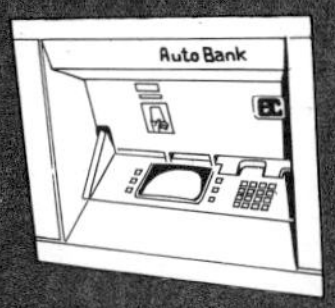

Banks are now open 24 hours a day for certain services.

A mint workshop, Bern, Switzerland 1486

Power over money

Minting money needs few resources, but strict controls. Images shown on coins and later on banknotes came to symbolise this power. The right to mint money was often a royal prerogative. Having a powerful image shown on coins meant that penalties for tampering or forging money could be 'treasonable'. An example of changing politics was the replacing of 'heraldic' designs (i.e. the tyrants) by the 'democratic' goddess Athena and her owl, at the time the tyrants were expelled from Athens in 510 B.C. Coins often had mint marks identifying the moneyer or the time it was minted – the date is not found until the late 16th century.

Brass Chinese coins, 1796-1820, with secret mint marks identifying which part of the Beijing mint made them.

This Anglo-Saxon gold and garnet pendant from Norfolk, about A.D. 620, is set with a Byzantine gold solidus. It may come from the same workshop as the Sutton-Hoo ship treasures.

Metal is money

Metal in many shapes and forms has always been popular as money. It is easy to measure, is universally accepted, and even small payments can be made accurately.
Silver was rare in ancient Egypt, and a hoard from El Amarna, dated about 1400 B.C., contained bars, rings and plate silver. Pliny (A.D. 23-79) reports that the Romans and the Etruscans used lumps of raw copper in the 5-4th centuries B.C. Iron in the shape of hoes, or other tools, was used until the 19th century in Sudan, and copper rings called 'Manillas', which could be worn as armlets, were widely used in west Africa until 1948.

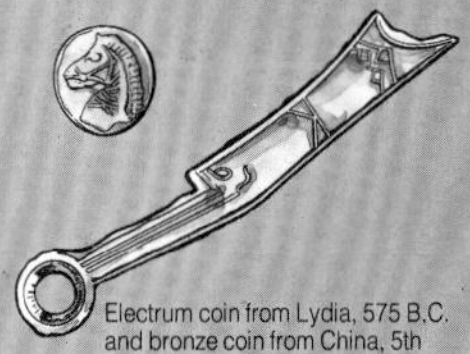

Electrum coin from Lydia, 575 B.C. and bronze coin from China, 5th century B.C.

The first coins

Greeks in western Turkey in about 600 B.C. were the first to stamp pieces of metal of fixed size and weight, to guarantee their value. The metals used were electrum, gold and silver. Within 100 years, this practice had spread widely. In about 425 B.C. Greek cities in Italy started to make small copper coins to replace silver ones – this was the first time a coin had a monetary value above its real value. The Chinese invented coins independently in the 5th century B.C.

By 1750 Cadiz and Seville were the main ports of entry for goods from South America.

Black slaves from Africa worked the silver mines in South America. The Spanish also converted treasure looted from the Incas.

Canton was the largest port in China. This picture shows the river lined with the warehouses of the Europeans.

All our coinage today is 'token', as it has no metallic value; however items such as this telephone token and sweet (from Italy) have been used as small change.

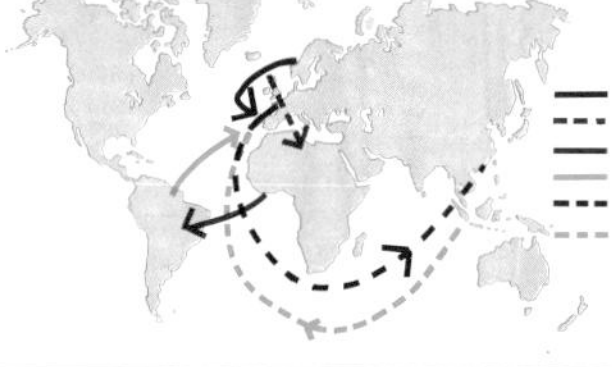

Iron bars	Sweden – Spain
Weapons	England – N.Africa
Slaves	Africa – S.America
Silver	S.America – Spain
Silver	Spain – Canton
Tea, Spices	Canton – Sweden

The luxury goods the Europeans bought were mostly tea, silks and porcelain.

Sterling – This word is thought to derive from 'Easterlings', the name given to Baltic merchants who settled in the City of London in the 15th century.

Assay – is the test done on coined money to ensure the correct purity and weight. In England it is carried out once a year, and is called 'The Trial of the Pyx'.

Capital from the Latin 'caput', head, was first used by Italian traders in the 12th century, referring to a firm's assets. 'Capitalist' is first recorded in Holland in 1633.

United States foreign stock
US residents in 1960 who were immigrants or one or both of whose parents were immigrants.

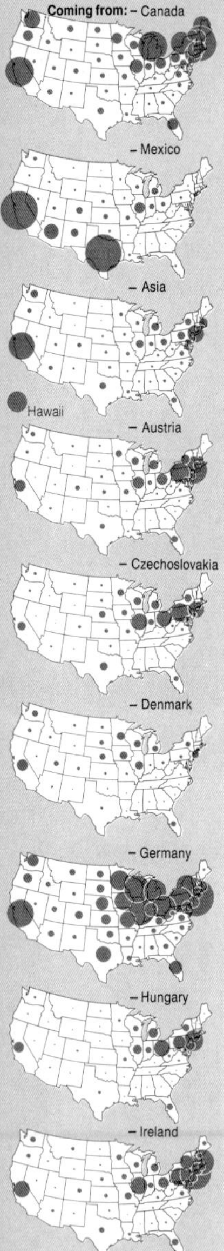

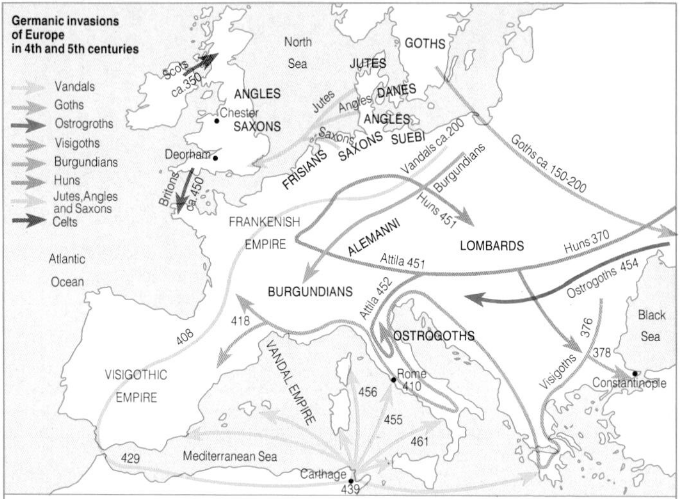

They did not want to stay

The movement of groups of people from place to place is as old as men and women themselves (see page 24). Most migrations are over relatively short distances but from time to time in history there have been periods of mass movements or emigrations. The spread of the Germanic tribes over Europe in the 4th and 5th centuries was one such movement and the emigration of millions of Europeans to North America in the 19th and early 20th centuries another.

Most emigration is caused by pressure of population in the home country; a shortage of land combined with rural poverty has been the driving force for many people seeking a fresh life and economic opportunities in a new land. The highly educated and skilled may also be attracted by the opportunities offered by another country. Religious or political persecution has been another potent reason for emigrating and many people have also been displaced by wars.

Guided tours through Europe

Europe was disrupted in the 4th and 5th centuries by the invasion of peoples from Central Asia. The Huns, who were nomadic tribes, originated in Mongolia, and having overthrown the Ostrogoths, invaded Greece and Italy under their leader Attila. This invasion triggered off a general movement of the germanic tribes with the Visigoths sacking Rome in 410 before moving into France. The movements of other groups such as the Vandals and the Burgundians can be followed on the map.

From the middle of the 5th century Britain was invaded by Angles, Saxons and Jutes from northern Europe. They met with fierce resistance but finally established control over the country after victories at Deorham in 577 and Chester in 616.

On continental Europe the number of invaders was too small to alter radically the nature of Romanized society. Germanic rulers took the place of Roman leaders. In Britain however the advent of different peoples began a new phase in the country's history.

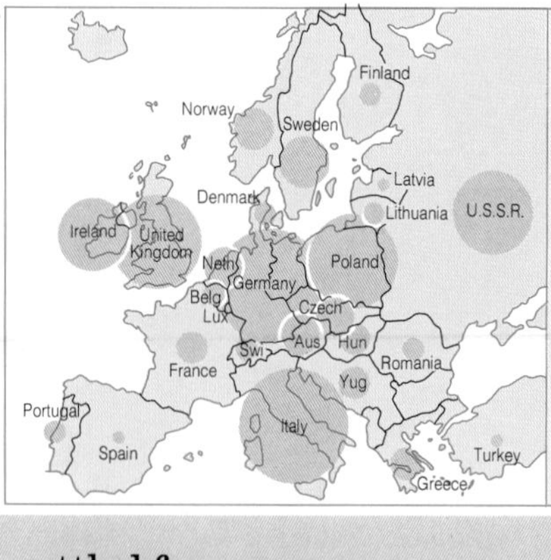

Next stop – United States

The United States has received more immigrants than any other country, and is sometimes called a 'nation of immigrants'. In the 18th and early 19th centuries there was a steady stream of arrivals from Europe, and until the abolition of the slave trade, many black people from Africa.

After 1850 the second great wave of immigrants arrived, with 2·5 million in the first decade. In the early stages there were many Irish and Germans, escaping from famine and economic depression. From 1870 an increasing number of other East Europeans arrived in America. The largest immigrant group however came from Italy in the latter part of the century and into the 20th century. Many European settlers started their own farm under the Homestead Act of 1862 and the industrial boom of the 1880s created many new jobs in the rapidly expanding towns.

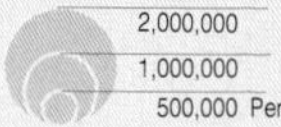

. . .and some settled far away

United States foreign stock
US residents in 1960 who were immigrants or one or both of whose parents were immigrants.

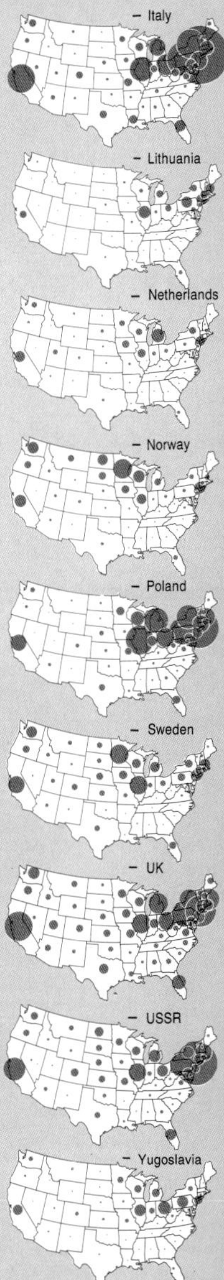

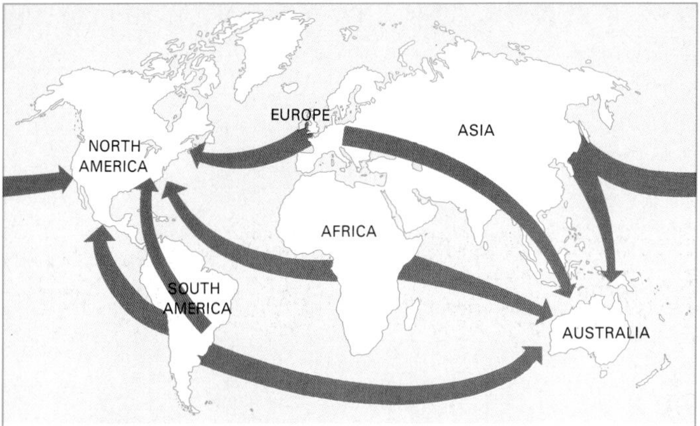

A mixed world

Most movements of people are over relatively short distances, usually within their own country. Since World War II there have been large movements within Europe and Asia in response to employment opportunities in Germany, France and Switzerland. Internal movements of people in many parts of Africa have often been the result of local or national wars. Starvation, too, has caused large scale migrations.
The population of the oil-rich states of the Middle East has also been swelled by Arab immigrants from Syria and Jordan who work in middle management. Workers from Pakistan and India have also moved in to work in more menial jobs on construction sites or in oilfields.
Movements between countries are now strictly controlled by immigration policies but skilled people have moved to North America and Australia. The immigration of many highy educated people has been termed 'brain drain'. Economic and political uncertainties in their country have encouraged many Argentinians to move to the United States and to other South American states.

Chinatown

A substantial Chinese community can be found in most large cities round the world. In some cases workers were brought to work on agricultural or construction projects. Cities such as Singapore have a very large Chinese element. A great number of Chinese have settled along the west coast of North America, and their influence stretches far beyond the ubiquitous Chinese restaurant. Many Chinese hold important jobs in commerce, politics and education.

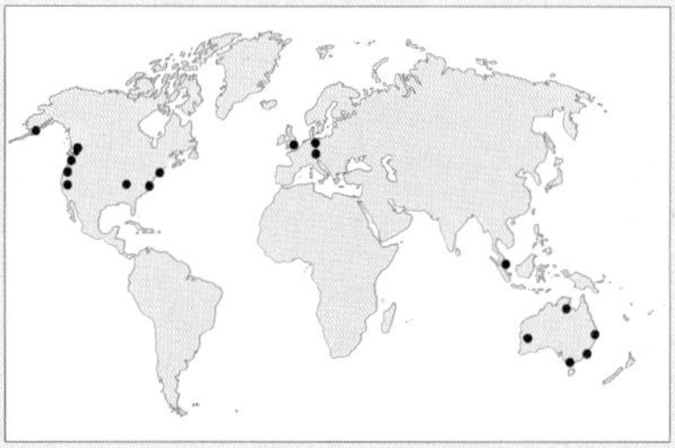

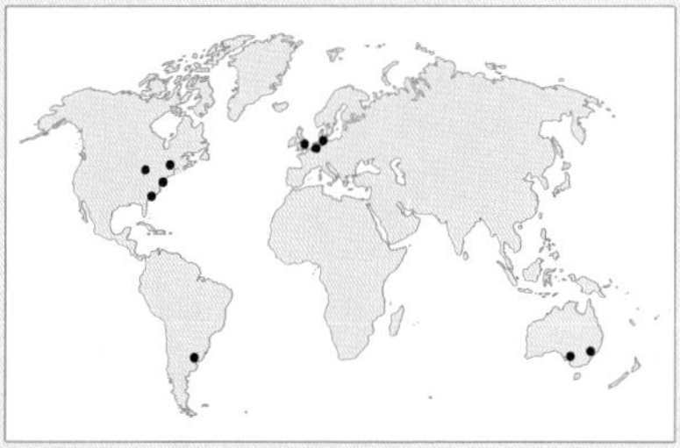

Little Italys

Pressure of population and poverty, particularly in the south of the country have encouraged many millions of Italians to move abroad. Many found a new life in North America particularly in the cities of the eastern seaboard. In Australia the Italians form the largest immigrant group after the British. Their numbers build up by chain migration. As well as settlers in Sydney and Melbourne there are many Italians working in Queensland and the orchards of New South Wales.

Chain migration. Individual sponsors help relatives and friends from their home village or area to emigrate. They in turn enable others to join the community.

Minorities

There is probably no state in the world that can claim a totally homogeneous population. Every country has minority groups differentiated by race, religion, culture or language; frequently by a combination of these social characteristics.
The drawing of political boundaries has not always taken into account cultural identities; the peace settlement following World War I satisfied some ethnic groups but still left many frustrated minorities.
There is frequently discrimination against minorities and discontent can explode into violence. The objective of the group may be for better treatment, for a state of their own, or sometimes for reunion with the nation from which they have been separated.

The Caucasus area

The Caucasus mountains run northwest to southeast between the Black Sea and the Caspian Sea. They straddle the Soviet Republics of Gruziya (Georgia), Armeniya (Armenia) and Azerbaydzhan (Azerbaijan). The region is a mosaic of different languages and ethnic groups. It has long been a battlefield, and this century has witnessed the uprooting and persecution of a number of these cultural entities.
The principal conflict is between the Muslim and Christian groups. There is a large enclave of Christian Armenians in Nagorno-Karabakh which is part of southern Azerbaydzhan, where the people are mainly Shia Muslims. In 1988 crowds in the Armenian capital of Yerevan protested about the way their compatriots are treated and demanded the return of the region to their control. The Azerbaydzhanis are equally determined to retain the area and friction between the two groups continues.

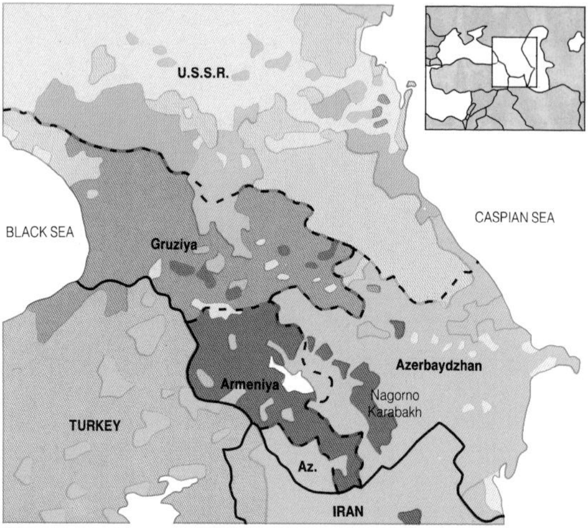

Georgian
Other Caucasian languages
Turkish (including Azeri)
Other Turkish languages
Kurdish
Other Iranian languages
Armenian
Russian (including Ukranian)
Kalmuck Mongolian

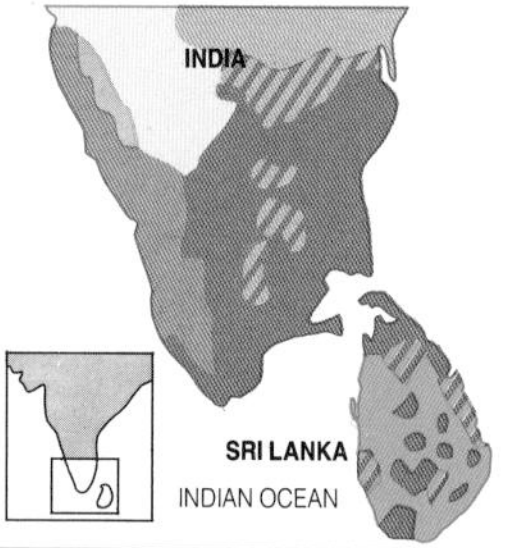

Sri Lanka – Southern India

The people of Sri Lanka are divided principally between the Sinhalese who are Buddhist and form 70% of the population and the Tamils who are Hindu. This minority group makes up 18% of the population and they live mostly in the drier northern and eastern parts of the country.
From the 7th century onwards the Tamils came in increasing numbers from South India, and in later centuries established Jaffna Kingdom, their own state separate from the Sinhalese kingdoms.
In addition to the long-settled Ceylon Tamils there are many Indian Tamils who came to work on the tea plantations in the 19th and 20th centuries and live in the centre of the island.
Tension between the Sinhalese and the Tamils increased after 1970 and agreements were made to repatriate some of the labourers and grant citizenship to the remainder, but these were not fully implemented. There has been considerable violence in recent years as the Ceylon Tamils in the north have been pressing for a separate state; Indian troops have been assisting the government to keep order.

Tamil
Malayalam
Telugu
Kannara
Other Dravidian
Sinhalese

Before James Cook

In 1770 James Cook landed in Botany Bay, south eastern Australia and claimed the territory for his country. It is estimated that in 1788, there were about 750 000 Aborigines, the indigenous inhabitants of Australia, living in 600 tribal groups each speaking their own dialect. At that time 200 separate languages existed of which half have now been lost.
The Aborigines tended to live in small isolated communities but followed a nomadic way of life with tribal areas of vast extent. Although there were no defined frontiers the Aboriginal population had a very special attachment to their land. Their way of life included fishing, hunting and gathering wild plants, and although they used stone tools, the manufacture of metal implements had no part in their lives.
Throughout the 19th century and into the present century the Aborigines were often treated badly; their lands were seized and many were killed.
The last Tasmanian aborigine died in 1876. Since 1930 this minority group has received better treatment though discrimination still exists.
Originally the policy was to settle the people in reserves based on tribal homelands. More recently the emphasis has been on assimilating the Aborigines and giving them full citizenship. Many now live in cities. In 1987 a treaty of understanding with the Aboriginal people was proposed and in 1988 further concessions were discussed.

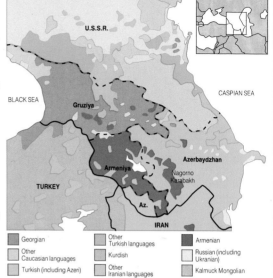

Just some of the 28 major aborigine language families

. . .and there were fewer of some
minority people

Some American Indian tribes

Once a majority

When Europeans arrived in North America there were already well established Indian groups living throughout the continent. Estimates of their numbers vary but there must have been at least 1 million. In the 1980 US census there were 201 311 Indian people; the majority living in California, Oklahoma and Arizona. When the country was settled conflicts between the colonists and the native peoples were inevitable as land was taken over and enclosed for farming. After 1850 some consideration was given to the Indian peoples and the question of tribal sovereignty over land was addressed. The government policy was to set up reservations, partly to prevent conflict between the whites and the Indians, and partly so that they could be taught western ways of agriculture.

The present policy of self determination allows the Indians to take responsibility for their own management. They have all the rights of American citizens but individuals are also subject to certain restrictions and concessions relating to land sales and taxes.

Indian viewpoint

At the time of the white man's arrival there were numerous tribal groups which had evolved life styles adapted to the use of available land resources. Most sustained themselves through hunting, gathering and sometimes agriculture. The abundance of fish in the northwest provided a basis of life for tribes such as the Haida whilst the Indians of the Great Plains based their economy on buffalo hunting. The white pioneers of the West impinged on the Indians' hunting grounds and they were pushed north and into the arid south-west, where they became cultivators. Archaeological remains show that in the southwest the Pueblo people were living in villages 6000 years ago.

Indian way of life

- Fishing
- Cultivation
- Hunting
- Wild plants
- Hunting and wild plants

China: – the largest minority nationalities

Population density

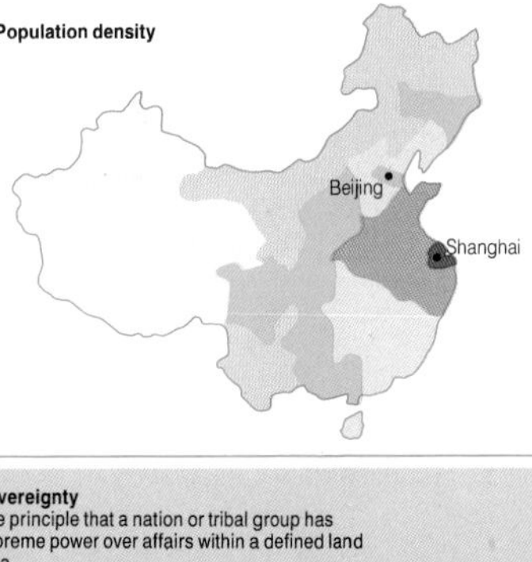

67 million but still in a minority

In China there are 57 ethnic minority groups numbering 67 million people. They form, however, only 7% of the total population, the remaining 93% are of the Han group.

These minority groups generally live in remote areas which are sparsely inhabited, and as the map shows they are to be found round the borders of the country. Indeed such groups as the Mongols have close links with neighbouring countries. The situation in Tibet is somewhat different as the country became effectively independent from China in 1911 but was reconquered in 1959. A revolt which was quelled led to the exile of the spiritual leader, the Dalai Lama together with many followers. There is considerable unrest there because of the suppression of Tibetan culture.

In order to foster minority culture, a number of books are published in the most widely spoken languages. In addition to Han Chinese, the local languages are taught in schools and many textbooks are now available in the minority languages.

Minorities and the one-child policy

The population of China was estimated to be 1 085 million in mid 1987. The rate of increase is 35 068 a day which means 12·8 million babies are born each year.

The changing birthrate shown in the diagram reflects political events. During the Cultural Revolution the birthrate rose, but since the 1970s there have been vigorous efforts to control the growth of the population.

Because of the strategic position of many of the minority groups it is thought desirable to increase the size of the local population. The minority groups, therefore, do not have to conform to the one-child policy and the population in their areas is growing more rapidly than in the rest of the country.

Birth rate per 1000 population

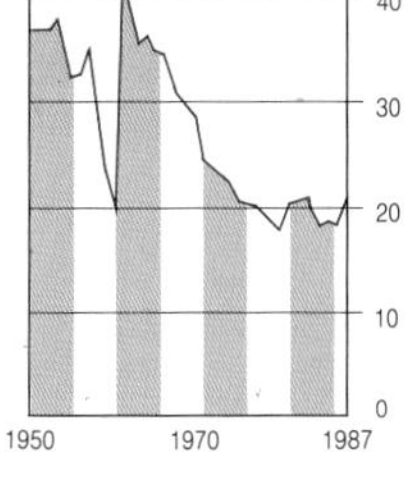

Population density per km²

- <10
- 10-100
- 100-200
- 200-300
- 300-1000
- >1000

Sovereignty
The principle that a nation or tribal group has supreme power over affairs within a defined land area.

1 km² equals 0·39 sq miles **1 sq mile equals 2·59 km²**

Refugee population by country and original state of the refugees – 1988

Country	Refugees	Origin
• Asia	5 668 400	
Iran	2 830 000	Afghanistan
Thailand	112 000	Laos, Cambodia, Vietnam
• Africa	3 975 900	
Somalia	840 000	Ethiopia
Sudan	807 000	Ethiopia, Uganda
Malawi	401 600	Mozambique
Zaire	320 000	Angola, Rwanda
Burundi	270 000	Rwanda

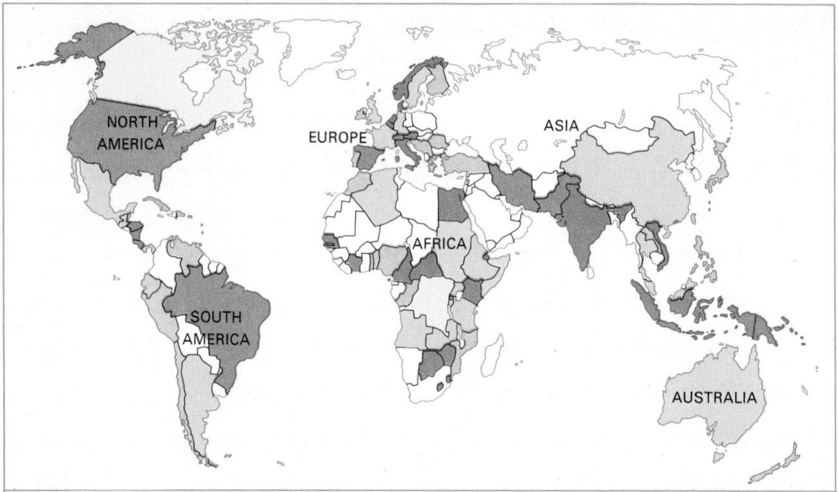

Refugees, 1985

- Less than 500 or no data
- 500–5000
- 5000–50 000
- 50 000–250 000
- 250 000–500 000
- 500 000–1 million
- Over 1 million

Refugees

During and after the World Wars there were many millions of people who were forced to leave their homelands. More recently there has been an escalation in the number of refugees and there are now at least 14 million stateless people, half of them children.

The 1951 United Nations Convention defines refugees as those who cannot live in their own country owing to a well-founded fear of persecution 'for reasons of race, religion, nationality, political association or social grouping'. At present those fleeing because of war are not included, though there are many who leave their home for this reason. Five million Afghans have fled since 1980 and there are many refugees in Africa, Central America and the Middle East.

In Africa the victims of poverty and starvation who hope to find better conditions are added to those homeless because of wars. It is estimated that as many as 300 million people could become environmental refugees next century if the predictions of some climate change scientists are correct.

The United Nations seeks to help refugees through UNHCR. The organization has two principal functions; to protect refugees and to seek longterm solutions to their problems. As well as trying to ensure that refugees are not forced to return home to face persecution, UNHCR endeavours to enable them to make a new start. There are 3 possible solutions; the preferred one is to encourage people to return voluntarily to their own country. Other possibilities include integration into the host country or resettlement in another country.

A region in turmoil

During much of the last decade political unrest and armed conflict have been widespread throughout Central America. Thousands of people have been forced to leave their homes, either to seek refuge elsewhere in their own countries or in neighbouring countries; often relying on hosts as poor as themselves.

Displaced people from Guatemala, Nicaragua and El Salvador have fled across international boundaries and many live in refugee camps. UNHCR's initiative is enabling some to become self-sufficient through rural resettlement projects, though these have aroused some criticism. In one such camp in Guatemala, everyone in the community has some activity; for the children it may be feeding the hens, for adults there is teaching or perhaps making shoes.

Where it is safe to do so, refugees are encouraged to return to returnee settlement zones in their own country and begin productive lives.

Some hope for peace in the area lies with the Esquipulas II plan submitted by Nobel prize winner President Arias of Costa Rica. However the economic problems facing the region are enormous.

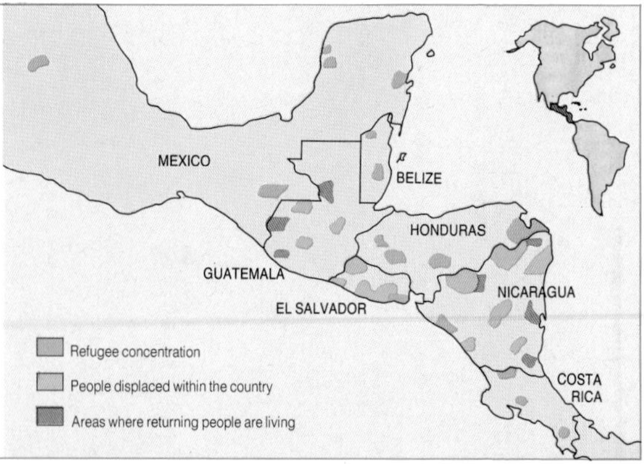

. . .and some could not stay

refugees

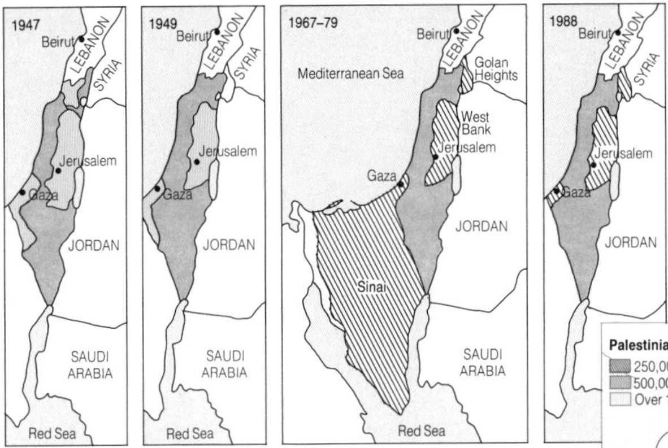

The world's conscience

There are about 4 million Palestinian Arabs, of whom about 700 000 became refugees after the 1948 war. About 40 per cent of Jordan's population are Palestinian. Some live in the 10 refugee camps administered by UNRWA, but many have drifted to Amman. In Jordan refugees are given citizenship and have the same economic and political rights as other citizens.

As a result of the 1974 Rabat Conference the PLO now represents the Palestinians. In 1988 the PLO, proclaimed an independent state of Palestine. They have also accepted the UN resolution 242 which implies recognition of the Jewish state of Israel.

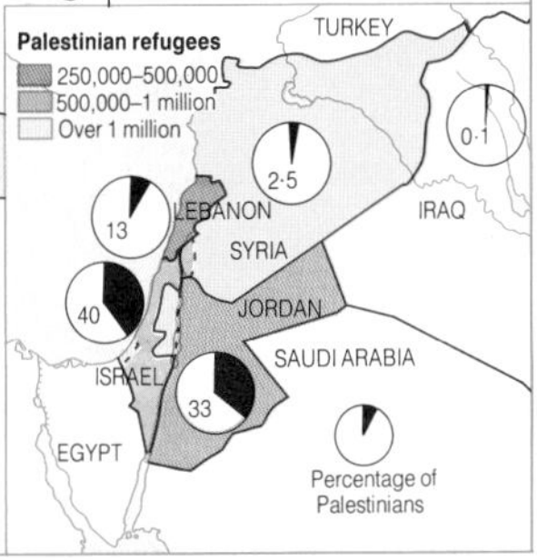

Arabs and Israelis

In 1947 the United Nations recommended that Palestine be divided into an Arab state and a Jewish State. (Map 1 ''New'' Palestine is shown in green; Israel is shaded). The Jews agreed but the Arabs thought the whole country should be theirs. War ensued. By 1949 the Jewish state of Israel held more territory than the UN had intended for it. The first wave of refugees fled to neighbouring countries.

After the Six Day War in 1967 Israel occupied the Gaza strip, the Sinai peninsula and the part of Jordan known as the West Bank. This led to more Arabs seeking asylum in neighbouring countries.

The Camp David Accords in 1977 led to an Israeli withdrawal from Sinai but plans for self government for the West Bank did not proceed. Throughout the 1970s and 1980s there have been frequent clashes.

1987 saw the beginning of the intifada, the Palestinian uprising of peaceful disobedience in the West Bank and Gaza Strip. The Palestinians claim to have renounced terrorism, and their new stance has given them greater international credibility.

Victims of war and drought

Racial and religious tensions in North East Africa have created a complex ebb and flow of homeless people. Somalia and Ethiopia fought briefly over the Ogaden area, and many of the Somali inhabitants have become refugees. There are still many refugees leaving Somalia for Ethiopia. In the north of Ethiopia, Eritrea and Tegre have been fighting for self-determination. As a result of war and recent droughts in northern Ethiopia, large numbers of refugees have crossed into Sudan and while many from Tegre have returned home there are many made homeless by the Eritrean war. There are also 100 000 refugees from Chad in western Sudan. In southern Sudan, an intermittent civil war has lasted over 20 years, and refugees have fled to Ethiopia and northern Sudan, where many suffered in the intense rainfall of August 1988 in the Khartoum area. Those who stay within their own country cannot be given the same protection as those who cross borders and become officially refugees.

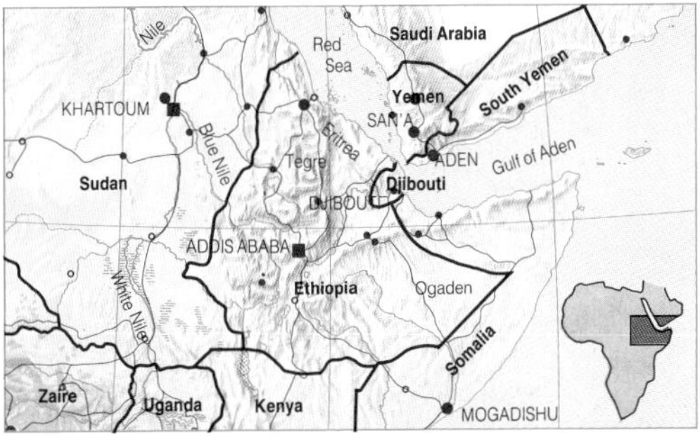

Locals versus refugees

During the 1980s Eastern Sudan suffered several years of poor harvests. Into this heavily populated area arrived refugees from the Eritrea and Ethiopia war. They live in settlements, and tensions and conflicts with the nationals are not uncommon.

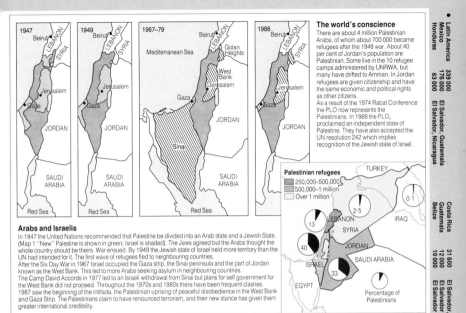

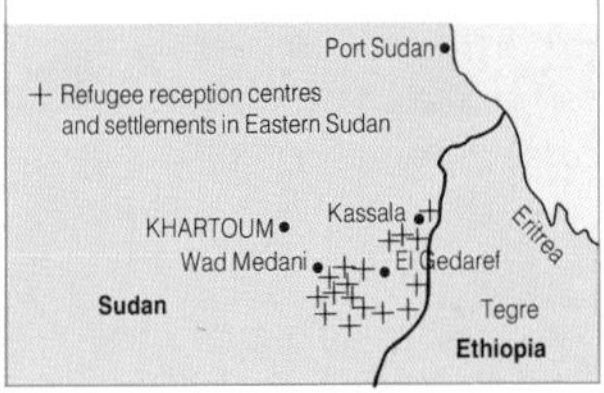

• **Latin America**	**339 000**	
Mexico	**176 000**	**El Salvador, Guatemala**
Honduras	**63 000**	**El Salvador, Nicaragua**
Costa Rica	**31 600**	**El Salvador, Nicaragua, Guatemala**
Guatemala	**12 000**	**El Salvador**
Belize	**10 000**	**El Salvador**
• **North America**	**1 353 000**	
United States	**1 000 000**	
Canada	**353 000**	
France	**179 000**	
F.R.Germany	**146 000**	

UNHCR – United Nations High Commission for Refugees

UNRWA – United Nations Relief and Works Agency for Palestine Refugees

PLO – Palestine Liberation Organization

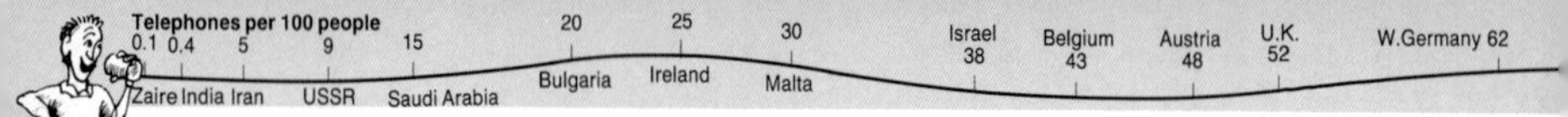

You cannot always rely on a carrier pigeon

Communication is the sharing of information by signs, speech, writing or by telecommunications. There is a broad distinction between personal communication by, for instance telephone or fax, and the mass media like newspapers, radio and television.

The earliest messages were probably in the form of clay tablets carried by professional runners who were employed by kings and rulers. In 500 BC the Greeks developed a communication system based on brick walls, each of which was visible from the next one. The tops of the walls were indented, each notch representing a letter. Fires lit in appropriate places could convey a message – a brief one! Carrier pigeons used in classical times, and in World War I, carried a longer message.

More recently the Royal Mail was delivered by horse-drawn coach. Trains, boats and planes now deliver mail world wide, but for short distances the professional courier, now on a motorcycle, is still the quickest.

Before the advent of telegraphy, messages could be sent by semaphore with the alphabet represented by the positions of wooden arms. Towers were built on hills, some of which retain the name Telegraph Hill. In 1837 messages were first sent by electric signals, and the Morse code was used to send messages between England and France in 1851. The code was a series of short and long signals representing letters of the alphabet. The telephone was invented by Alexander Bell in 1875 and enables speech to be carried as electric signals along a wire. Modern cordless phones use a transmitter and a radio receiver.

Messages were carried in a roll with the sender's seal attached to prevent tampering.

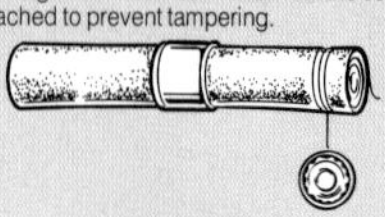

A message, which had to be enlarged to be read, could be attached to the leg of a homing pigeon.

The Penny Post (regardless of weight or distance) was introduced in Britain in 1840. The stamp first used was the Penny Black.

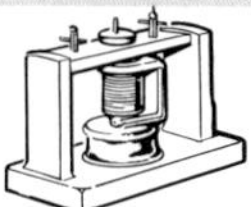

The telephone was invented in 1875 by Graham Bell.

To operate this telephone a handle had to be turned. This connected the caller with the operator.

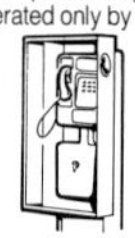

A modern public telephone kiosk. Some are operated only by a special card.

Most modern telephones use push-button dialling.

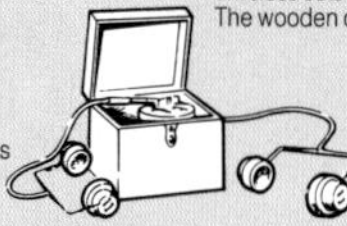

The earliest 'wireless' sets used a thin wire like a cat's whisker connected to a crystal. They were known as 'cats' or crystal sets.

Wireless sets of the 1930's were more convenient to use. The wooden casings were often elaborately decorated.

A modern radio uses very high frequency wavelengths for good quality reception.

The personal stereo is very compact and can be used anywhere.

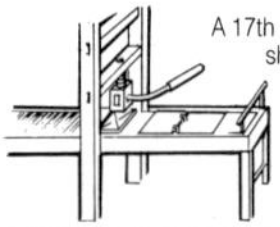

A 17th century printing press shows the platen which is brought down on to the paper.

The Daily Courant.

The Daily Courant 1707. The first daily newspaper to appear in Britain.

Newspapers from round the world can be bought on many news stands. Local editions may be published.

Morse Code

A •—
B —•••
C —•—•
D —••
E •
F ••—•
G ——•
H ••••
I ••
J •———
K —•—
L •—••
M ——
N —•
O ———
P •——•
Q ——•—
R •—•

International ship signals

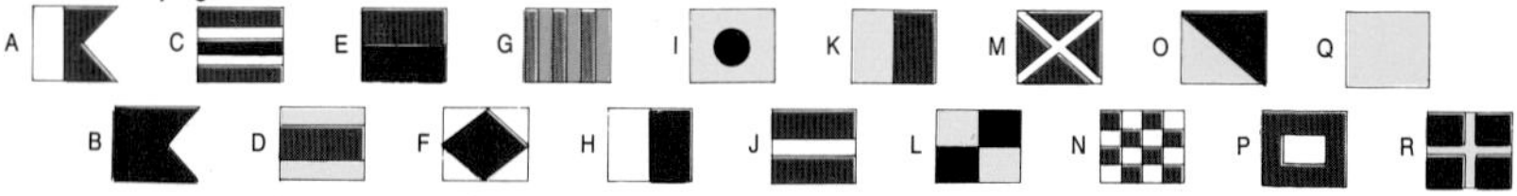

International semaphore alphabet

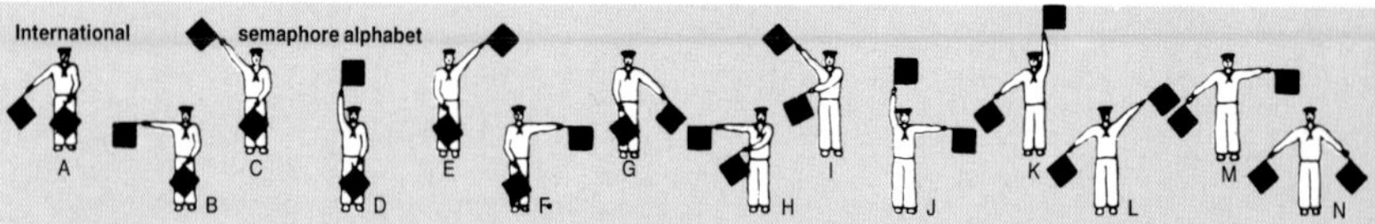

Ship signals: lights and shapes

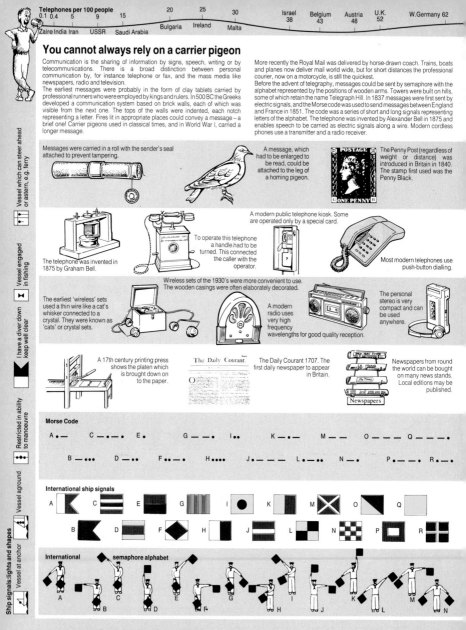

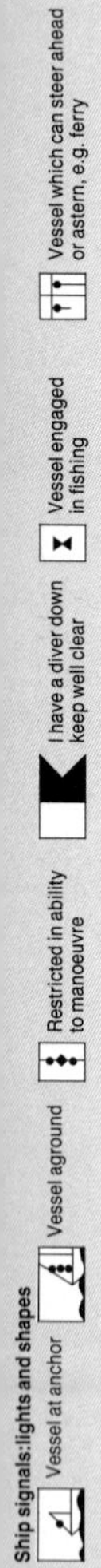

. . .and we sent messages

the development of mass media

The use of radio waves for communication was pioneered by Marconi in 1895, and in 1901 he succeeded in transmitting a signal across the Atlantic. Sound can be transmitted at low frequencies, whereas pictures require higher frequencies.
The first practical demonstration of television was made by Baird in 1926 from the top of the London store of Selfridges. In 1937 the BBC started a service from Alexandra Palace in London. Both the technology of television and the style of programming have changed greatly. The use of the transistor from 1960 led to a huge reduction in the size of the electric circuits needed in TVs.
The communications satellite Telstar went into orbit in 1962 to provide telephone and TV links between Europe and America. Many more Intelsats now relay telephone pictures round the world. Recent developments include Direct Broadcasting Satellites (DBS). Optic fibre cables which carry channels, and video both allow the individual to select from a wide range of entertainment.

Newspaper production and circulation

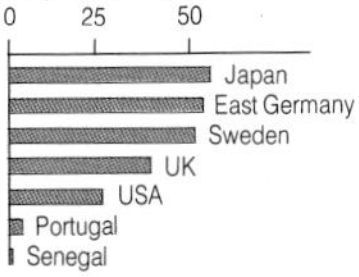

The first newspaper may have circulated in Ancient Rome in 59 BC. Newspapers in their modern form originated in 17th century Europe. In 1788 the Times was founded. The first tabloid was published in 1903 by Lord Northcliffe. Colour printing and multi-section papers are commonplace today. The heaviest paper was a 1987 edition of the Sunday New York Times, it weighed 6.35 kg, 14 lb.

British postboxes were painted green until 1874 when the colour was changed to red. The monarch's initials are on the letterbox.

Foreign mail is usually carried by airmail.

Postmarks may be used for advertising or for carrying messages:

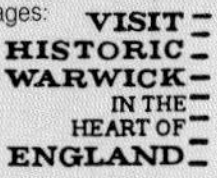

Pictorial stamps cover many themes from natural history to commemorative issues.

The cordless phone can be carried anywhere. It uses radio signals to transmit and receive.

Information may also be sent down a telephone line and after decoding, it can be displayed on a screen.

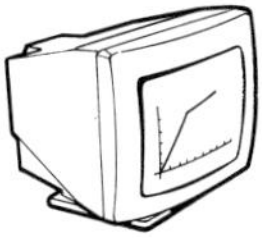

The fax machine allows text and graphic information to be quickly and easily relayed.

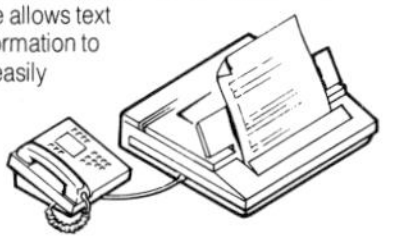

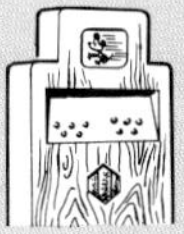

Television sets in the 1930's had very small screens compared with the size of the whole set.

There are three systems used for encoding colour picture information for transmission: U.S. system – used in U.S.A., Canada, Mexico and Japan; PAL (phase alteration line) used in most of Western Europe; and SECAM (système electronique couleur avec mémoire) used in France, E.Europe and USSR.

Different types of aerial are needed to receive TV signals. Dish aerials can receive satellite transmissions.

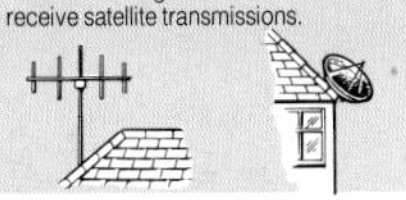

Modern newspaper production

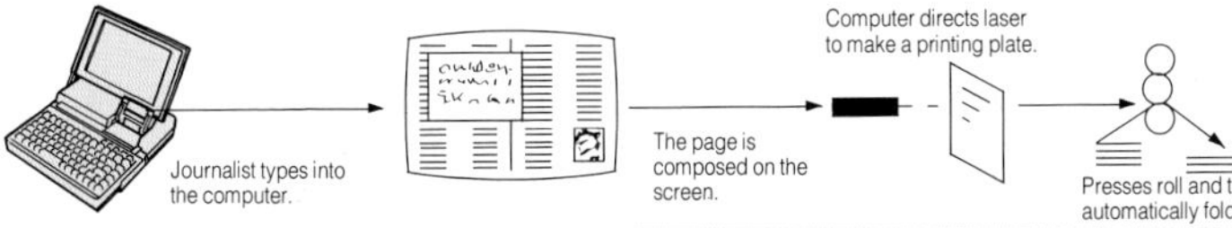

S •••   U ••—   W •— —   Y —••—   1 •— — — —   3 •••— —   5 •••••   7 — —•••   9 — — — — •

T —   V •••—   X —••—   Z — —••   2 ••— — —   4 ••••—   6 —••••   8 — — —••

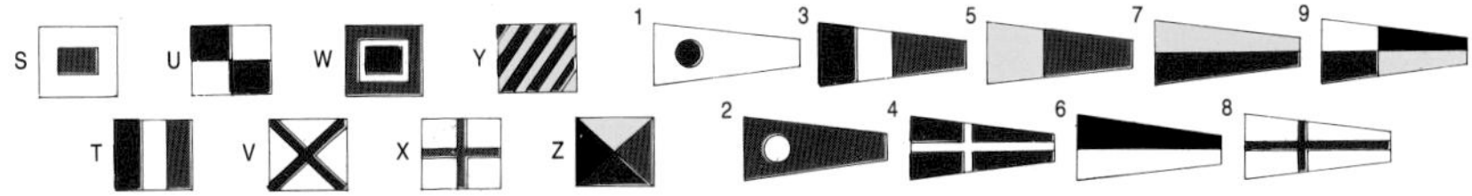

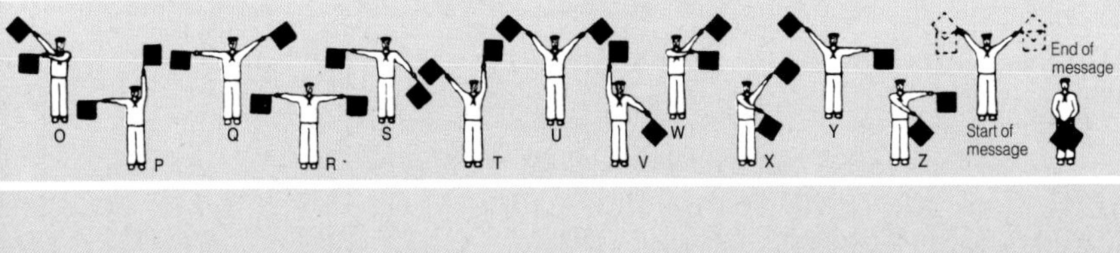

Ship signals sound:

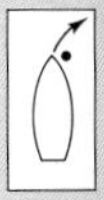

I am altering my course to starboard

I am altering my course to port

I am operating a stern propulsion

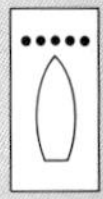

I doubt whether you are taking sufficient avoiding action

Victory 1765

Swedish East Indiaman 1786

Submarine.

Santa Maria, 1492. Ship sailed by Christopher Columbus.

Sailing ship with stern rudder. 13th century

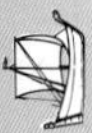

Foochow junk. China

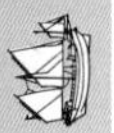

Oseburg. Norway 10th century

Phoenician ship 800 BC

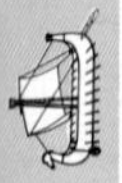

One horse power or more?

In the earliest days men would have used sledges and rafts would have been used for the movement of goods and people. When the horse, the camel and other animals were trained people could travel farther. The wheel, which was in use in Mesopotamia in 4000 BC, led to the development of many vehicles; from the most basic wagon to the mail coach and the Ferrari car.

The earliest sea traders were the Phoenicians with their carvel-built boats which

Assyrian chariot from Mesopotamia 200 BC. Spoked wheels had replaced solid ones.

16th century coach. This vehicle without springs must have been very uncomfortable.

Types of suspension. The leather straps and the C shaped spring, introduced at the end of the 17th century.

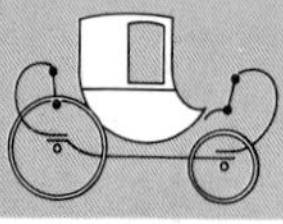

Montgolfier hot-air balloon of 1783. 2 Frenchmen made a successful flight of 8 km, 5 miles.

Giffards first steam driven airship of 1852. The speed was 8 kph, 5 mph.

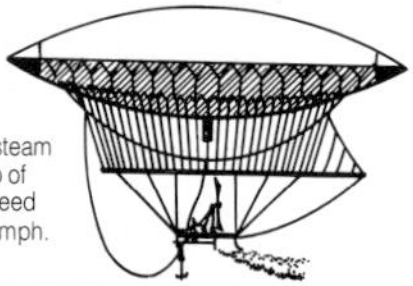

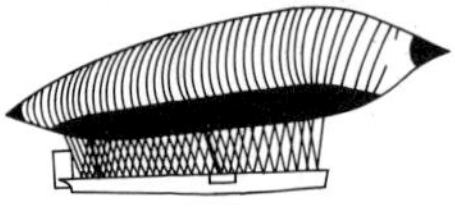

'La France', 1884. The airship was powered by battery electric motors and could be steered.

A Fiat car of 1900. One of the first small cars to be built. Notice the similarity to the Victoria in row 1.

Model T Ford, 1927. Also known as the Tin Lizzie. 15 million of these cars were made.

Cadillac, 1936. A typical large streamlined American car of the 1930s. It had overdrive and synchromesh gears.

Mini, 1959, designed by Issigonis. This compact car featured a transversely mounted engine.

Petrolette. First production motorcycle. It had direct rearwheel drive and a 1488cc engine.

Triumph, 1914. The engine was placed in the centre of the frame. It could develop 2·5 horsepower.

Lambretta scooter, 1968. The small motorized scooter is cheap and popular. Its modest speed and safe ride are useful in towns.

Hobbyhorse, 1818. A crossbar with 2 wheels. The rider used his feet for propulsion.

Pennyfarthing, 1872. The penny and the farthing were a large and a small coin. The large front wheel was turned by the rider.

Safety bicycle, 1940. The chain drive is enclosed, and the machine had a single gear and free wheel.

Rocket, 1829. Stephenson's engine won the prize for the best locomotive on the Manchester/Liverpool line.

American locomotive of 1850. The tall chimney and cow-catcher grid in the front are familiar.

'Castle' class British locomotive of 1920. The Great Western Railway used these for pulling expresses to Bristol and the west.

Bleriot's monoplane, 1909. The first cross channel flight took 37 minutes.

Vickers Vimy, 1919. Alcock and Brown made the first non-stop crossing of the Atlantic in 16 hours.

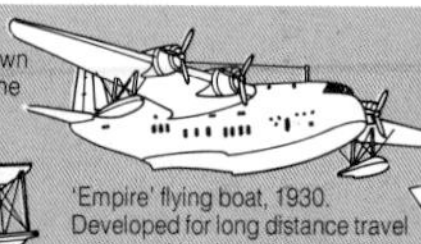

'Empire' flying boat, 1930. Developed for long distance travel over sea. Used on routes to North America.

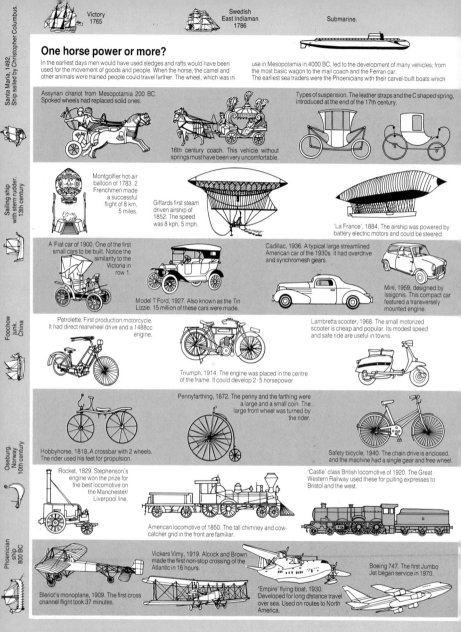

Boeing 747. The first Jumbo Jet began service in 1970.

. . .and we travelled

from carriage to space shuttle

Great Britain
1843

Queen Mary
1934

suited Mediterranean conditions. In northern countries hulls were of clinker construction. The Portuguese caravels of the 16th century enabled sailors to cross the oceans (see page 38).
At the beginning of the 19th century people were travelling no faster than they were in Roman times but then the steam engine and the railways revolutionized transport. For the first time many people could be transported at once. In the last 90 years cars, lorries and aeroplanes have made travel and transport much easier and quicker. Astronauts in the space shuttle travel round the earth many times as they carry out experiments or take equipment into space.

An English mail coach of 1874. These fast coaches were protected against highwaymen.

A horse-drawn omnibus of 1851. 24 passengers were carried and some sat back to back on the roof.

A Victoria of about 1870, used by well off people in the US and Britain. A hood could be raised if needed.

Zeppelin, 1900. This airship was driven by two 16 horsepower engines. Cells containing hydrogen added bouyancy.

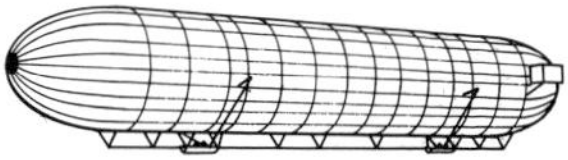

Hindenburg, 1937. This airship operated transatlantic commercial flights until she crashed in 1937 killing all aboard.

Hot-air balloon. An average-sized balloon may be enclosed by 836 sq. metres of nylon material.

Triumph TR5. A sporting car with a 'Surrey' roof.

Volvo 740 SE, 1989. Many modern cars incorporate numerous safety features.

McLaren-Honda Turbocharged 1½ litre MP4/4, 1988. This Formula 1 car dominated the 1988 season.

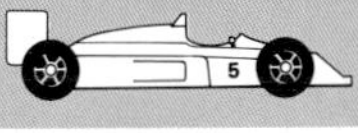

Articulated lorry. The trailer may be disconnected allowing the operator flexibility.

Harley Davidson. The American 'easy-rider' of the 1970's. The comfortable tourer.

Kawasaki. Modern Japanese sports bike. An aerodynamic bodywork design.

Three-wheeled vehicle. Popular in the 1950s. It was cheap to run and offered more protection than a motorcycle.

Racing cycle, 1980. For maximum speed these bicycles have narrow-profile wheels, 10-speeds and a very light frame.

Folding bicycle. Very light and can be folded down in about 45 seconds and carried in a canvas carrier.

Mountain bicycle. With relatively wide wheels, up to 15 gears and a very sturdy frame these bicycles can go almost anywhere.

Mount Pilatus Line, Switzerland, 1960. A rack and pinion system is used to raise the cab on the steepest track in the world.

Diesel-electric trains are fast and economical to run.

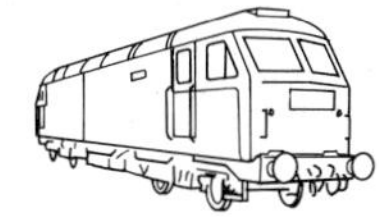

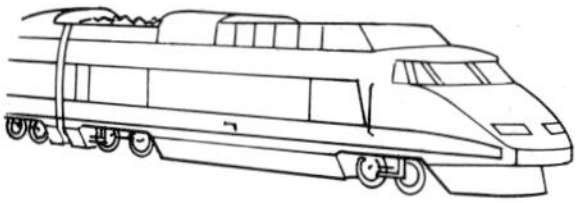

TGV France. Train à Grande Vitesse. The fastest train in Europe.

Helicopter. First made in 1919 and now widely used for military purposes, also police and rescue services.

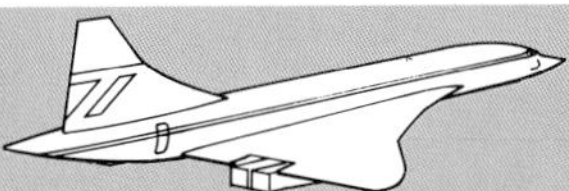

Anglo-French Concorde, 1969. Supersonic passenger aircraft. When the plane is flying subsonic the nose can be lowered.

Space shuttle. The first reusable spacecraft. When fuelled for launch it weighs 4·4 million lbs, 2 million kg.

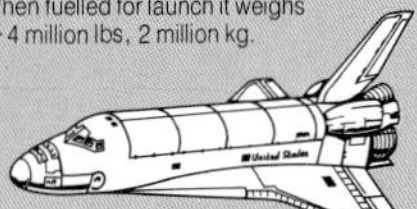

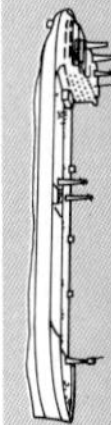

Tanker

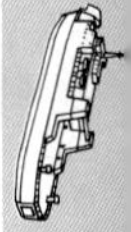

Car Ferry

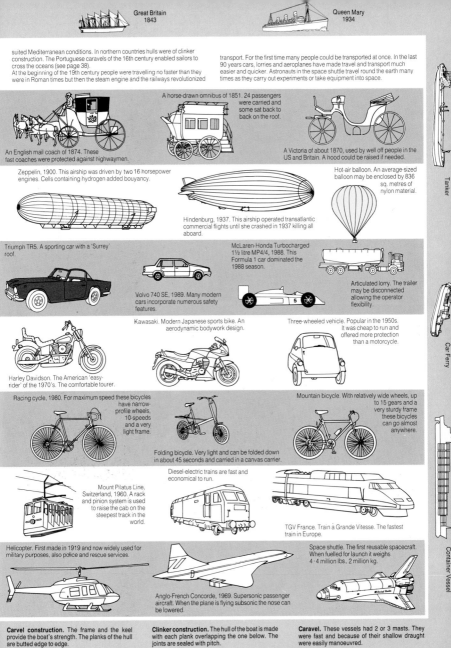

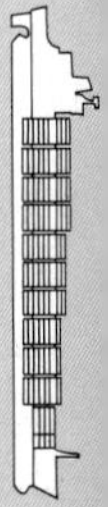

Container Vessel

Carvel construction. The frame and the keel provide the boat's strength. The planks of the hull are butted edge to edge.

Clinker construction. The hull of the boat is made with each plank overlapping the one below. The joints are sealed with pitch.

Caravel. These vessels had 2 or 3 masts. They were fast and because of their shallow draught were easily manoeuvred.

The fastest Pacific crossing by ship was by the container ship 'Sea-Land commerce' which set out from Yokohama on 30th June 1973 and took 6 days, 1 hour and 27 minutes to reach Long Beach, Cal. Average speed 61·6 km/hour.

Heathrow Airport, London, handles the largest amount of international traffic: in 1987 34,742,051 people. Chicago's O'Hare Field Airport is the world's busiest, in 1986 54,770,673 passengers passed through it.

Walsinghasm, Norfolk, UK, ranked with Rome, Jerusalem and Santiago de Compostela (Spain) as a place of pilgrimage in the Middle Ages. Today, after a 19th century revival, Anglican and Catholic shrines to 'Our Lady of Walsingham' are visited by many people.

A cabin on the 'Hindenburg' airship, built 1936, cost $750 (shared $450) for the 43 hour flight from New Jersey to Frankfurt.

The first ever scheduled passenger air service was run between Friedrichshafen and major German cities from 1910. It was by Zeppelin airships, which cruised at 32 knots and carried 24 passengers.

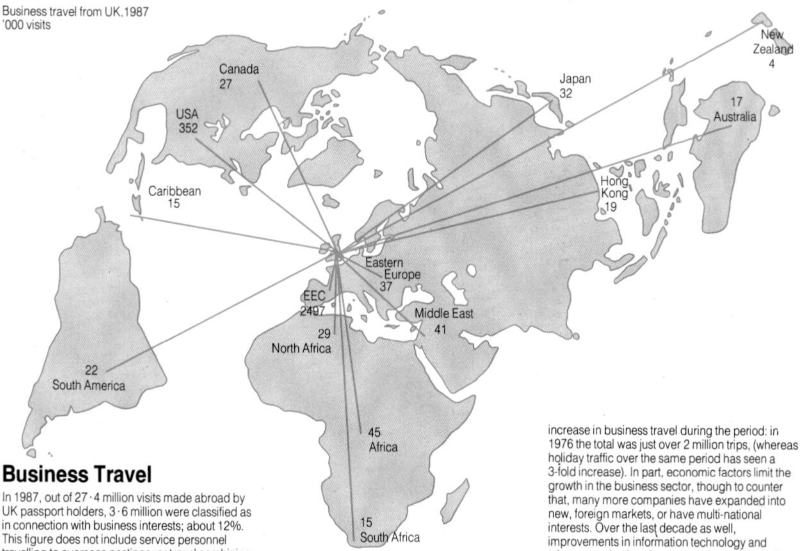

Business Travel

In 1987, out of 27·4 million visits made abroad by UK passport holders, 3·6 million were classified as in connection with business interests; about 12%. This figure does not include service personnel travelling to overseas postings, or travel combining business and holiday; so the true total may be reckoned at about 15%. Of this total, 2·5 million trips were to other EEC countries, in descending order France (0·6 million), Germany (0·5), Netherlands (0·3), Belgium, Ireland, Italy and Spain (0·2) each, all others 0·3 million. Thus the EEC accounts for about 66% of business travel. Another 10% travel, or 387,000 trips, was to non-EEC Europe, slightly more than the 379,000 trips to North America (of which 352,000 or 93%, was to the US). Changing patterns over the last decade illustrate economic changes: trips to the Middle East stood at 41,000 in 1987, compared to 69,000 in 1976. In 1987 Japan and East Europe had similar volumes of travel from the UK (respectively 32,000 and 37,000 trips), whereas in 1976 the same figures were 8,000 and 36,000. The 4-fold increase to Japan is twice as large as the overall increase in business travel during the period: in 1976 the total was just over 2 million trips, (whereas holiday traffic over the same period has seen a 3-fold increase). In part, economic factors limit the growth in the business sector, though to counter that, many more companies have expanded into new, foreign markets, or have multi-national interests. Over the last decade as well, improvements in information technology and telecommunications have made it possible for far more sophisticated contacts to be made between business parties, mitigating the need for some to travel.

One expanding part of the business sector is the conference 'industry', which is eagerly courted by all countries, as it brings in high earnings and can use hotel facilities outside the peak holiday season. The UK has about 10% of the international conference trade.

Religion

Mecca is the holiest city for Muslims, the birthplace of Muhammed in 570 AD who founded Islam. The month of pilgrimage, or 'hajj', is June-July, and in 1988 attracted 1,379,556 people, though the total number visiting Mecca was over 2 million. Another great pilgrimage is to the Hindu festival of 'Kumbh Mela', which takes place every 12 years at the confluence of the rivers Ganges and Yamuna. First recorded in 644 AD, in 1988, over 15 million people came to bathe in the holy river, following the example of monks and holy men from many different sects of Hinduism.

Nature

Many people are attracted to the natural wonders of the world, such as the Grand Canyon, where the Colorado River has cut a gorge up to 1.6 km, 1 mile deep. Active volcanoes, such as Etna in Sicily, or the hot springs and mud in Iceland or North Island New Zealand, are also favourite tourist destinations. Deserts form another sparsely populated, harsh environment, and places such as Ayers Rock in Australia, at 30 m, 100 ft high and 9.6 km, 6 miles circumference the largest single rock in the world, are increasingly visited.

Politics

Red Square is a modern focus for a modern type of pilgrimage. Lenin is seen as the founding father of Communism, and his mausoleum is a massive red granite monument overlooking Red Square. His body is on show in a mausoleum underground, and the upper part of the building is the tribune where Soviet leaders watch the May Day parade. In South Dakota, USA, 20 m, 60 ft statues of Washington, Jefferson, Lincoln and Theodore Roosevelt are carved out on the side of Mount Rushmore. The sculptor was Gutzon Borglum.

In 1960, it is estimated that 90% of door-to-door travel time for an air trip was spent in the air. By 1980 this had dropped to just under 60%.

Lourdes is today the most famous place of Christian pilgrimage, and since Bernadette Soubirous saw a vision of the Virgin in 1858, over 60 'miracle' cures have occurred, as well as hundreds of other recorded cases.

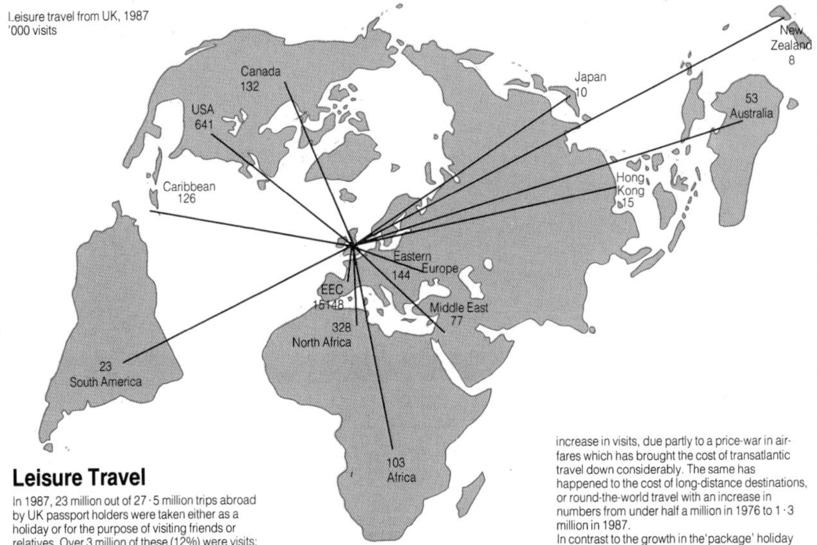

Leisure Travel

In 1987, 23 million out of 27·5 million trips abroad by UK passport holders were taken either as a holiday or for the purpose of visiting friends or relatives. Over 3 million of these (12%) were visits; 7·5 million (33%) were independently arranged holidays; but nearly 12 million (over 55%) were inclusive or 'package' holidays.

Of these package trips, 4·4 million were to Spain, and other popular destinations include Greece (1·5 million), France (1·3 million), Yugoslavia (0·6 million), Italy (0·6 million), Austria (0·5 million) and north Africa (0·3 million). Package holidays developed in the 1950s as the cost of air travel came down, enabling many people to get quickly to warm sunny beaches. Spain developed resorts in the 1950s and 1960s, and is still by far the most popular destination for Britons. The tourist industry in Greece developed later, during the 1970s; only 250,000 package holidays were taken there in 1976.

Visits from the UK to America topped 800,000 in 1987, 25% to friends and relatives, the rest as holidays. The last decade has seen a 3-fold increase in visits, due partly to a price-war in air-fares which has brought the cost of transatlantic travel down considerably. The same has happened to the cost of long-distance destinations, or round-the-world travel with an increase in numbers from under half a million in 1976 to 1·3 million in 1987.

In contrast to the growth in the 'package' holiday over the last 30 years, the trend today is towards a greater variety of options and destinations, such as 'adventure' tours, and longer holidays not spent in one location. As most airports and airlines are at full capacity during peak travel periods, many travellers are taking holidays outside the traditional July/August time – this has been reflected in the growing popularity of Australasia and the Far East; in 1976 7,000 people holidayed in Australia, this had risen 6-fold to 46,000 in 1987.

Culture

From the 17th century, aristocrats would complete their education by going on a 'Grand Tour' of Europe to see the sights of ancient Greece and Rome.

Today, cultural monuments and sites are one of several attractions to a country. In particular, people from north America visit the 'Old World', often in search of their roots; and the great temple complexes in S.E.Asia, the palaces of India and the ruins of early American civilization are foci for tourists.

Sun and Sand

Despite medical warnings about exposure to strong sunlight, holidays in hot, sunny spots are the most popular choice today. Sea bathing was originally thought to be good for health, similar to visiting a spa to 'take the waters'. A visit to the sea offered the chance to relax and do nothing in very different surroundings, and became popular with factory workers by the turn of the century. The first 'charter' flight to the south of France was arranged by Cook's in 1939.

Cult Pilgrimage

Since Elvis Presley died on the 16th August 1977, his home 'Gracelands' has become a memorial to his life and music. Rooms at the house display instruments, costumes and memorabilia, and many thousands of people visit annually.

There are other places which have become popular as centres for modern cults. Stonehenge, in Wiltshire, a neolithic stone monument, attracts many who wish to watch the sunrise at the summer solstice on June 21st.

The longest railway journey is The Trans-Siberian route from Moscow to Nakhodka. It takes 8 days to make the 5778 mile trip, and Moscow time is kept on the train, though this may be 7 hours out from local time.

The first steam only crossing of the Atlantic was made by the 'Royal William' in 1833. When she left Pictou in Nova Scotia, she was carrying nearly half her 1,370 ton weight in coal.

In 1986, 11,563,000 US citizens travelled abroad. Of these, 5,126,000 visited western Europe and the Mediterranean, and 3,800,000 Central America and the Caribbean.

- 1957 Sputnik 1, USSR, first satellite
- 1958 Explorer 1, USA, first American satellite
- 1961 Vostok 1, USSR, first manned flight
- 1962 Mercury-Atlas, USA, first American to orbit
- 1965 Mariner 4, USA, close photos of Mars
- 1966 Gemini 8, USA, first docking of 2 spacecraft
- 1969 Apollo 11, USA, first landing of human on moon
- 1971 Salyut 1, USSR, first manned space station
- 1972 Pioneer 10, USA, leaves solar system
- 1973 Skylab 1, USA, first American space station
- 1974 Helios 1, USA-FRG, closest approach to Sun
- 1975 Viking 1, USA, first soft landing on Mars
- 1977 Salyut 6, USSR, space station visited by 27 astronauts
- 1977 Voyager 1 and 2, USA, mission to Jupiter and Saturn
- 1979 Ariadne 1, Europe
- 1981 STS 1, USA, first trip with space shuttle 'Columbia'
- 1983 STS 3, USA, Challenger carries Tracking and Data Relay Satellite

Yuri Gagarin's descent module on April 12, 1961, at the end of the first manned space flight

Apollo 15 splashdown using two of its three parachutes only. August 7, 1971

We return. . . but how

Equal only to the problems of launching a crewed spacecraft are those of bringing the men and women home safely. As the orbiter descends it encounters an increasingly dense atmosphere, and at a height of 130 km, 80 miles friction slows the craft from 27 400 to 13000 kph, 17000-8000 mph. (Deceleration is about 2G.) This is the fiery stage of reentry, maximum temperature 2500°C., and coincides with a 12 minute communication blackout. Further reductions in speed are achieved by high-banking S-turns. The vehicles must then be landed in a predictable place.

The Gemini Apollo and Mercury capsules of the 1960's and 70's were protected by a material that burnt away. The landings were aided by spectacular drogue parachutes that allowed the command module to descend gently to the sea.

In the 1980's space exploration has been revolutionized by the development of the reusable space shuttle. It is launched as a rocket, orbits as a spacecraft and returns as a glider to a runway. Touchdown speed is in excess of 320 km per hour, 200 mph, and as the craft has no power only one attempt is possible.

Protection against the temperatures at reentry is provided by 31 000 tiles of almost pure silica which dissipate the heat but remain intact. The tiles vary in thickness from 1 cm, ·5 inch to 9 cm, 3·5 inches. They are of varying sizes and each is shaped to fit round the orbiter's aluminium outer skin.

America now has three shuttles, Columbia, Discovery and Atlantis. The USSR has similar craft such as Buran which made its initial journey fired by the launch vehicle Energia in November 1988. The Soviets also appear to have developed a much smaller space plane which can be launched coventionally and will land on a normal runway.

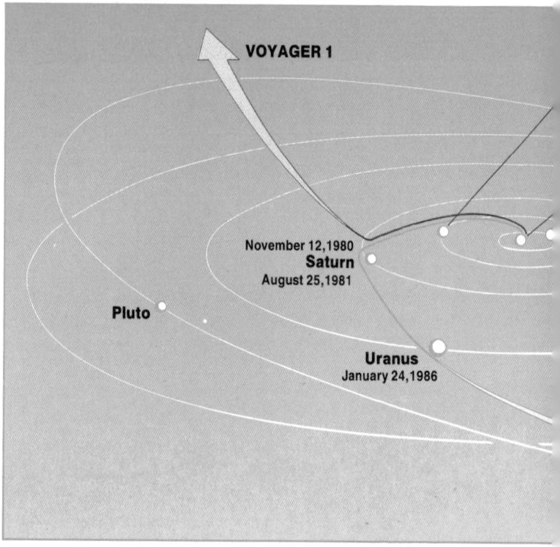

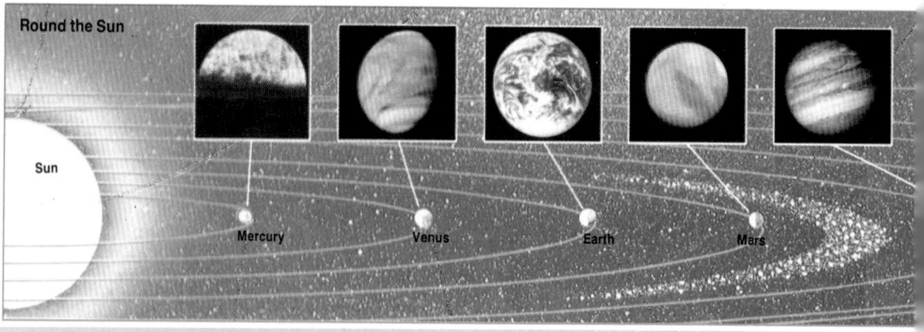

. . .and we conquered Space?

space voyages and planets

- 1984 Soyuz T10, USSR, astronauts longest time in space, 236 days
- 1985 Spacelab 3 USA, 15 experiments in material performance, fluid mechanics, and life sciences
- 1986 Voyager 2 USA, Uranus probe, photographs taken
- 1988 Buran, USSR, pilotless shuttle launched by Energia
- 1989 Atlantis USA. probe to map Venus

Shuttle Challenger landing, November 6, 1985

Aiming at the stars

The 'space age' is a phrase which has been much used in the last 35 years. Since 1957 when Sputnik 1 was launched by the USSR, hundreds of space craft have been sent into the sky. Most satellites originally had military purposes such as surveillance but later the same techniques were used for civil telecommunications, for search and rescue, and for natural resource investigations.

Knowledge and skill in space technology have increased tremendously. In the 1960's the major challenge was to place a person on the moon's surface and in the 1970's the space laboratories were an important part of space exploration. The American space shuttle programme in the 1980's suffered a serious setback after the Challenger disaster of 1986.

The Voyager project has met with more success. Two space craft, Voyager 1 and Voyager 2 were launched in 1977 to explore the fringes of our planetary system. In 1986 the Voyager 2 was able to send high-quality photographs of Uranus and its moons, although the radio signals took almost 3 hours to reach Earth.

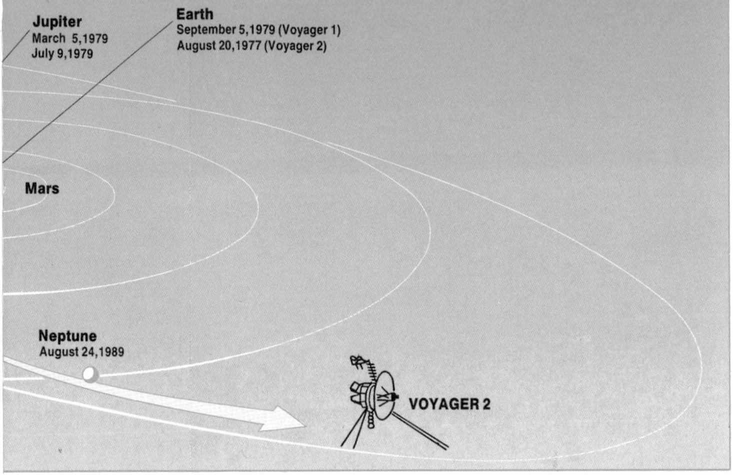

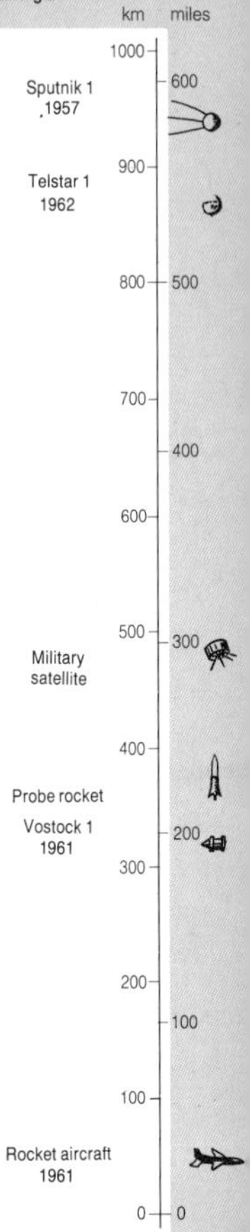

G is the earth's normal acceleration. As the orbiter slows down astronauts are subject to deceleration forces expressed in terms of G. The moon astronauts had to undergo 8G during reentry because of the speed of the vehicle. In the Gemini and similar capsules the crews were subject to 5G. The deceleration of current spacecraft is 2G.

Geostationary orbit. The orbit of a satellite which is moving at a speed and height which keeps it always above the same place on earth.

1820 Bellingshausen sights the Antarctic continent
1831 Biscoe sights Enderby Land
1841 John Ross penetrates the pack ice
1895 Bull lands on Antarctica
1902 Nordenskjöld discovers Antarctic peninsula
1902 Scott, Wilson and Shackleton attempt to reach the South Pole
1909-12 Mawson reaches the South Magnetic Pole
1911 Amundsen first at the South Pole
1912 Scott reaches Pole and dies on return journey

Profiles of the major whale species as they breach the surface.

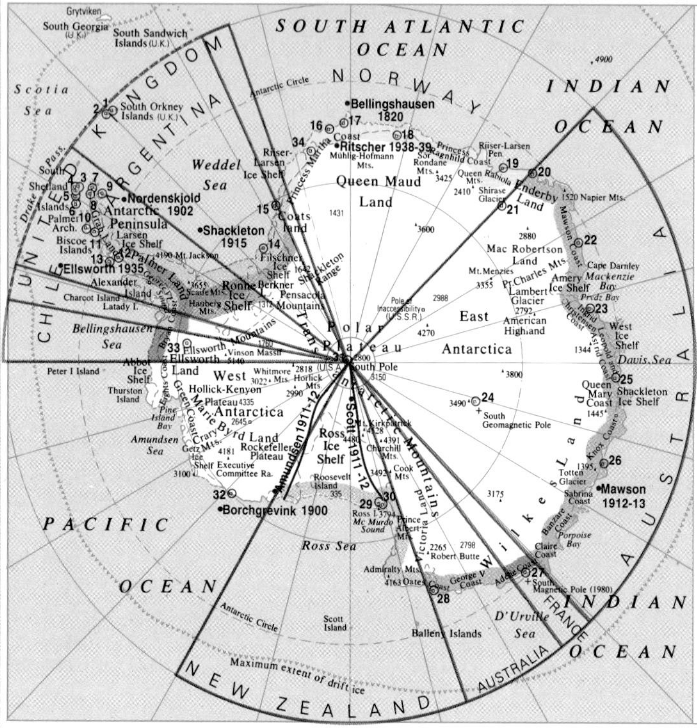

The Antarctic Treaty of 1959 placed the region under international management. There are 18 countries, some with territorial claims, who make decisions. A further 14 countries have acceded to the treaty.

Territorial claim
Disputed claim
• Major explorations
Exploration route
◎ Permanent research station

1 Orcadas
2 Signy
3 Arctowski
4 Bellinghausen
5 Teniente Marsh
6 Arturo Prat
7 Esperanza
8 Bernado O'Higgins
9 Marambio
10 Palmer
11 Faraday
12 San Martin
13 Rothera
14 Belgrano II
15 Halley Bay
16 Georg von Neumeyer
17 Sanae
18 Novolazarevskaya
19 Syowo
20 Molodezhnaya
21 Mizuho
22 Mawson
23 Davis
24 Vostok
25 Mirny
26 Casey
27 Dumont d'Urville
28 Leningradskaya
29 Scott
30 Mc Murdo
31 Amundsen-Scott
32 Russkaya
33 Siple
34 Wasa

Antarctica: yesterday, today and tomorrow

Antarctica, now a frozen continent, once supported semi-tropical forests and land animals while huge reptiles swam in the seas. Evidence for this scenario some 70 million years ago, comes from fossils of ferns, trees and animal fragments. Antarctica was once at the heart of the super continent and as the other continents broke away Antarctica drifted to its present position over the South Pole.

This continent is the fifth largest and covers 14 million km², 5.4 million sq. miles. It is the size of the United States and Mexico together. The 2 ice sheets of Greater and Lesser Antarctica are separated by the Trans-Antarctic Mountains, from which huge glaciers flow. About 2% of the surface is ice-free, and these Dry Valleys are thought to be the nearest equivalent on earth to the planet Mars.

The first sighting of the Antarctic continent was by a Russian explorer in 1820. Subsequently several expeditions visited the area. The early 20th century was the heyday of exploration, culminating in Amundsen's dash to the South Pole in 1911. Scott arrived there a month later, but he and his party died tragically on the return journey.

The succeeding years were marked by international rivalries, partly resolved by the Antarctic Treaty of 1959 which provided for the demilitarisation of the continent.

Scientists of many nations now work in Antarctica and their research covers climate, oceans, glaciers and geology as well as biological investigations.

The extreme temperatures make this last wilderness a fragile ecosystem, for which the 1982 convention CCAMLR was an important safeguard. The environmental group Greenpeace has done much to focus public attention on the dangers of human impact. Their aim now is to press for the establishment of a World Park within the framework of CCAMLR.

Cross-section of Antarctica

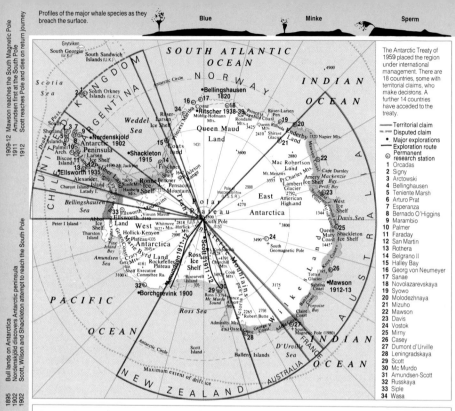

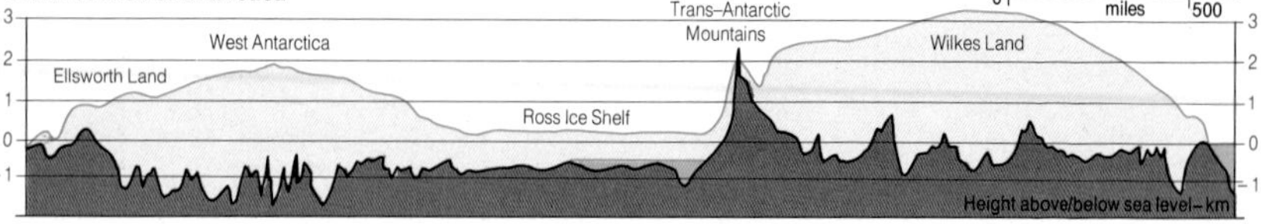

. . .and we explored the seventh continent

Antarctica-ice cap under the ozone hole

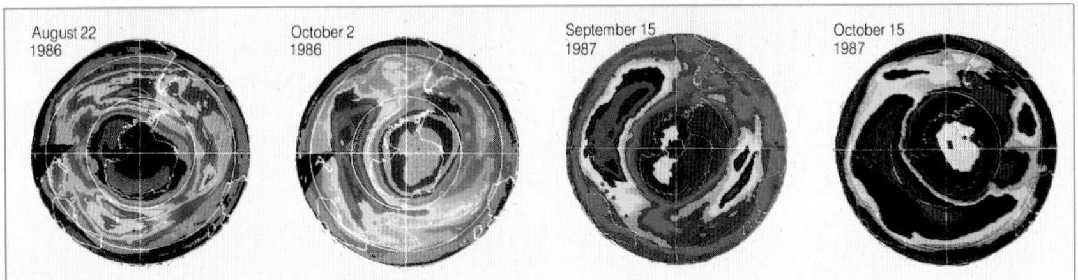

Cutting holes in our ozone umbrella

The two pairs of TOMS images show the 'hole' in the ozone over Antarctic (outlined in white). The hole is at its largest in the Antarctic spring and has been growing each year since its discovery in 1981. By October 1986 it was the size of the USA. Grey and purple portray the hole in the 1986 pair and by October 1987 the increased size is visible as a deep blue, purple, black and pink area.
The earth is protected from the damaging effects of radiation by the ozone layer, which is destroyed by chlorine released by CFCs. The 1989 Helsinki Declaration recommends a total ban on these gases by 2000.

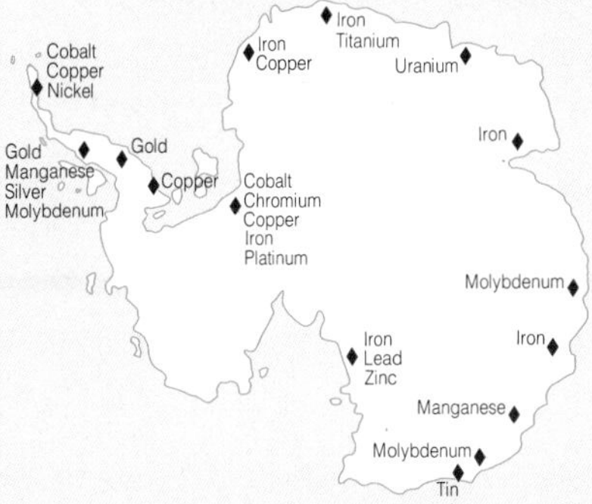

A rich continent

The geological history of the continent suggests that it may contain mineral deposits similar to areas such as South Africa, to which it was once joined. There are thought to be large coal deposits in the Trans-Antarctic Mountains; possibly the world's largest.
The Dufek Massif in the Pensacola Mountains covers an area of 50 000 km², 19 500 sq miles. The magma in this formation has cooled slowly so that the minerals have separated into layers. Such formations elsewhere are very rich in mineral deposits.
It is possible that substantial reserves of oil and gas underlie the Ross Sea Shelf, and test drillings in 1973 confirmed the existence of potential off-shore deposits. Exploitation would be difficult and expensive.
The existence of these minerals is largely conjecture, based on geological similarities, together with a few test drills. The reserves are by no means proven. The Antarctic Treaty made no provision for minerals, and although they are not at present exploited, most governments believe that mining is inevitable in Antarctica. Proposals were drawn up in 1988 setting out a legal framework to control exploration and exploitation, and establishing a procedure to assess environmental impact.
Commercial whaling has now ceased in the Southern Ocean, but factory ships (see page 52) catch and process huge quantities of fin fish. There are fears that stocks will be overfished. The small shrimp-like krill (Norwegian for 'small fry') is also being caught, mainly for use in animal feed.

The human impact

The scientific work carried out in Antarctica is of immense value. However the building, servicing and running of the stations, as well as their research generates waste products and can adversely affect the surroundings. Few research stations have yet solved the problems of waste disposal. Currently there is debate as to whether larger aircraft requiring more sophisticated ground facilities and runways are appropriate.

Pollution

It is unknown how much environmental damage was done when the Bahia Paraiso ran aground near the Antarctic Peninsula in January 1989. Diesel fuel leaked forming a slick several centimetres thick. Oil contaminated the coastline causing birds and penguins to become sodden. There were particular fears for a nearby Adélie penguin rookery containing some 10 000 breeding pairs. The prawn-like krill too were affected. The oil spill was the first major ecological disaster in Antarctica but with increasing shipping and human activity will it be the last?

Reproduced from news photograph.

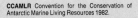

CCAMLR Convention for the Conservation of Antarctic Marine Living Resources 1982.

TOMS Total Ozone Mapping Spectrometer on board the polar orbiting satellite Nimbus. Images made at NASA Goddard Space Flight Center.

CFCs Chlorofluorocarbons are used in some aerosols, also in refrigerators and air conditioning units.

Why a patchwork of nations?

Towards the end of the 15th century the world as it was known to the ancients suddenly expanded. Europeans with stronger, larger ships and improved navigational instruments could explore new regions and return to them with reasonable accuracy. This expansion on a near global scale was similar in many ways to that which, many centuries earlier had brought the migrant tribes from Asia to Europe.

When the European nations began to carve up the world between them, the local populations were rapidly overwhelmed by the superior force of the new arrivals. The new colonial powers all had developed institutions at home necessary for the administration of large political units, and these were used to take over wide areas containing disparate tribal groups, at the cost of the local systems. In Australia, the indigenous population was originally ignored, but imported diseases killed large numbers of the Aborigines, and when the new arrivals became more land hungry, the surviving Aborigines were forced to live in designated areas.

Likewise the 'Indians' discovered by Columbus were, within a few decades, almost entirely eliminated by Spanish savagery. They were replaced by Negro slaves, kidnapped on the west coast of Africa. The Spanish also brought with them diseases such as smallpox to which the Indians had no natural resistance.

Local interests and simple economies were subordinated to the economic needs and interests of the imperial power. When the colonies became independent the national boundaries were often based on colonial administrative divisions, not on local conditions and needs. Independence, in many cases, seemed to bring simply a change in personnel. But it soon became clear that Western-style political institutions could not be grafted onto more pluralistic Third World societies and many of the newly independent states fell to military rule. They now have the task of evolving their own systems which can take into account local needs, but also enable them to survive in a world dominated by Western values.

Within Europe the Industrial Revolution led to a shifting of power from nations like France and Austria whose economies were largely dependent on agriculture to the nations in which industry was now the dominant factor. USA became one of the strongest and Japan, due largely to American intervention, also opened up to trade with other nations.

In 1914, 84% of the world's land area was subject to European influence. Some countries came under direct colonial rule, while others, such as the ex-colonies of North and South America, had strong financial and economic ties with their previous mother country. After the Second World War, these vast European empires crumbled: most of the new states which emerged after 1945 had previously been colonies of one of the Western European countries. As new nations were being created, environmental issues, concern about nuclear war and terrorism, the problems of food supply and energy showed the interdependence of all nations. Co-operation was clearly in the national interests of large and small countries.

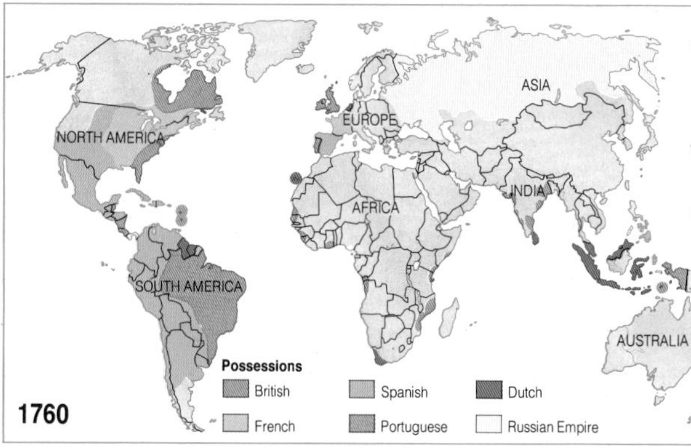

European trading empires

The decline of the Spanish, Portuguese and Dutch empires in the seventeenth century gave France and Britain the chance to strengthen their own empires. Commercial wars were fought to capture trade controlled by the declining empires, since trade was clearly the basis of power and the means of expansion. Britain and France both had trading posts in Canada and Britain had eleven colonies in North America but it was the Sugar Islands, which depended on slave labour, that provided Britain with most of her wealth and hence her power.

Nationalism in Europe

The nineteenth century saw a further weakening of the Spanish Empire. Most of her colonies in South America became independent between 1810 and 1824, although they kept strong links with their mother country. France and Britain continued to build up their empires but in 1850, much of the world was still unclaimed by the European powers. The Congress of Vienna (1815) tried to create a balance of territorial interests between the old European multi-national empires but it ignores the feelings of the nationalist minorities. Nationalist risings throughout Europe were suppressed, but Greece won her independence (1821), encouraging the nationalist feelings of other Balkan states, which weakened the Turkish Ottoman empire. Belgium became independent in 1830.

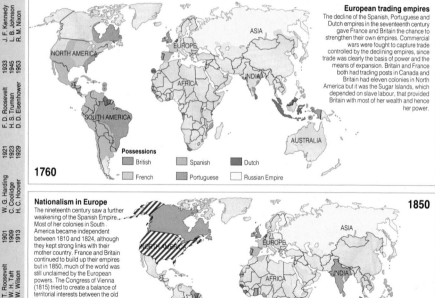

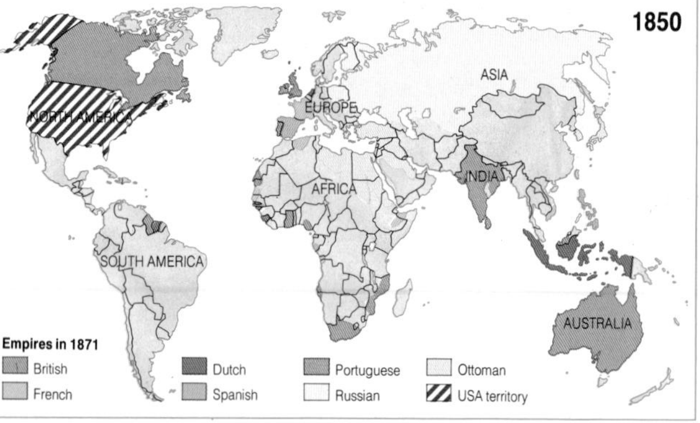

Russia and the Soviet Union Czars and Empresses since 1700								
Peter I 1682–1725	Anne 1730–1740	Peter III 1762	Alexander I 1801–1825	Alexander III 1881–1894
Catherine 1725–1727	Ivan IV 1740–1741	Catherine 1762–1796	Nicholas I 1825–1855	Nicholas II 1894–1917
Peter II 1727–1730	Elizabeth 1741–1762	Paul 1796–1801	Alexander II 1855–1881	Russian Revolution 1917

The world is colonized

In 1897, a quarter of the earth's surface was ruled by Britain, containing a quarter of the world's population (about 372 million people). The 'scramble for Africa' had brought another huge area of the world under European colonial rule and in 1894-5, Japan defeated China and joined the Western powers and Russia in the 'scramble for China'. More than 5 million people now lived under imperial rule. In Europe the unification of Italy (1870) and the creation of the German state (1871) had changed the balance of power and Germany had emerged as a new, and extremely powerful, force.

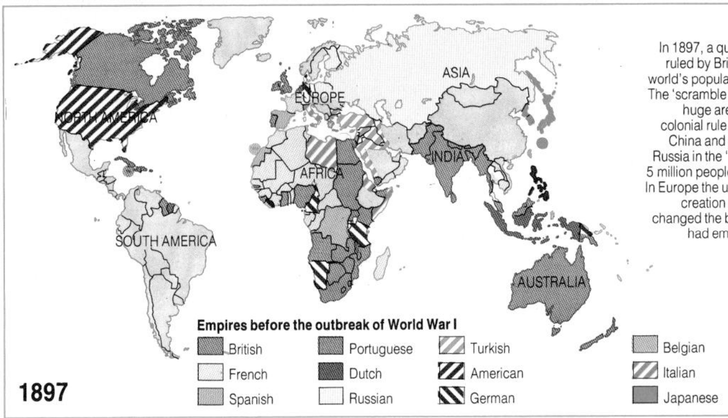

1897

1920

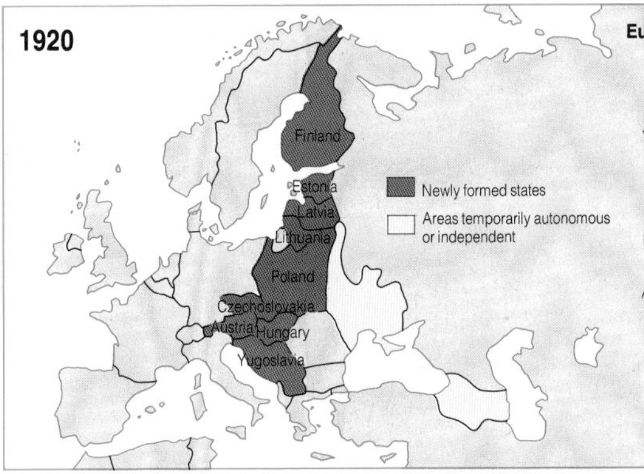

European political problems post World War I

At the end of the First World War, the map of Europe was virtually redrawn. The peace settlement at Versailles (1919) asserted the principle of national self-determination and broke up the German, Austro-Hungarian and Russian empires into small nation states. The republic of Yugoslavia was created from former Austrian states, and a new state of Czechoslovakia came into being. Romania made significant gains, while Mustapha Kemal Ataturk managed to strengthen Turkey. The new states contained significant national minorities and the settlement created a huge refugee problem and a widespread feeling of injustice. The defeated powers never accepted the validity of the imposed settlement and the new states felt continually threatened by their stronger neighbours. Authoritarian governments emerged in Germany, Italy and Russia. In Germany, Hitler promised to restore the self-respect of a defeated nation and to bring the German minorities in the new states back under German authority. Defying the Treaty of Versailles, he began to strengthen his army and the young nations were powerless to resist. Germany occupied Austria and Czechoslovakia without encountering any opposition. It soon became evident that this was only a beginning, and that war was inevitable.

The world today

The 20th century has seen the inevitable process of decolonisation. In 1926, there were more than 80 separate colonies and dependencies covering ⅓ of the world's population and land surface area. The Second World War encouraged anti-colonial nationalistic feelings and was followed by the rapid break up of the British and French empires. The two powers which emerged from the Second World War, Russia and the U.S.A., have been charged with a form of neo-colonialism, where they exert, military influence and economic control over other countries. All nations have to learn to find a balance between national interest and collaboration in the interests of global security.

1989

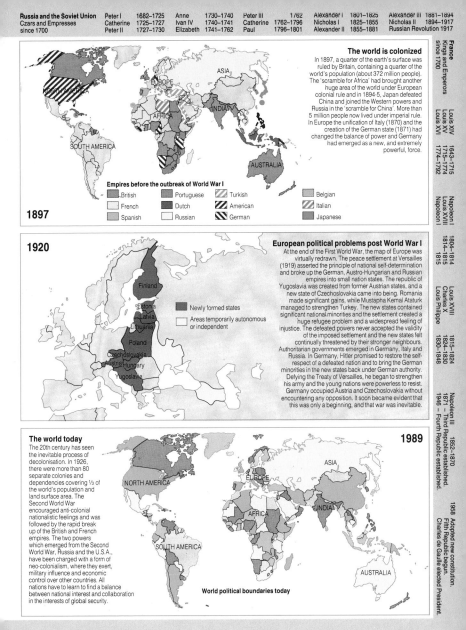

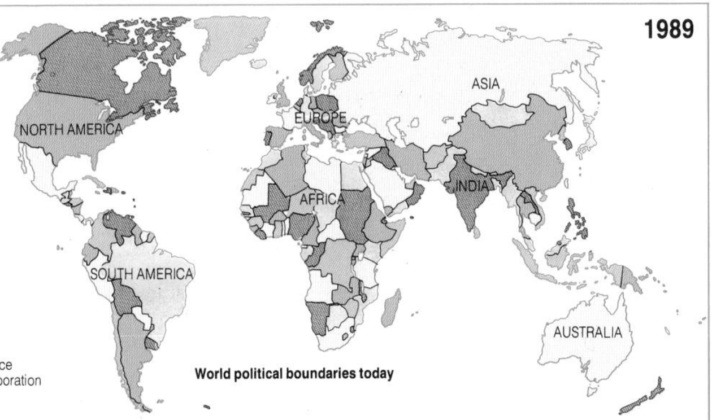

World political boundaries today

France
Kings and Emperors since 1700

Louis XIV	1643–1715
Louis XV	1715–1774
Louis XVI	1774–1792
Napoleon I	1804–1814
Louis XVIII	1814–1815
Napoleon I	1815
Louis XVIII	1815–1824
Charles X	1824–1830
Louis Philippe	1830–1848
Napoleon III	1852–1870

1871 – Third Republic established.
1946 – Fourth Republic established.
1958 Adopted new constitution. Fifth Republic begun. Charles de Gaulle elected President.

Comparing superpower forces. Source: The Economist – The two sides' official figures.

- **Total population** — NATO 561 948 000; Warsaw Pact 380 737 000
- **Ground troops** — NATO: NATO figures 2.2m, WP figures 3.0m; Warsaw Pact: WP figures 3.2m, NATO figures 3.1m
- **Tanks** — NATO: NATO figures 22 200, WP figures 30 700; Warsaw Pact: WP figures 59 500, NATO figures 57 500
- **Missile launchers up to 500km range** — NATO: NATO figures 88, WP figures 136; Warsaw Pact: WP figures 1600, NATO figures 1450
- **Combat aircraft** — NATO: NATO figures 3975, WP figures 7125; Warsaw Pact: WP figures 7875, NATO figures 8250

Military balancing act

The 'military' industry is second only in importance to the petroleum industry in the world today. The major powers thankfully avoid confrontation but both the US and the USSR have large numbers of forces abroad and local disputes easily become wars.

Spending on defence is seen as spending on security and even non-aggressive, neutral Sweden spends 3.4% of its national product on defence. Governments in both East and West need to buy security and have built up arsenals of nuclear and conventional weapons. With the far greater strength of the Soviet armed forces the West has been reluctant to cut its defences.

The Soviet Union needs to invest in its economy and in the West money is needed for health, education and 'green' issues among others.

There are signs that trust may be growing between the superpowers. The Soviet Union has announced arms reductions and if reforms continue NATO could cut its forces too. Nevertheless, East and West are still very far apart politically and both blocs will wish to retain minimum security.

Defence spending

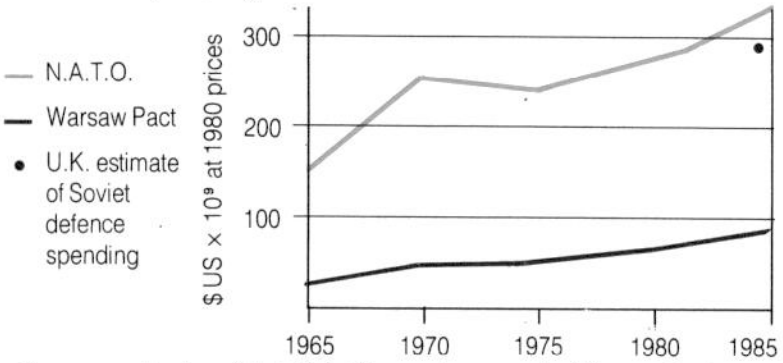

Figures are the two sides' official figures, however the Warsaw Pact's are regarded as an underestimate. For instance, the Soviet figure for spending in 1986 is given as $ US 47 × 10^9, the British estimate of 1986 Soviet spending is $ US 293 × 10^9. Source: IISS Military Balance 1987-88

Trade balances

The pattern of world trade is extremely complex. The old industrial countries are competing with the newly industrialized ones such as Brazil and Korea. The success of Japan as a trading nation is phenomenal. On the other hand the less-developed countries who are producers of raw materials are striving to industrialize, and alter their trading patterns. The poorest nations' share of the world market is only 1%.

A major threat to Third World countries is the 'new protectionism', a policy of erecting barriers which subtly restrict trade. Currency crises, oil crises, debt crises and the world recession have created demands for protection. Restrictions on imports of textiles, cars and particularly on steel aim to maintain employment and protect incomes in the importing countries. Developing countries, too, erect trade barriers to protect their emerging 'high tech' industries.

Oil production and prices

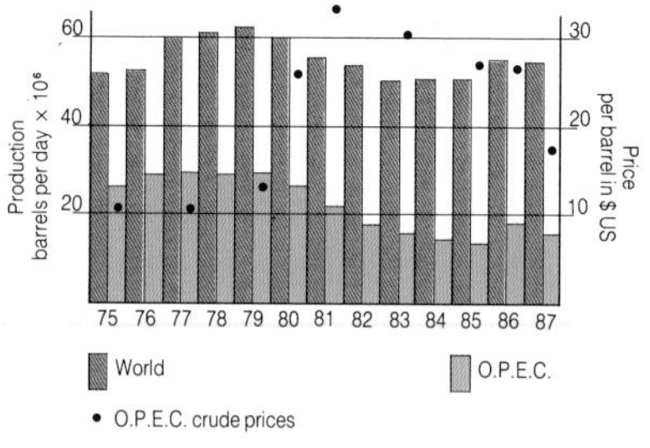

Source: Institute of Petroleum

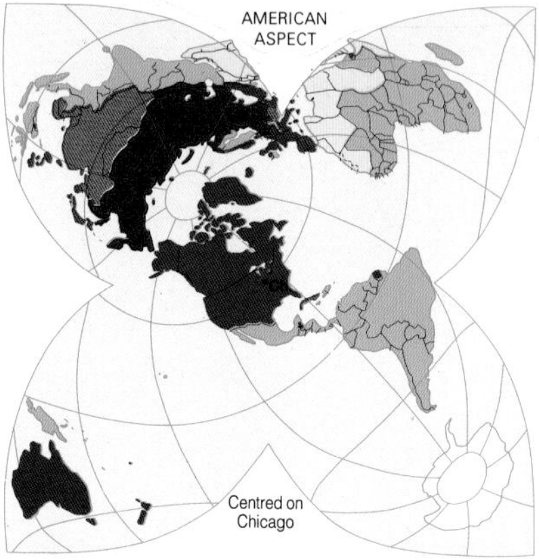

Military blocs

The North Atlantic Treaty Organisation (NATO) was formed in 1949 as an alliance between a group of European countries and the US and Canada. The purpose of the organization was to provide a form of collective security against possible Soviet aggression. The areas covered by NATO are divided into 3 Commands, the Atlantic, Europe and the North Sea.

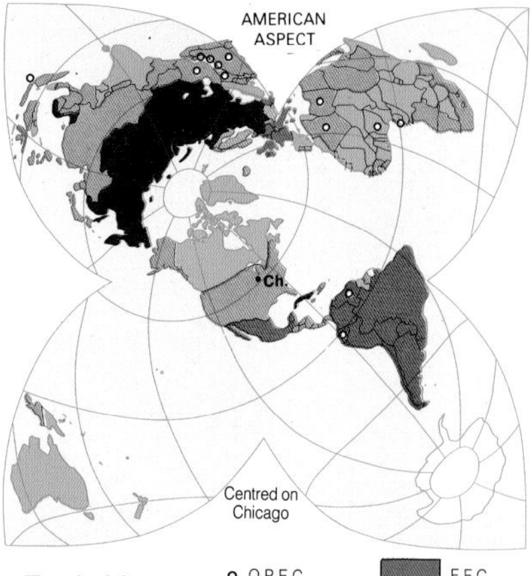

Trade blocs

More than one third of world trade takes place between the countries of the European Community. In 1987 the Single European Act was passed. It aims to remove frontier controls, technical barriers and fiscal barriers so that a single unified market will exist by 1992. However, there are many obstacles still to overcome.

. . .and we were weak standing alone

military and trade blocs

World groups showing some economic indicators 1982

Group	Population 000	Gdp $mn	Imports $mn
EEC	271 706	2 370 158	617 277
EFTA	41 604	393 629	115 259
Comecon (Europe)	380 737	1 339 948	154 731
OAU	467 515	292 107	73 984
OAS	605 469	3 812 019	346 811
Arab League	176 519	398 074	138 140
OPEC	350 069	635 302	154 746
OECD	813 636	7 622 726	1 267 326

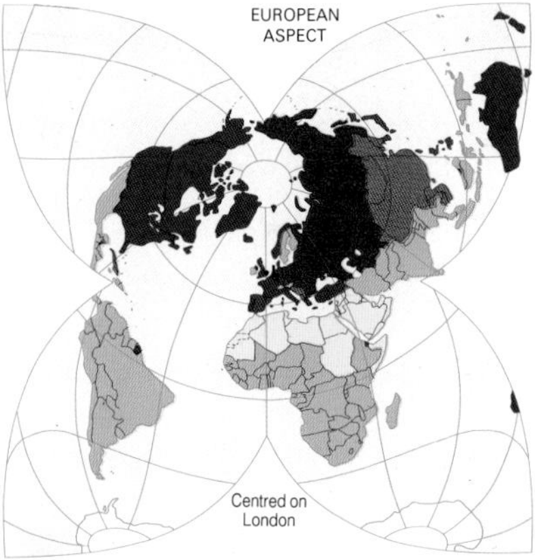

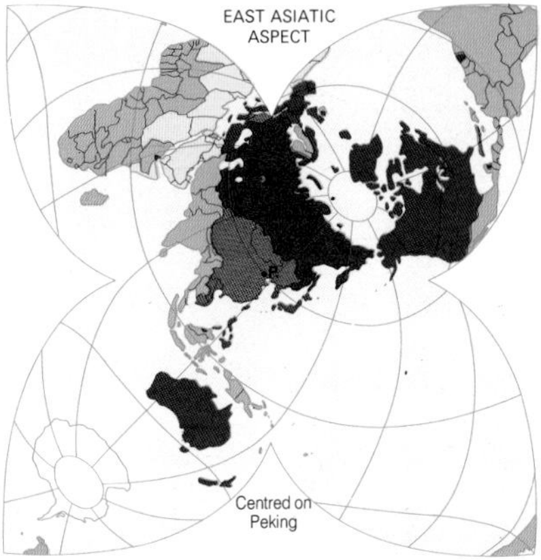

N.A.T.O., A.N.Z.U.S. | Warsaw Pact | Other Communist States | Arab League | Other states

The Warsaw Treaty of Friendship, Cooperation and Military Assistance, known as the Warsaw Pact, was signed in 1955 and gave the communist countries of Europe a unified military command. It was formed as a counterpart to NATO and has its headquarters in Moscow. The signatories were the Soviet Union, Albania, Bulgaria, Czechoslovakia, East Germany, Hungary, Poland and Romania.

The League of Arab States was formed in 1945 and included the Arab States which were then independent; others have joined as they became independent. In 1979 Egypt was expelled but it has now become a member once more. The purpose of the League is to promote cooperation between members and protect national sovereignty.

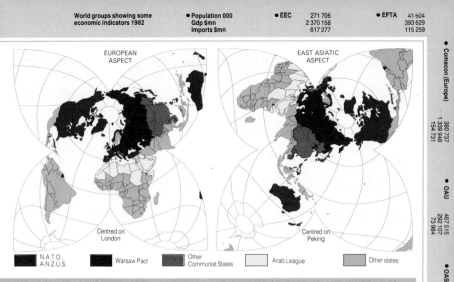

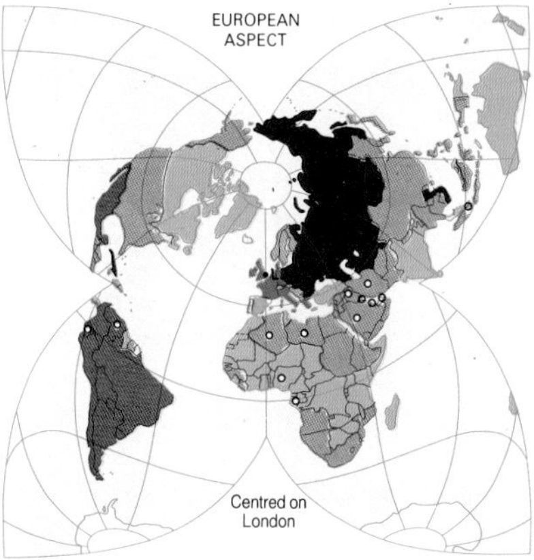

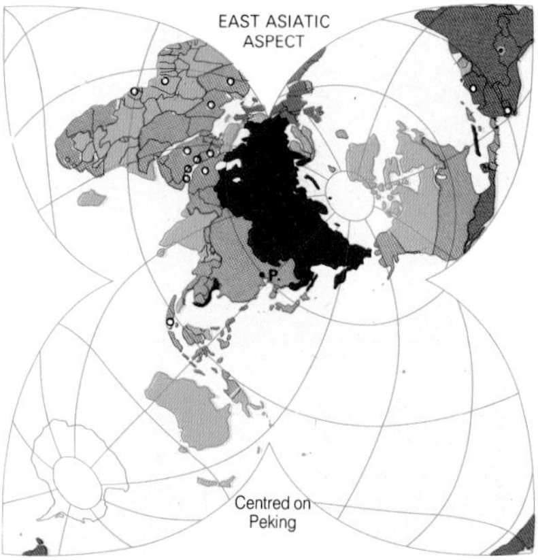

E.E.C. assoc Commonwealth | E.F.T.A. | L.A.I.A. | Comecon | Other countries

There are a number of other associations whose objectives include economic cooperation and the coordination of social, and sometimes political policies. Alliances such as the Organization of African Unity (OAU) cover specific areas. OPEC comprises thirteen nations heavily dependent on oil exports. The Organization for Economic Cooperation and Development (OECD) is an association of 24 countries, which fosters cooperation among industrialized nations and also coordinates their efforts to aid developing countries. The Commonwealth is an association of nations with British connections, straddling the North South divide and bringing together countries with a wide variety of policies, economies and cultures.

GDP Gross domestic product. The total income of the people of the country before allowing for wear and tear on machinery etc.

OPEC. Organization of Petroleum Exporting Countries. Founded in 1960 so that the main producers could have more control over pricing and production.

Maghreb Permanent Consultative Committee was founded in 1964 to coordinate economic affairs and improve cooperation between Tunisia, Morocco, Algeria and Libya.

German Agency for Technical Cooperation GTZ (Germany)
International Agency for Agricultural Development (Italy)
Ministry for Development Cooperation (Netherlands)
Ministère de la Cooperation (France)
Agence de Cooperation Culturelle et Technique (France)
Ministry of Economic Cooperation BMZ (Germany)
United Nations
Organisation for Economic Cooperation and Development
International Coooperation for Socio-Economic Development
International and national aid agencies

Vienna Institute for Development (Austria)
Administration Générale de la Cooperation au Developpement (Belgium)
Danish International Development Agency (Denmark)
Directorate for Agricultural Assistance to Developing Countries (Netherlands)
Norwegian Agency for International Development (Norway)
Swedish International Development Authority (Sweden)

Support- the international response

When the United Nations was formed in 1945 it had a membership of 51 countries. There are now 158 members and its role has broadened from peacekeeping into many other areas that affect Third World countries. The International Monetary Fund was initiated in 1944 and part of its role is to finance short term payments to the less developed countries. Complex financing arrangements channel aid through agencies such as FAO, UNESCO and WHO.

The UN has promoted and financed health care schemes and education; the work of UNICEF is particularly related to the needs of children.

. . .and the individual response

Individuals too, respond to the needs of the others through their contributions to a wide variety of non governmental organizations, usually charities.Some of these may benefit the disadvantaged in their own country, many are concerned with supporting the Third World.

Among the oldest is the Red Cross which was founded by the Geneva Convention of 1864 to care for war casualties.

The deep personal concern of individuals to the predicament of others can be seen from the response to the Band Aid appeal in 1984. In contrast to the established charities this appeal resulted from the concern of one man, the pop-star Bob Geldof. Through this organization money was raised from all over the world. Food and equipment was sent to help the famine areas in Ethiopia in the short term. Long-term projects and rehabilitation schemes in 7 of the most drought affected African countries were also started.

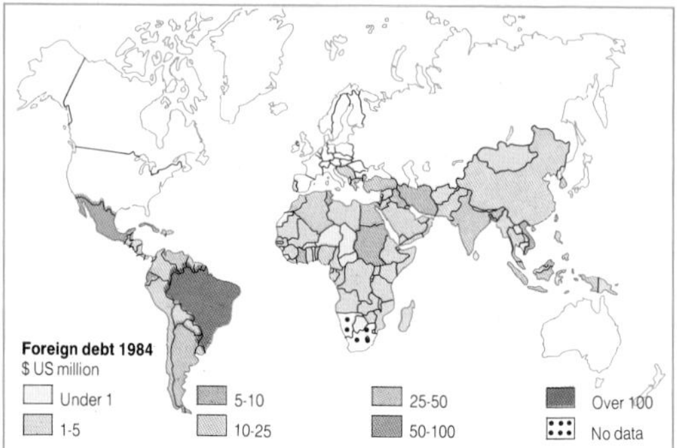

Development and debt

After the dramatic rise in oil prices in 1973 many Third World countries borrowed heavily from the World Bank to finance development. The burden of debt is heaviest for some of the African countries, though Brazil had the greatest debt in 1984.

The problem has been exacerbated by high interest rates and sometimes by the new protectionist policies in the developed countries. Long term aid is of most benefit to developing countries but only if it is accompanied by informed management.

We all need support

Depending on the family

It is within the family that the child is given security and support and learns to give this help to the old, the handicapped and those in need. In some Third World countries a large family is considered necessary to work and to support elderly parents, but there is often insufficient food for all.

Depending on the community

Self help within a community can achieve remarkable results. In Burkino Faso the Naam programme covers 700 village groups. Through the cooperation of village members dams have been built and grain stores constructed. The success of these projects lies in the willingness of the whole community to tackle local problems.

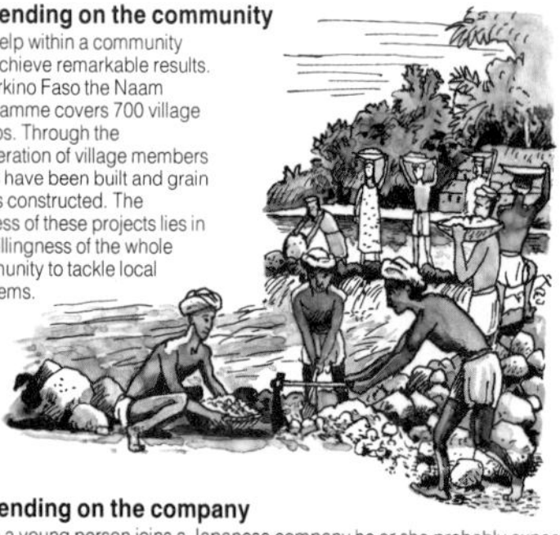

Depending on the company

When a young person joins a Japanese company he or she probably expects to spend his or her whole working life with the same firm. The company takes a very paternal attitude to its employees and makes provision for their medical needs and sometimes for their housing and even holidays. Workers are supported throughout their working life and beyond. In return for this care the company expects total commitment from its employees.

Depending on society

Government is best equipped to provide some of the needs of the community. In some countries many social services are publicly funded and provide a wide range of medical treatments and health care. Provision is also made for the elderly and handicapped. Charities funded by individual contributions also play a part in caring for those in need.

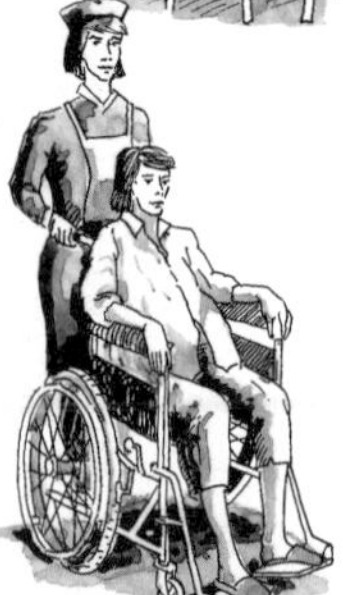

Some non government organizations concerned with overseas aid

Christian Aid
Deutsche Welthungerhilfe
Euroaction- ACORD
Help the Aged
Irish Concern
Mani Tese
Medicins sans Frontières
Methodist World Development Programme
Norwegian Refugee Council
NOVIB
Oxfam
Oxfam America
Oxfam Belgique
Peace Corps
Quaker Peace Corps and Service
Rädda Barnen
Save the Children Fund
Swissaid
Tear Fund
Third World First
Trocaire
Voluntary Service Overseas
War on Want
Water Aid

Supporters and supported

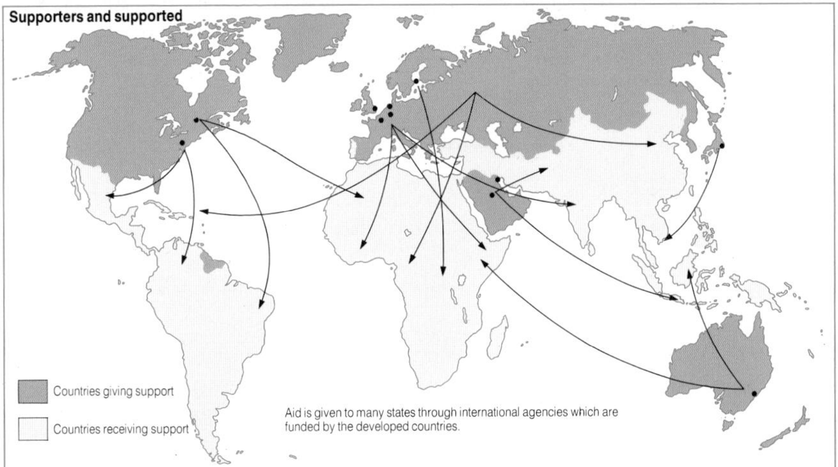

The imbalance of wealth between the developed countries and the developing countries is great. It is estimated that the one fifth of the population that live in the First World have 66% of the world's income, while only 20% goes to the 75% who live in the Third World. Much of this is absorbed by government administration or may even be used in military activities.

Economic aid may be channelled through international agencies such as the World Bank. Governments also make loans or give aid through their official development agencies often cofinancing with banks and other lending agencies. Multilateral or bilateral arrangements are common.

During the past two decades funds have been directed towards meeting the needs of poor people. Education, population planning, agricultural development and small industrial projects have taken priority. This bottom up development should improve the general economy of recipient countries.

The colours on the map show what group a country has been placed in on the basis of its GNP per capita, and in some instances its distinguishing economic characteristics. Countries with populations of more than 1 million are included.

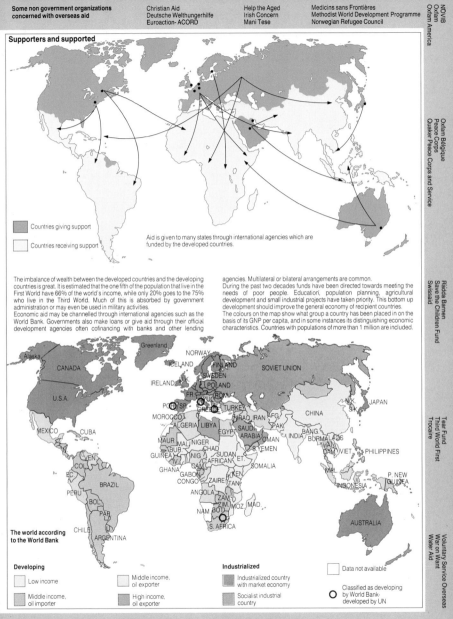

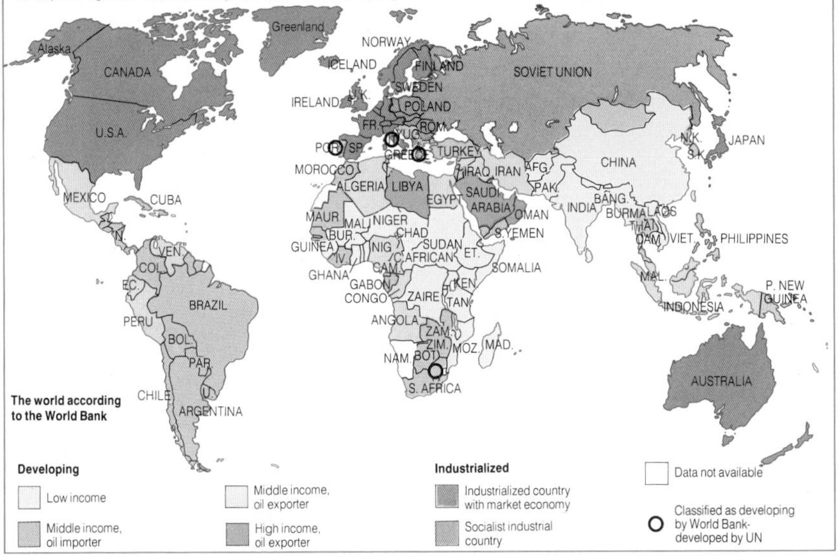

WHO World Health Organization
FAO Food and Agriculture Organization
UNESCO United Nations Education, Scientific and Cultural Organization
UNICEF United Nations Childrens Emergency Fund

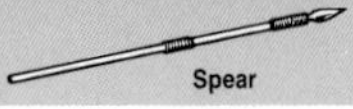
Spear

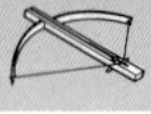
Bow and arrow

Sword

Cannon

Countries with more than 1 million killed in the 20th century

USSR	22 409 000
Poland	6 865 000
Germany	6 401 000
China	6 278 000
Vietnam	2 988 000
Korea	2 890 000
Austria	2 305 000
France	2 305 000
Cambodia	2 150 000
Japan	2 031 000
Nigeria	2 006 000
Turkey	1 673 000
Yugoslavia	1 593 000
Bangladesh	1 500 000
United Kingdom	1 465 000
Spain	1 201 000
Italy	1 170 000
Romania	1 003 000

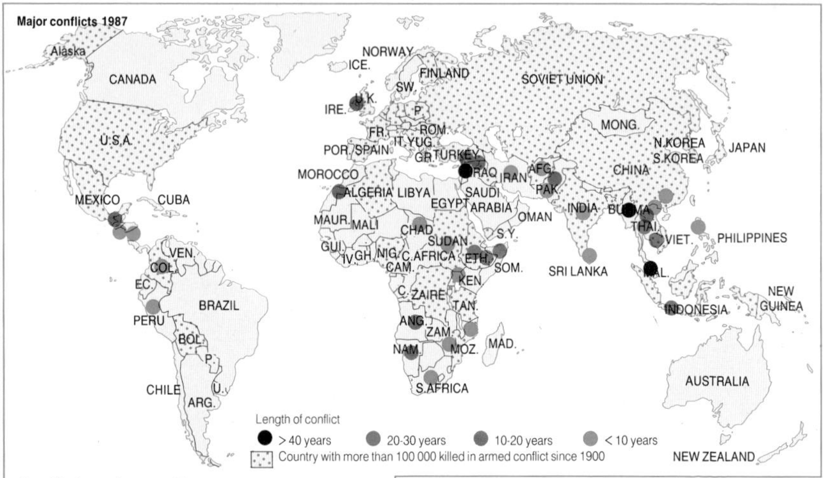

Conflicts and casualties

Compared with the preceding centuries, the 20th century has been marked by a high death rate from wars. More powerful weapons together with motorized transport and the aeroplane have meant that the effects of violence are more widespread.
So far this century, there have been about 240 wars, that is, conflicts with more than 1000 people killed in a year. This includes 13 major wars. Casualties this century number more than 100 million people; 12 times more than the number killed in the 19th century. Apart from the World Wars some of the worst conflicts have been civil wars such as the Biafran war in Nigeria in 1967-70.
In 1987, twenty-two wars were going on, more than in any one year recorded so far.

Do we never learn?

Until the end of World War II, Europe was the principal battlefield, though in the latter stages of this conflict the Far East saw heavy fighting by land and sea. Since 1945 most wars have taken place in the Third World, and though some, such as the Iran-Iraq war, have been between neighbouring states, many have been civil wars.
Traditionally, wars were the main way by which a country sought to extend its territory or influence, or a nation tried to gain independence from a colonial power. In recent times power struggles within nations have increased and are now major causes of wars. However, local wars frequently involve superpower intervention and the supply of arms to adversaries.

Victims – soldiers

Until the 20th century wars were largely a series of pitched battles between professional armies and navies. The battle of Hastings was one such. The victims of these conflicts were the soldiers and sailors, while civilians were expected to supply provisions and keep clear.

Victims – soldiers and civilians

During World War I it is estimated that at least 19 million people were killed, of whom about 5% were civilians. The death toll of World War II was 40 million and a far greater percentage of these, some 20 million, were civilians. In addition 6 million Jews died in concentration camps.

Victims

The bombing of Hiroshima in 1945 caused total devastation to the city, and the death of 130 000 people, almost all civilians. In the Iran-Iraq war 40 years later, civilians, mostly women and children were again victims. Chemical weapons, including nerve gas caused horrifying injuries.

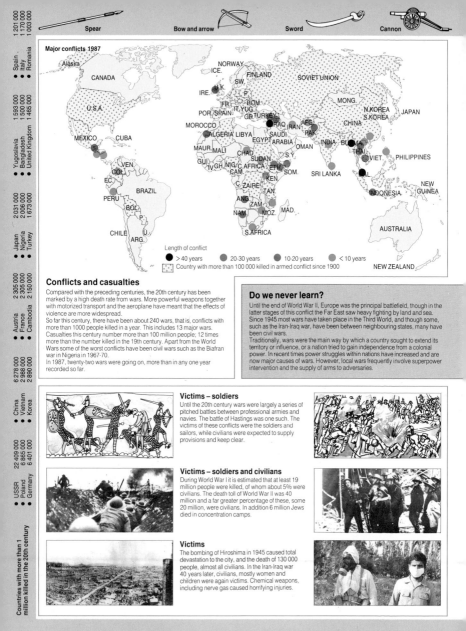

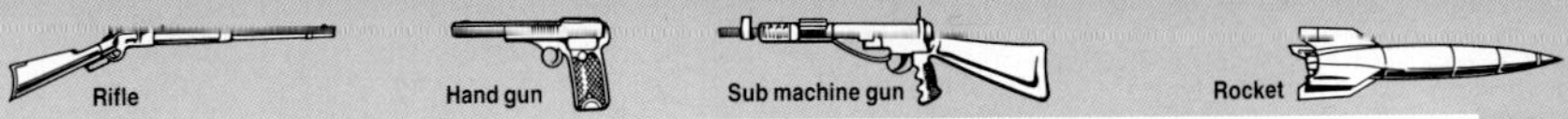

Weapons and education

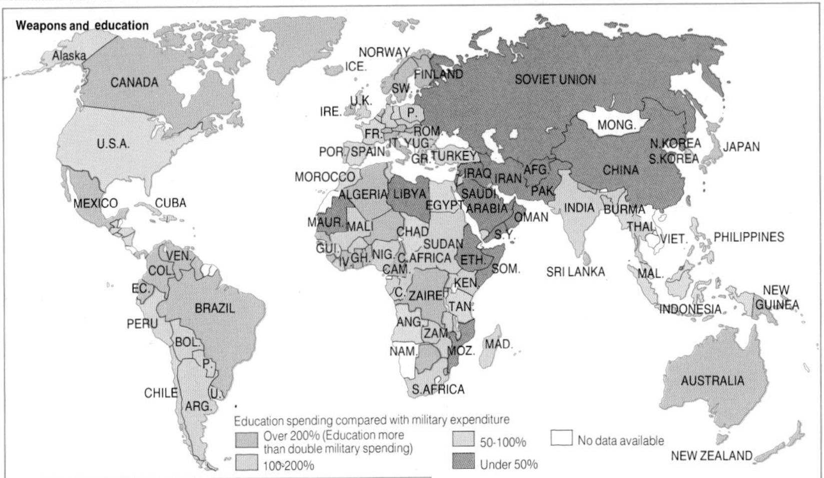

Books or bullets

Resources spent for military purposes are not available for civilian use. While the developed countries, taken as a whole, spend more on military equipment than on health or education, the imbalance in developing countries is even more serious. Again, taken as a whole, their military expenditure exceeds the total for health and education.

If only a proportion of arms expenditure was diverted to social improvements, it would be possible to reduce hunger and provide better medical care and education. Teachers and books could be provided for the world's many illiterate people, and for the children for whom neither primary nor secondary education is at present available. The cost of 1 minute of arms spending could provide 2000 classrooms for 60 000 children.

The global trade in arms is huge and exports to developing countries from the North are at a high level. A small but increasing share of the arms trade is from such countries as Israel and Brazil, to other countries in the South, while several countries do big business in arms reexports.

Conventional weapons of ever-greater destructive force are being developed, but all-out nuclear war is the greatest threat to humanity. The technology of nuclear weaponry is now in the hands of some countries with unstable governments and its use by terrorists is another threat. The greatest fear is of an escalation of a local war into a nuclear one, which might then involve the superpowers.

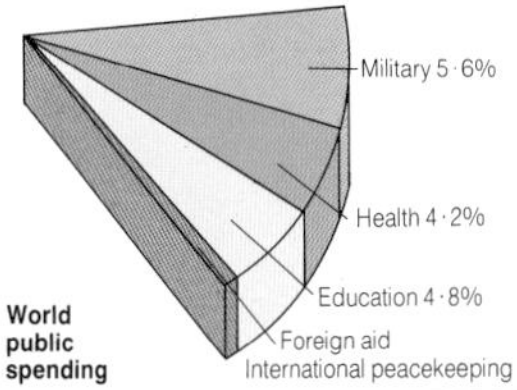

The cost of wars

1986 was officially the International Year of Peace. In that year the World's armed forces totalled more than 27 million, and in that year world military spending was about US $825 billion. The developed countries spent, in total, about 4 times as much as the developing countries in total. For the world this represented 5·6% of public expenditure.

In terms of spending per person some of the oil-rich countries of the Middle East spend a much greater proportion of their public expenditure on military weaponry.

Military spending – as a percentage of public spending

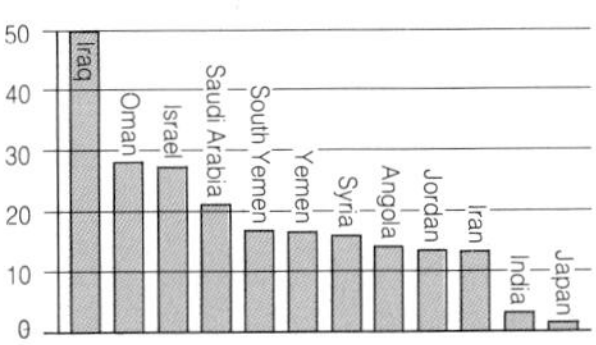

Striving for peace

After the destruction of Hiroshima and Nagasaki, efforts were made to secure total disarmament. This proved impossible and by the 1960s arms limitation had become the objective.

Negotiations have had only a limited effect and the superpowers have built up arsenals of lethal weapons including many with nuclear capability. The 1987 INF Treaty between the USA and the USSR was signed. The agreement was the first to eliminate medium range nuclear weapons, and some progress was made on cutting back on other nuclear arms.

Disarmament and arms control

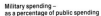

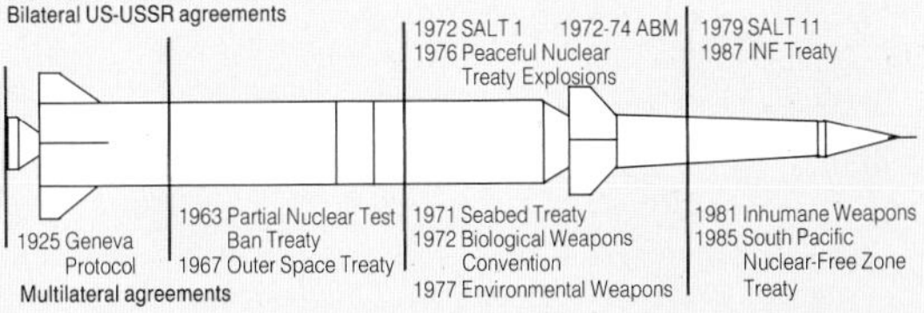

Population distribution

Some parts of the world are very thickly populated, while elsewhere there are huge areas which are virtually uninhabited. Geographical factors play a large part in explaining these wide variations.
Many people can live and grow their own food where there is flat fertile land along a river valley, especially if the climate is favourable and several crops can be grown during the year. Where the climate is too dry as it is in the desert areas of Africa, or too cold as it is in high northern and southern latitudes, and in mountainous areas, few people can live. Those who make their homes in these harsh environments have evolved specialized ways of life to cope with the problems of survival.
Historical as well as geographical reasons help to account for the concentration of people in Europe, Japan and the eastern seaboard of North America. Industrialization and exchange of manufactured goods for food are linked with the growth of cities, particularly of large ports.
But, as the map shows, there are wide variations not only in the distribution of people throughout the world, but also in their appearance. Yet even within one ethnic group there may be many subgroups, each with its own culture and often with its own unique physical characteristics.

Here is the total population on Earth represented by 100 people

Of them there are :

11 from Europe
6 from Soviet Union
6 from North America
22 from China
16 from India
16 from the rest of Asia
4 from the Middle East
8 from Latin America and the West Indies
10 from Africa
1 from Oceania

And among them:

36 are less than 15 years old
6 are more than 65 years old
20 will have a radio
5 have a car
7 have a television set
60 do not have clean drinking water in or near their house
25 are undernourished
20 are very poor
8 are unemployed

Prosperity – more to it than just luck?

Most, but not all, high-income countries are in the temperate lands. The majority of low-income countries, which make up 47% of the world's area, are in the tropical regions of Africa and Asia. The density of population is high in some of the wealthier countries and the level of health care is high in all of them. Families are generally small and most babies will grow to adulthood and many will reach a ripe old age. But in low-income countries a lot of children will die young. There are many babies born, but the level of infant mortality and maternal death is high. It is in these tropical areas that the most serious consequences of soil erosion and deforestation are apparent. The pressure on the land caused by an increasing population, means that land that is really unsuitable for farming is being used. In consequence the topsoil is eroded and further problems are caused (see page 76). The reckless destruction of large areas of tropical rainforest is another serious problem. Already much of the forest in West Africa has been cut down and the rainforests, in South America and southeast Asia are under threat (see page 48).
Whether you are reading this book cover to cover like a novel, or dipping into it at random, much has been happening in the world during the time you have been holding the book. Children have died and the environment has been damaged, but there have also been good things.

Events:	Number born	Number of children dying of starvation
every minute	240	27
every hour	14 400	1620

. . .a moment on Earth

Number of women dying in childbirth	Number of new refugees	Area of devastated rainforest	Area of top soil destroyed	Military expenditure $
1	**4**	**252 m²** **(2 712 sq ft)**	**204 m²** **(2 196 sq ft)**	**1 902 587**

60	**240**	**15 120 m²** **(162 756 sq ft)**	**12 240 m²** **(131 755 sq ft)**	**114 155 220**

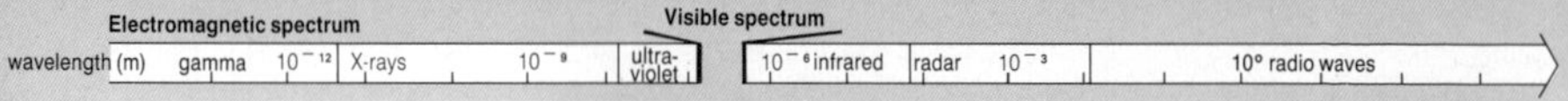

Satellites and remote sensing

Since the launching of Sputnik 1 in 1957, several thousand satellites have orbited the Earth. Early on it was realized that satellites were very good platforms for scanning the earth's surface and monitoring the atmosphere.

This scanning has been performed with a variety of instruments in a great portion of the electro-magnetic spectrum, such as ultra-violet, visible light, infra-red, thermal and radar. Both the scanning and the end-result can be received as photographic film in black and white or in colour, or as digitally stored data. The remote sensing technique has given interesting results for scientists and planners as well as for the general public. Current uses of satellite data are for weather observations and mapping of Earth resources such as topographic maps and monitoring the environment.

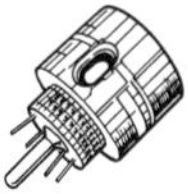

Meteosat

Meteosat is a system with geostationary satellites covering almost the whole Earth. The satellite seemingly is hanging in the space over the same area since it has a height over the equator of 36,000 km, 22,000 mi, and moves in the same direction as our planet.

The Meteosat picture (right) shows the hemisphere which is scanned continuously by the satellite. Africa is in the centre of the picture

Landsat

Landsat 1 was launched in 1972 by NASA (National Aeronautics and Space Administration) as the first remote sensing satellite.

Landsat 1, 2 and 3 were satellites at an altitude of 917 km, 570 miles. The resolution was 80 m, 262 ft for the MSS (Multi Spectral Scanner). The total globe was covered every 18 days. Landsat 4 and 5 are also polar orbiting satellites at an altitude of 705 km. Besides the MSS sensor they had a TM (Thematic Mapper) sensor. The resolution for the TM sensor was 30 m, 100 ft.

Stockholm, Sweden

The image to the right is a Thematic Mapper image of the Stockholm area presented in "false colours". It is not a photograph but a combination of data from different wave length bands coded into special colours. Black is water, brown is forest, red is chlorophyll-rich vegetation and all shades of blue are bare ground, such as bedrock, gravel pits and urban areas. This picture was made in May 1985.

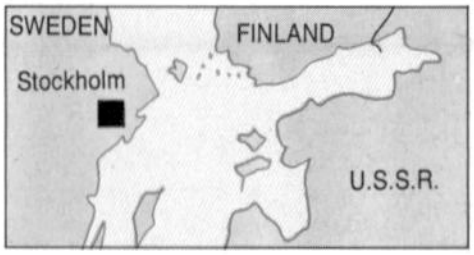

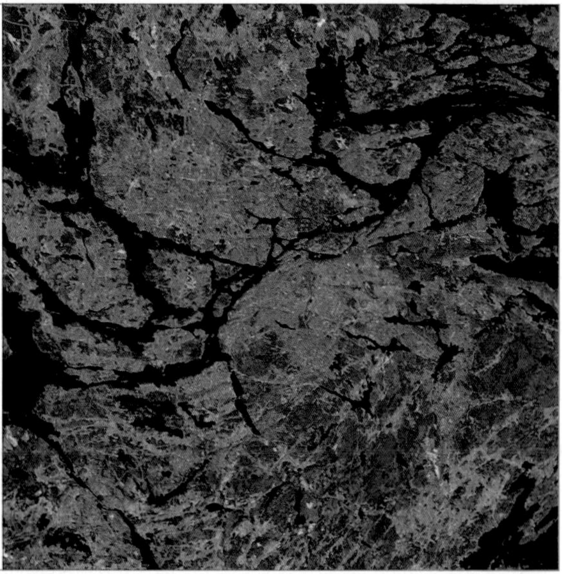

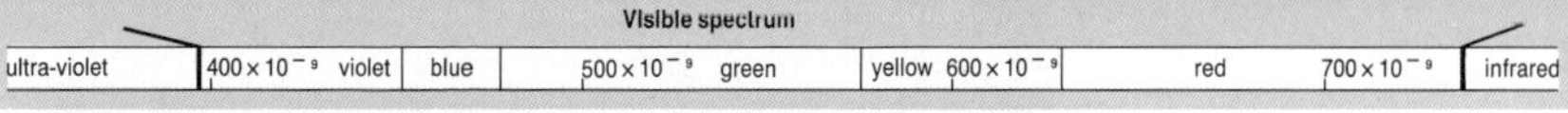

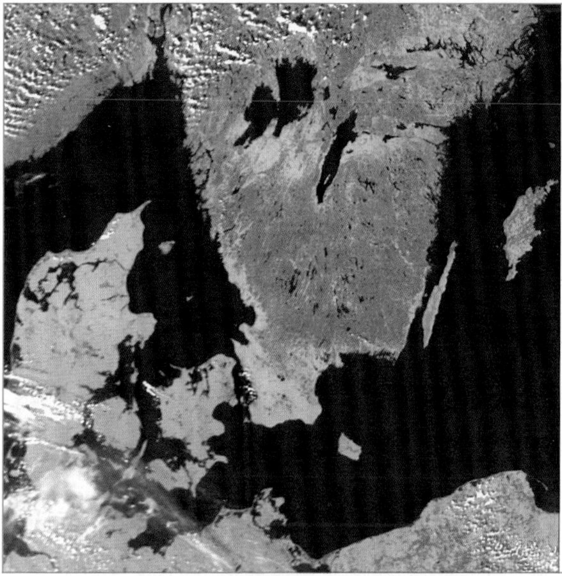

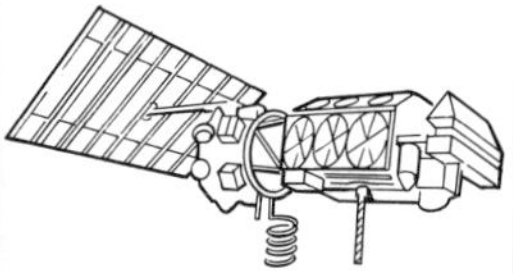

NOAA polar orbiting satellite

The satellite orbits the poles at an altitude of 1860 km, 1155 miles. There are at least two polar orbiting NOAA (National Oceanic and Atmospheric Administration) satellites maintained, viewing both the night and day sides of earth twice a day. They can transmit a wide range of environmental data: earth's cloud cover, surface temperature, atmospheric temperature and humidity, water-ice moisture boundaries, distribution of ozone, heat budget and cloud altitude.

Southern Scandinavia

This image was made in January 1989. The digital signals from the satellite were coded into colours when producing this picture. You can see Denmark in the lower left corner and Stockholm in the upper right. Arable land is light brown, bedrock and urban areas reddish-brown and the clouds white.

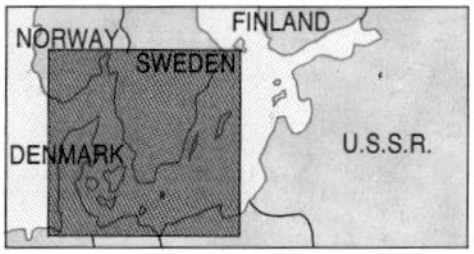

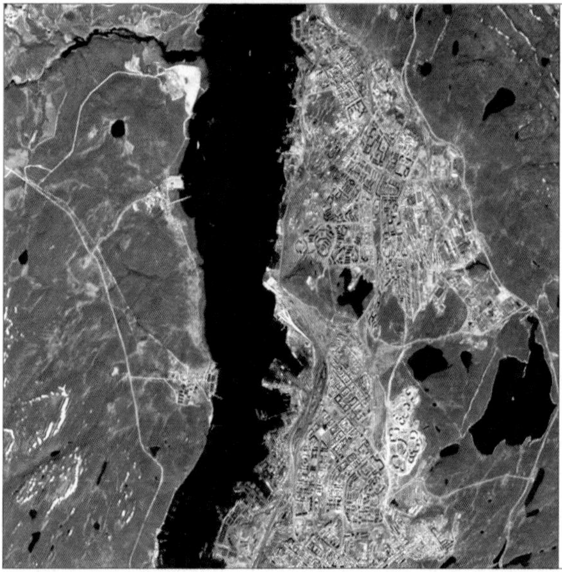

Spot

The French Spot satellite is the first commercial earth observation satellite. It is a polar orbiting satellite at an altitude of 832 km, 517 miles. The Spot sensors have a resolution of 20 m, 60 ft, in multispectral mode and 10 m, 33 ft in panchromatic mode. A new feature for Spot was the possibility of stereoscopic registration. With the ability to produce 3 dimensional information Spot has become a useful tool for topographic mapping as well as for detecting changes and monitoring the environment.

Murmansk, USSR

The Spot image to the left shows the city of Murmansk with its big port. The picture is a combination of images in natural colours and black and white. The resolution is so good that roads, residential blocks and some features in the port can be seen. The scale is approx. 1:25 000 and the resolution is 20 m, 60 ft, so that objects with a length of 20 m can be detected.

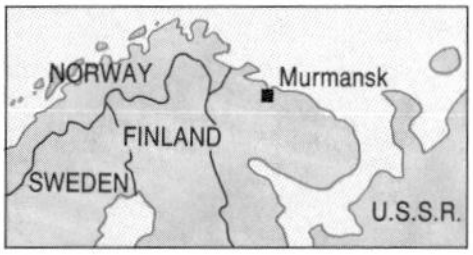

Our World from other aspects

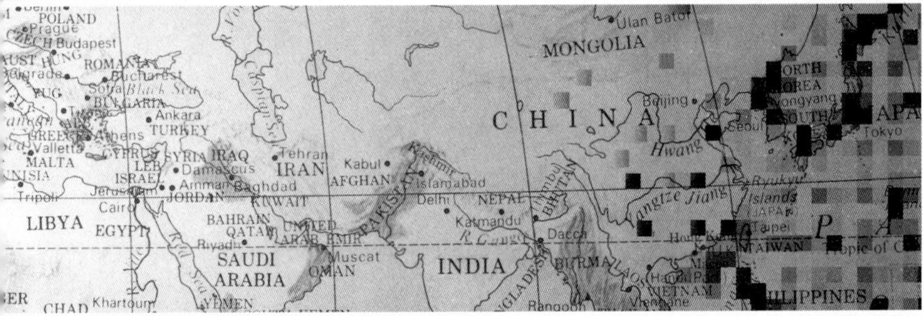

POLAND
Prague
Budapest
Bucharest
Black Sea
Ankara
TURKEY
Athens
Valletta
MALTA
SYRIA
IRAQ
Tehran
IRAN
Kabul
Damascus
Amman
Baghdad
JORDAN
Cairo
LIBYA
EGYPT
BAHRAIN
SAUDI
ARABIA
Muscat
OMAN
Islamabad
Delhi
NEPAL
Katmandu
INDIA
Dacca
Ulan Bator
MONGOLIA
C H I N A
Beijing
Seoul
Tokyo
Taipei
CHAD
Khartoum
YEMEN
Rangoon
PHILIPPINES

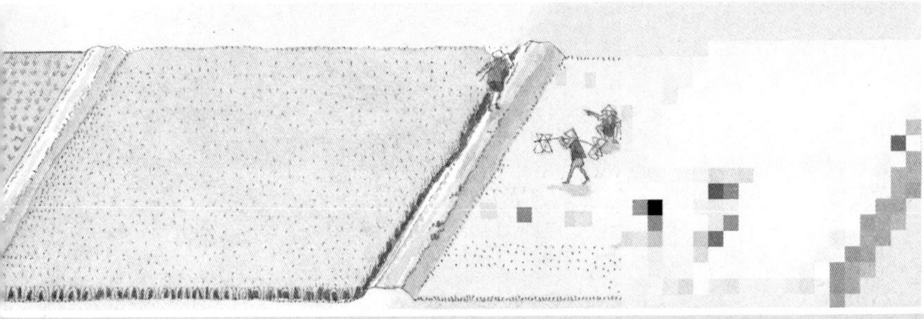

Alfred Nobel (1833–1896)

Alfred Nobel was one of the most imaginative of the 19th century inventors. Nobel built up a fortune from his invention of dynamite in 1867 and a smokeless gunpowder in 1889; but as he had no children to whom he could leave it he established a foundation for the Nobel Prizes.

These consist of five annual awards (for physics, chemistry, physiology or medicine, literature, and peace). In 1969 the Swedish National Bank financed a sixth prize for economics in the memory of Alfred Nobel.

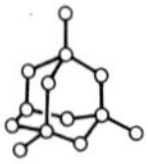

Physics

Year	Name	Discovery	Lifespan	Country
1901	Röntgen, W.C.	Rays named after him	1845–1923	Germany
1902	Lorentz, H.A.	Influence of magnetism upon radiation	1853–1928	Netherlands
	Zeeman, P.	phenomena	1865–1943	Netherlands
1903	Becquerel, A.H.	Spontaneous radioactivity	1852–1908	France
	Curie, P.	Radiation phenomena discovered by	1859–1906	France
	and his wife, Curie, M.	Professor H. Becquerel	1867–1934	France
1904	Rayleigh, Lord J.W.	Densities of gases and discovery of argon	1842–1919	UK
1905	Lenard, P.E.	Work on Cathode rays	1862–1947	Germany
1906	Thomson, Sir J.	Work on conduction of electricity by gases	1856–1940	UK
1907	Michelson, A.	Optical precision instruments	1852–1931	USA
1908	Lippmann, G.	Reproducing colours photographically	1845–1921	France
1909	Marconi, G.	Development of wireless telegraphy	1874–1937	Italy
	Braun, C.F.		1850–1918	Germany
1910	Van der Waals, J.D.	Work on the equation of state for gases and liquids	1837–1923	Netherlands
1911	Wien, W.	Laws governing the radiation of heat	1864–1928	Germany
1912	Dalén, N.G.	Automatic regulators for illuminating lighthouses and buoys	1869–1937	Sweden
1913	Kamerlingh-Onnes, H.	Properties of matter at low temperatures; led to production of liquid helium	1853–1926	Netherlands
1914	Von Laue, M.	Diffraction of X-rays by crystals	1879–1960	Germany
1915	Bragg, Sir W.H.	Analysis of crystal structure by X-rays	1862–1942	UK
	and son, Bragg, Sir W.L.		1890–1971	UK
1917 (awarded 1918)	Barkla, C.G.	Work on Röntgen radiation	1877–1944	UK
1918 (awarded 1919)	Planck, M.K.	Energy quanta	1858–1947	Germany
1919	Stark, J.	Döppler effect in canal rays	1874–1957	Germany
1920	Guillaume, C.E.	Anomalies in nickel steel alloys	1861–1938	Switzerland
1921 (awarded 1922)	Einstein, A.	Law of the photo-electric effect and Theoretical Physics	1879–1955	Germany
1922	Bohr, N.	Structure of atoms and radiation from them	1885–1962	Denmark
1923	Millikan, R.A.	Elementary charge of electricity and the photo-electric effect	1868–1953	USA
1924 (awarded 1925)	Siegbahn, K.M.	X-ray spectroscopy	1886–1978	Sweden
1925	Franck, J.	Laws governing the impact of an electron	1882–1964	Germany
(awarded 1926)	Hertz, G.	upon an atom	1887–1975	Germany
1926	Perrin, J.B.	Sedimentation equilibrium	1870–1942	France
1927	Compton, A.H.	An effect named after him	1892–1962	USA
	Wilson, C.T.	Paths of electrically-charged particles visible by condensation of vapour	1869–1959	UK
1928 (awarded 1929)	Richardson, Sir O.W.	Thermionic phenomenon and the law named after him	1879–1959	UK
1929	De Broglie, Prince L-V.	Wave nature of electrons	1892–1987	France
1930	Raman, Sir C.V.	Scattering of light and effect named after him	1888–1970	India
1932 (awarded 1933)	Heisenberg, W.	Creation of quantum mechanics	1901–1976	Germany
1933	Schrödinger, E.	New productive forms of atomic theory	1877–1961	Austria
	Dirac, P.A.		1902–1984	UK
1935	Chadwick, Sir J.	The Neutron	1891–1974	UK
1936	Hess, V.F.	Cosmic radiation	1883–1964	Austria
	Anderson, C.D.	The Positron	1905-	USA

Year	Name	Discovery	Lifespan	Country
1937	Davisson, C.J.	Diffraction of electrons by crystals	1881–1958	USA
	Thomson, Sir G.P.		1892–1975	UK
1938	Fermi, E.	Nuclear reactions brought about by slow neutrons	1901–1954	Italy
1939	Lawrence. E.O.	Cyclotron and with regard to artificial radioactive elements	1901–1958	USA
1943 (awarded 1944)	Stern, O.	Molecular ray method and the magnetic moment of the proton	1888–1969	USA
1944	Rabi, I.	Resonance method for recording the magnetic properties of atomic nuclei	1898–	USA
1945	Pauli, W.	The Exclusion Principle or Pauli Principle	1900–1958	Austria
1946	Bridgman, P.W.	High pressure physics	1882–1961	USA
1947	Appleton, Sir E.V.	Appleton layer in the upper atmosphere	1892–1965	UK
1948	Blackett, Lord P.M.	Wilson cloud chamber, nuclear physics and cosmic radiation	1897–1974	UK
1949	Yukawa, H.	Prediction of existence of mesons	1907–1981	Japan
1950	Powell, C.F.	Photographic method of studying nuclear processes	1903–1969	UK
1951	Cockcroft, Sir J.D.	Transmutation of atomic nuclei by	1897–1967	UK
	Walton, E.T.	artificially accelerated atomic particles	1903–	Ireland
1952	Block, F.	New methods for nuclear magnetic precision	1905–1983	USA
	Purcell, E.M.	measurements	1912–	USA
1953	Zeruike, F.	Phase contrast microscope	1888–1966	Netherlands
1954	Born, M.	Statistical interpretation of the wave function	1882–1970	UK
	Bothe, W.	Coincidence method	1891–1957	Germany
1955	Lamb, W.E.	Fine structure of the hydrogen spectrum	1913–	USA
	Kusch, P.	Magnetic moment of the electron	1911–	USA
1956	Shockley, W.	Semiconductors and discovery of the	1910–	USA
	Bardeen, J.	transistor effect	1908–	USA
	Brattain, W.H.		1902–	USA
1957	Yang, C.N.	Parity laws and elementary particles	1922–	China
	Lee, T-D.			China
1958	Cerenkov, P.A.	Interpretation of the Cerenkov effect	1904–	USSR
	Frank, I.M.		1908–	USSR
	Tamm, I.J.		1895 – 1971	USSR
1959	Segrè, E.G.	The Antiproton	1905–	USA
	Chamberlain, O.		1920–	USA
1960	Glaser, D.	The bubble chamber	1926–	
1961	Hofstadter, R.	Electron scattering in atomic nuclei and the structure of the nucleons	1915–	USA
	Mössbauer, R.L.	Resonance absorption of gamma radiation	1929–	USA
1962	Landau, L.D.	Theories for condensed matter especially liquid helium	1908 – 1968	USSR
1963	Wigner, E.P.	Application of fundamental symmetery principles	1902–	USA
	Goeppert-Mayer, M.	Nuclear shell structure	1906 – 1972	USA
	Jensen, J. Hans D.		1907 – 1973	Germany
1964	Townes, C.H.	Quantum electronics led to construction of	1915–	USA
	Basov, N.G.	oscillators and amplifiers on the master-	1922–	USSR
	Prochorov, A.M.	laser principle	1916–	USSR
1965	Tomonaga, S.	Quantum electrodynamics	1906 – 1979	Japan
	Schwinger, J.		1918–	USA
	Feynman, R.		1918–	USA
1966	Kastler, A.	Optical methods for studying hertzian resonances in atoms	1902 – 1984	France
1967	Bethe, Hans A.	Energy production in stars	1906–	USA
1968	Alvarez, L.W.	Technique of using hydrogen bubble chamber and data analysis	1911–	USA
1969	Gell-Mann, M.	Classification of elementary particles and their interactions	1929–	USA
1970	Alfven, H.	Magneto-hydrodynamics and applications in different parts of plasma physics	1908–	Sweden
	Néel, L.	Antiferromagnetism and ferrimagnetism and its applications in solid state physics	1904–	France
1971	Gabor, D.	Holographic method and its development	1900–	UK
1972	Bardeen, J.	Developed theory of superconductivity – the	1908–	USA
	Cooper, L.N.	BCS theory	1930–	USA
	Schrieffer, J.R.		1931–	USA
1973	Esaki, L.	Tunnelling phenomena in semiconductors and	1925–	Japan
	Giaever, I.	superconductors	1929–	USA
	Josephson, B.D.	Theoretical predictions of the properties of a supercurrent through a tunnel barrier – the Josephson effects	1940–	UK
1974	Ryle, Sir M.	Research in radio astrophysics	1918–1984	UK
	Hewish, A.		1924–	UK

Year	Name	Discovery	Lifespan	Country
1975	Bohr, A.	Connection between collective motion and particle motion in atomic nuclei and theory development of the structure of the atomic nucleus	1922–	Denmark
	Mottelson, B.		1926–	Denmark
	Rainwater, J.		1917–1986	USA
1976	Richter, B.	A new kind of heavy elementary particle	1931–	USA
	Ting, S.C.		1936–	USA
1977	Anderson, P.W.	Electronic structure of magnetic and disordered systems	1923–	USA
	Mott, Sir N.F.		1905–	UK
	Van Vleck, J.H.		1899–1980	USA
1978	Kapitsa, P.L.	In the area of low-temperature physics	1894–1984	USSR
	Penzias, A.	Cosmic microwave background radiation	1936–	USA
	Wilson, R.W.		1933–	USA
1979	Glashow, S.L.	Prediction of the weak neutral current	1932–	USA
	Salam, A.		1926–	Pakistan
	Weinberg, S.		1933–	USA
1980	Cronin, J.W.	Violation of fundamental symmetry principles in the delay of neutral K-mesons	1931–	USA
	Fitch, V.L.		1923–	USA
1981	Bloembergen, N.	Laser spectroscopy	1920–	USA
	Schawlow, A.L.		1921–	USA
	Siegbahn. K.M.	Development of high-resolution electron spectroscope	1918–	Sweden
1982	Wilson, K.G.	Theory for critical phenomena in connection with phase transitions	1936–	USA
1983	Chandrasekhar, S.	Theoretical studies of physical processes to the structure and evaluation of the stars	1910–	USA
	Foster, W.A.	Nuclear reaction in the formation of the chemical elements in the universe	1911–	USA
1984	Rubbia, C.	Field particles W and Z, communicators of weak interaction	1934–	Italy
	Van der Meer, S.		1925–	Netherlands
1985	Von Kutzing, K.	The quantified Hall effect	1943–	Fed.Rep. of Germany
1986	Ruska, E.	Design of the first electron microscope	1906–	Fed.Rep. of Germany
	Binning, G.	Design of the scanning tunnelling microscope	1947–	Fed.Rep. of Germany
	Rohrer, H.		1933	Switzerland
1987	Bednorz, J.G.	Superconductivity in ceramic materials	1950–	Switzerland
	Muller, K.A.		1927–	Switzerland
1988	Lederman, L.M.	Neutrino beam method. Demonstration of doublet structure of leptons through discovery of muon neutrino	1922–	USA
	Schwartz, M.		1932–	USA
	Steinberger, J.		1921–	USA
1989	Ramsey, N.F.	Invention of separated oscillatory fields method	1915–	USA
	Dehmelt, H.G.	Development of ion trap technique	1922–	USA
	Paul, W.		1913–	Fed.Rep.of Germany

Chemistry

Year	Name	Discovery	Lifespan	Country
1901	Van't Hoff, J.H.	Laws of chemical dynamics and osmotic pressure in solutions	1852–1911	Netherlands
1902	Fischer, H.E.	Work on sugar and purine synthesis	1852–1919	Germany
1903	Arrhenius, S.A.	Electrolytic theory of dissociation	1859–1927	Sweden
1904	Ramsay, Sir W.	The inert gaseous elements in air and their place in the periodic system	1852–1916	UK
1905	Von Baeyer, J.F.	Work on organic dyes and hydroaromatic compounds	1835–1917	Germany
1906	Moissan, H.	Investigation and isolation of the element flourine and for the adoption in the service of science of the electric furnace named after him	1852–1907	France
1907	Buchner, E.	Biochemical researches and discovery of cell-free fermentation	1860–1917	Germany
1908	Rutherford, Lord E.	Disintegration of the elements and the chemistry of radioactive substances	1871–1937	UK
1909	Ostwald, W.	Work on catalysis and investigations into the fundamental principles governing chemical equilibria and rates of reaction	1853–1932	Germany
1910	Wallach, O.	Services to organic chemistry and the chemical industry by work in the field of alicyclic compounds	1847–1931	Germany
1911	Curie, M.	The elements radium and polonium by the isolation of radium, the study of the nature and compounds of this remarkable element	1867–1934	France

Year	Name	Discovery	Lifespan	Country
1912	Grignard, V.	The Grignard reagent	1871–1935	France
	Sabatier, P.	The method of hydrogenating organic compounds in the presence of finely disintegrated metals	1854–1941	France
1913	Werner, A.	Work on the linkage of atoms in molecules which opened up new fields of research in inorganic chemistry	1866–1919	Switzerland
1914 (awarded 1915)	Richards, T.W.	Accurate determination of the atomic weight of a large number of chemical elements	1868–1928	USA
1915	Willstätter, R.M.	Research on plant pigments especially chlorophyll	1872–1942	Germany
1918 (awarded 1919)	Haber, F.	Synthesis of ammonia from its elements	1868–1934	Germany
1920 (awarded 1921)	Nernst, W.H.	Work in thermochemistry	1864–1941	Germany
1921 (awarded 1922)	Soddy, F.	Contributions to the knowledge of the chemistry of radioactive substances. Investigations into the origin and nature of isotopes	1877–1956	UK
1922	Aston, F.W.	Isotopes in a large number of non-radioactive elements by his mass spectro-graph and the enunciation of the whole-number rule	1877–1915	UK
1923	Pregl, F.	Invention of the method of micro-analysis of organic substances	1869–1930	Austria
1925 (awarded 1926)	Zsigmondy, R.A.	Demonstration of the heterogenous nature of colloid solutions and methods used	1865–1929	Germany
1926	Sveldberg, T.	Work on disperse systems	1884–1971	Sweden
1927 (awarded 1928)	Weiland, H.O.	Investigations of constitution of the bile acids and related substances	1877–1957	Germany
1928	Windaus, A.O.	Research into the constitution of the sterols and their connection with the vitamins	1876–1959	Germany
1929	Harden, Sir A.	Investigations on the fermentation of sugar	1865–1940	UK
	Von Euler-Chelpin, H.K.	and fermentative enzymes	1873–1964	Sweden
1930	Fischer, H.	Research into the constitution of haemin and chlorophyll, especially for the synthesis of haemin	1881–1945	Germany
1931	Bosch, C.	Contributions to the invention and development	1874–1940	Germany
	Bergius, F.	of chemical high pressure methods	1884–1949	Germany
1932	Langmuir, I.	Discoveries and investigations in surface chemistry	1881–1957	USA
1934	Urey, H.C.	Heavy hydrogen	1893–1981	USA
1935	Joliot, F.	Synthesis of new radioactive elements	1900–1958	France
		and his wife Joliot-Curie, I.	1897–1956	France
1936	Debye, P.J.	Investigations on dipole moments and on the diffraction of X-rays and electrons in gases	1884–1966	Netherlands
1937	Haworth, Sir W.N.	Investigations on carbohydrates and vitamin C	1883–1950	UK
	Karber, P.	Investigations on carotenoids, flavins and vitamins A and B	1889–1971	Switzerland
1938 (awarded 1939)	Kuhn, R.	Work on carotenoids and vitamins	1900–1967	Germany
1939 (Forced to decline award but later received diploma)	Butenandt, A.F.	Work on sex hormones	1903–	Germany
	Ruzicka, L.	Work on polymethylenes and higher terpenes	1887–1976	Switzerland
1943 (awarded 1944)	De Hevesy, G.	Work on the use of isotopes as traces in the study of chemical processes	1885–1966	Hungary
1944 (awarded 1945)	Hahn, O.	The fission of heavy nuclei	1879–1968	Germany
1945	Virtanen, A.I.	Research and inventions in agricultural and nutrition chemistry especially for the fodder preservation method	1895–1973	Finland
1946	Sumner, J.B.	That enzymes can be crystallized	1887–1955	USA
	Northrop, J.H.	Preparation of enzymes and virus proteins	1891–	USA
	Stanley, W.M.	in a pure form	1904–1971	USA
1947	Robinson, Sir R.	Investigations on plant products of biological importance, especially alkaloids	1886–1975	UK
1948	Tiselius, A.W.	Research on electrophoresis and adsorption analysis, especially discoveries concerning the complex nature of serum proteins	1902–1971	Sweden
1949	Giauque, W.F.	Work on the behaviour of substances at extremely low temperatures	1895–1982	USA
1950	Diels, O.P.	Discovery and development of the diene	1876–1954	Germany
	Alder, K.	synthesis	1902–1958	Germany
1951	McMillan, E.M.	The chemistry of the transuranium elements	1907–	USA
	Seaborg, G.T.		1912–	USA
1952	Martin, A.J.	Invention of partition chromatography	1910–	UK
	Synge, R.L.		1914–	UK

Year	Name	Discovery	Lifespan	Country
1953	Staudinger, H.	In the field of macromolecular chemistry	1881–1965	Germany
1954	Pauling, L.C.	Research into the nature of chemical bond and its application to the elucidation of the structure of complex substances	1901–	USA
1955	Du Vigneaud, V.	The first synthesis of a polypeptide hormone	1901–1978	USA
1956	Hinshelwood, Sir, C.N.	Researches into the mechanism of chemical reactions	1897–1967	UK
	Semenov, N.N.		1896–1986	USSR
1957	Todd, Lord A.R.	Work on nucleotides and nucleotide co-enzymes	1907–	UK
1958	Sanger, F.	Work on the structure of proteins, especially insulin	1918–	UK
1959	Heyrovsky, J.	Discovery and development of the polaro–graphic methods of analysis	1890–1967	Czecho-slovakia
1960	Libby, W.F.	Method of using carbon-14 for age determination in archaeology, geology and geophysics	1908–1980	USA
1961	Calvin, M.	Research on carbon dioxide assimilation in plants	1911–	USA
1962	Perutz, M.F.	Study of the structures of globular proteins	1914–	UK
	Kendrew, Sir J.C.		1917–	UK
1963	Ziegler, K.	The field of the chemistry and technology of high polymers	1898–1973	Germany
	Natta, G.		1903–1979	Italy
1964	Hodgkin, D.C.	By X-ray techniques the structures of important biochemical substances	1910–	UK
1965	Woodward, R.B.	Outstanding achievements in the art of organic synthesis	1917–1979	USA
1966	Mulliken, R.S.	Work concerning chemical bonds and the electronic structure of molecules by the molecular orbital method	1896–1986	USA
1967	Eigen, M.	Studies of extremely fast chemical reactions, effected by disturbing the equi-librium by means of very short pulses of energy	1927–	Fed. Rep. of Germany
	Norrish, R.G.		1897–1978	UK
	Porter, Sir G.		1920–	UK
1968	Onsager, L.	The reciprocal relations bearing his name, which are fundamental for the thermodynamics of irreversible processes	1903–1976	USA
1969	Barton, Sir D.H.	Contributions to the development of the concept of conformation and its application in chemistry	1918–	UK
	Hassel, O.		1897–1981	Norway
1970	Leloir, L.F.	Sugar nucleotides and their role in the biosynthesis of carbohydrates	1906–	Argentina
1971	Herzberg, G.	Contributions to the knowledge of electronic structure and geometry of molecules, particularly free radicals	1904–	Canada
1972	Anfinsen, C.B.	Work on ribonuclease, especially the connection between the amino acid sequence and the biologically active conformation	1916	USA
	Moore, S.	Understanding of the connection between chemical structure and catalytic activity of the active centre of ribonuclease molecule	1913–1982	USA
	Stein, W.H.		1911–1980	USA
1973	Fischer, E.O.	Independent work on the chemistry of the organo-metallic so called sandwich compounds	1918–	Fed. Rep. of Germany
	Wilkinson, Sir G.		1921–	UK
1974	Flory, P.J.	Achievements, theoretical and experi-mental, in the physical chemistry of macromolecules	1910–1985	USA
1975	Cornforth, Sir J.W.	Work on the stereochemistry of enzyme–catalyzed reactions	1917–	Australia
	Prelog, V.	Research into stereochemistry of organic molecules and reactions	1906–	Switzerland
1976	Lipscomb, W.N.	Studies on the structure of boranes illuminating problems of chemical bonding	1919–	USA
1977	Prigogine, I.	Contributions to non-equilibrium thermo-dynamics, particularly the theory of dissipative structures	1917–	Belgium
1978	Mitchell, P.D.	Understanding of biological energy transfer through the formulation of the chemiosmotic theory	1920–	UK
1979	Brown, H.C.	Development of the use of boron- and phosphorus-containing compounds into important reagents in organic synthesis	1912–	USA
	Wittig, G.		1897–	Fed. Rep. of Germany
1980	Berg, P.	Studies of the biochemistry of nucleic acids, particularly recombinant DNA	1926–	USA
	Gilbert, W.	Contributions concerning the determination of base sequences in nucleic acids	1932–	USA
	Sanger, F.		1918–	UK
1981	Fukui, K.	Independent theories concerning the course of chemical reactions	1918–	Japan
	Hoffmann, R.		1937–	USA

Year	Name	Discovery	Lifespan	Country
1982	Klug, A.	Development of crystallographic electron microscopy and structural elucidation of biologically important nucleic acid-protein complexes	1926–	UK
1983	Taube, H.	Work on the mechanisms of electron transfer reactions, especially in metal complexes	1915–	USA
1984	Merrifield, R.B.	Development of methodology for chemical synthesis on a solid matrix	1921–	USA
1985	Hauptman, H.A.	Development of direct methods for the determination of crystal structures	1917–	USA
	Karle, J.		1918–	USA
1986	Herschbach, D.R.	Contributions in the dynamics of chemical elementary processes	1932–	USA
	Lee, Y.T.		1936–	USA
	Polanyi, J.C.		1929–	Canada
1987	Cram, D.J.	Development and use of molecules with structure-specific interactions of high selectivity	1919–	USA
	Lehn, J-M.		1939–	France
	Pedersen, C.J.		1904–	USA
1988	Deisenhofer, J.	Determination of the three-dimensional structure of a photosynthetic reaction centre	1943–	Fed. Rep. of Germany
	Huber R.		1937–	"
	Michel, H.		1948–	"
1989	Altman, S.	Discovery of catalytic properties of RNA	1939–	USA
	Cech, T.R.		1947–	USA

Physiology or Medicine

Year	Name	Discovery	Lifespan	Country
1901	Von Behring, E.A.	Work on serum therapy, especially its application against diphtheria	1854–1917	Germany
1902	Ross, Sir R.	Work on malaria, showing how it enters the organism	1857–1932	UK.
1903	Finsen, N.R.	Contribution to the treatment of diseases, especially lupus vulgaris, with concentrated light radiation	1860–1904	Denmark
1904	Pavlov, I.P.	Work on the physiology of digestion	1849–1936	Russia
1905	Koch, R.	Investigations and discoveries in relation to tuberculosis	1843–1910	Germany
1906	Golgi, C.	Work on the structure of the nervous system	1843–1926	Italy
	Ramon Y Cajal, S.		1852–1934	Spain
1907	Laverau, C.L.	Work on the role played by protozoa in causing diseases	1845–1922	France
1908	Mecnikov, I.I.	Work on immunity	1845–1916	Russia
	Ehrlich, P.		1854–1915	Germany
1909	Kocher, E.T.	Work on the physiology, pathology and surgery of the thyroid gland	1841–1917	Switzerland
1910	Kossel, A.	Work on proteins, including the nucleic substances, increasing the knowledge of cell chemistry	1853–1927	Germany
1911	Gullstrand, A.	Work on the dioptrics of the eye	1862–1930	Sweden
1912	Carrel, A.	Work on vascular suture and the trans–plantation of blood-vessels and organs	1873–1944	France
1913	Richet, C.R.	Work on anaphylaxis	1850–1935	France
1914	Bárány, R.	Work on the physiology and pathology of the vestibular apparatus	1876–1936	Austria
1919 (awarded 1920)	Bordet, J.	Relating to immunity	1870–1961	Belgium
1920	Krogh, S.A.	The capillary motor regulating mechanism	1874–1949	Denmark
1922	Hill, Sir A.V.	Relating to the production of heat in the muscle	1886–1977	UK
(awarded 1923)	Meyerhof, O.F.	Fixed relationship between the consumption of oxygen and the metabolism of lactic acid in the muscle	1884–1951	Germany
1923	Banting, Sir F.G.	Insulin	1891–1941	Canada
	Macleod, J.J.		1876–1935	Canada
1924	Einthoven, W.	The mechanism of the electrocardiogram	1860–1927	Netherlands
1926 (awarded 1927)	Fibiger, J.A.	The Spiroptera carcinoma	1867–1928	Denmark
1927	Wagner-Jauregg, J.	The therapeutic value of malaria inoculation in the treatment of dementia paralytica	1857–1940	Austria
1928	Nicolle, C.J.	Work on typhus	1866–1936	France
1929	Eijkman, C.	The antineuritic vitamin	1858–1930	Netherlands
	Hopkins, Sir F.G.	The growth-stimulating vitamins	1861–1947	UK
1930	Landsteiner, K.	The human blood groups	1868–1943	Austria
1931	Warburg, O.H.	The nature and mode of action of the respiratory enzyme	1883–1970	Germany

Year	Name	Discovery	Lifespan	Country
1932	Sherrington, Sir C.S.	Regarding the functions of neurons	1857–1952	UK
	Adrian, Lord E.D.		1889–1977	UK
1933	Morgan, T.H.	The role played by the chromosome in heredity	1866–1945	USA
1934	Whipple, G.H.		1878–1976	USA
	Minot, G.R.	Liver therapy in cases of anaemia	1885–1950	USA
	Murphy, W.P.		1892–	USA
1935	Spemann, H.	The organizer effect in embryonic development	1869–1941	Germany
1936	Dale, Sir H.H.	Relating to chemical transmission of	1875–1968	UK
	Loewi, O.	nerve impulses	1873–1961	Austria
1937	Szent-Györgyi von Nagyrapolt, A.	In connection with the biological combustion processes, special reference to vitamin C and the catalysis of fumaric acid	1893–1986	Hungary
1938 (awarded 1939)	Heymans, C.J.	The role played by the sinus and aortic mechanisms in the regulation of respiration	1892–1968	Belgium
1939 (had to decline the award, but later received the diploma and the medal)	Domagk, G.	The antibacterial effects of prontosil	1895–1964	Germany
1943	Dam, H.C.	Vitamin K	1895–1976	Denmark
	Doisy, E.A.	The chemical nature of vitamin K	1893–1986	USA
(awarded 1944)				
1944	Erlanger, J.	Relating to the highly differentiated	1874–1965	USA
	Gasser, H.S.	functions of single nerve fibres	1888–1963	USA
1945	Fleming, Sir A.	Penicillin and its curative effect in	1881–1955	UK
	Chain, Sir E.B.	various infectious diseases	1906–1979	UK
	Florey, Lord H.W.		1898–1968	UK
1946	Muller, H.J.	The production of mutations by means of X-ray irradiation	1890–1967	USA
1947	Cori, C.F.	The course of catalytic conversion of	1896–1984	USA
	and his wife, Cori, G.T.	glycogen	1896–1957	USA
	Houssay, B.A.	The part played by the hormone of the anterior pituitary lobe in the metabolism of sugar	1887–1971	Argentina
1948	Muller, P.H.	The high efficiency of DDT as a contact poison against several arthropods	1899–1965	Switzerland
1949	Hess, W.R.	The functional organization of the inter-brain as a co-ordinator of the activities of the internal organs	1881–1973	Switzerland
	Moniz, A.C.	The therapeutic value of leucotomy in certain psychoses	1874–1955	Portugal
1950	Kendall, E.C.	Relating to the hormones of the adrenal	1886–1972	USA
	Reichstein, T.	cortex, their structure and biological	1897–	Switzerland
	Hench, P.S.	effects	1896–1965	USA
1951	Theiler, M.	Yellow fever and how to combat it	1899–1972	USA
1952	Waksman, S.A.	Streptomycin, the first antibiotic effective against tuberculosis.	1888–1973	USA
1953	Krebs, Sir H.A.	The citric acid cycle	1900–1981	UK
	Lipmann, F.A.	Co-enzyme A and its importance for intermediary metabolism	1899–1986	USA
1954	Enders, J.F.	The ability of poliomyelitis viruses to	1897–1985	USA
	Weller, T.H.	grow in cultures of various types of tissue	1915–	USA
	Robbins, F.C.		1916–	USA
1955	Theorell, A.H.	The nature and mode of action of oxidation enzymes	1903–1982	Sweden
1956	Cournand, A.F.	Heart catherization and pathological	1895–	USA
	Forssmann, W.	changes in the circulatory system	1904–1979	Germany
	Richards, D.W.		1895–1973	Germany
1957	Bovet, D.	Synthetic compounds that inhibit the action of certain body substances, especially their action on the vascular system and the skeletal muscles	1907–	Italy
1958	Beadle, G.W.	That genes act by regulating definite	1903–	USA
	Tatum, E.L.	chemical events	1909–1975	USA
	Lederberg, J.	Genetic recombination and the organization of the genetic material of bacteria	1925–	USA
1959	Ochoa, S.	The mechanisms in the biological synthesis	1905–	USA
	Kornberg, A.	of ribonucleic acid and deoxiribonucleic acid	1918–	USA
1960	Burnet, Sir F.M.	Acquired immunological tolerance	1899–1985	Austria
	Medawar, Sir P.B.		1915–	UK
1961	Von Békésy, G.	The physical mechanism of stimulation within the cochlea	1899–	USA

Year	Name	Discovery	Lifespan	Country
1962	Crick, F.H.	The molecular structure of nuclear acids	1916–	UK
	Watson, J.D.	and its significance for information	1928–	USA
	Wilkins, M.H.	transfer to living material	1916	UK
1963	Eccles, Sir. J.C.	The ionic mechanisms involved in excitation	1903–	Australia
	Hodgkin, Sir A.L.	and inhibition in the peripheral and central	1914–	UK
	Huxley, Sir A.F.	proportions of the nerve cell membrane	1917–	UK
1964	Block, K.	The mechanism and regulation of the	1912–	USA
	Lynen, F.	cholesterol and fatty acid metabolism	1911–1979	Germany
1965	Jacob, F.	Genetic control of enzyme and virus	1920–	France
	Lwoff, A	synthesis	1902–	France
	Monod, J.		1910–1976	France
1966	Rlous, P.	Tumour-inducing viruses	1879–1970	USA
	Huggins, C.B.	Hormonal treatment of prostatic cancer	1901–	USA
1967	Granit, R.	The primary physiological and chemical	1900–	Sweden
	Hartline, H.K.	visual processes in the eye.	1903–1983	USA
	Wald, G.		1906–	USA
1968	Holley, R.W.	Their interpretation of the genetic code	1922–	USA
	Khorana, H.	and its function in protein synthesis	1922–	USA
	Nirenberg, M.W.		1927–	USA
1969	Delbrück, M.	The replication mechanism and the genetic	1906–	USA
	Hershey, A.D.	structure of viruses	1908–	USA
	Luria, S.E.		1912–	USA
1970	Katz,Sir B.	The humoral transmitters in the nerve	1911–	UK
	Von Euler, U.	terminals and the mechanism for their	1905–1983	Sweden
	Axelrod, J.	storage, release and inactivation	1912–	USA
1971	Sutherland, Earl W.	The mechanism of the action of hormones	1915–1974	USA
1972	Edelman, G.M.	The chemical structure of antibodies	1929–	USA
	Porter. R.R.		1917–1985	UK
1973	Von Frisch, K.	Organization and elicitation of	1886–1982	Fed. Rep. of
		individual and social behaviour patterns		Germany
	Lorenz, K.		1903–	Austria
	Tinbergen, N.		1907–	UK
1974	Claude, A.	Structural and functional organization of	1899–1983	Belgium
	De Duve, C.	the cell	1917–	Belgium
	Palade, G.E.		1912–	USA
1975	Baltimore, D.	The interaction between tumour viruses and	1938–	USA
	Dulbecco, R.	the genetic material of the cell	1914–	USA
	Temin, H.M.		1934–	USA
1976	Blumberg, B.	New mechanisms for the origin and	1925–	USA
	Gajdusek, D.C.	dissemination of infectious diseases	1923–	USA
1977	Guillemin, R.	The peptide hormone production of the brain	1924–	USA
	Schally, A.V.		1926–	USA
	Yalow, R.	Development of radioimmunoassays of	1921–	USA
		peptide hormones		
1978	Arber, W.	Restriction enzymes and their application	1929–	Switzerland
	Nathans, D.	to problems of molecular genetics.	1928–	USA
	Smith. H.O.		1931–	USA
1979	Cormack, A.M.	Development of computer-assisted tomography	1924–	USA
	Hounsfield, Sir G.N.		1919–	UK
1980	Benacerraf, B.	Genetically determined structures on the	1920–	USA
	Dausset, J.	cell surface that regulate immunological	1916–	France
	Snell, G.D.	reactions	1903–	USA
1981	Sperry, R.W.	The functional specialization of the	1913–	USA
		cerebral hemispheres		
	Hubel, D.H.	Information processing in the visual system	1926–	USA
	Wiesel, T.N.		1924–	Sweden
1982	Bergström, S.K.	Prostaglandins and related biologically	1916–	Sweden
	Samuelsson, B.I.	active substances	1934–	Sweden
	Vane, Sir R.		1927–	UK
1983	McClintock, B.	Mobile genetic elements	1902–	USA
1984	Jerne, N.K.	Theories concerning the specificity in	1911–	Denmark
	Köhler, G.J.	development and control of the immune	1946–	Fed. Rep of
	Milstein, C.	system and the principle for production		Germany
		of monoclonal antibodies	1927–	UK
1985	Brown, M.S.	The regulation of cholesterol metabolism	1941–	USA
	Goldstein, J.I.		1940–	USA
1986	Cohen, S.	Growth factors	1922–	USA
	Levi-Montakini, R.		1909–	Italy
1987	Tonegawa, S.	The genetic principle for generation	1939–	Japan
		of antibody diversity		
1988	Black, Sir J.W.	For important principles for drug treatment	1924–	UK
	Elion, G.B.		1918–	USA
	Hitchings, G.H.		1905–	USA
1989	Bishop, J.M.	Discovery of the cellular origin	1936–	USA
	Varmus, H.E.	of retroviral oncogenes	1939–	USA

Literature

Year	Name	Lifespan	Country
1901	Prudhomme, S.	1839–1907	France
1902	Mommsen, T.	1817–1903	Fed. Rep. Germany
1903	Bjornson, B.	1832–1910	Norway
1904	Mistral, F.	1830–1914	France
	Echegaray, J.	1833–1916	Spain
1905	Sienkiewicz, H.	1846–1916	Poland
1906	Carducci, G.	1835–1907	Italy
1907	Kipling, R.	1865–1936	UK
1908	Eucken, R.C.	1846–1926	Germany
1909	Lagerlof, S.	1858–1940	Sweden
1910	Heyse, P.	1830–1914	Germany
1911	Maeterlinck, M.	1862–1949	Belgium
1912	Hauptmann, G.	1862–1946	Germany
1913	Tagore, R.	1861–1941	India
1915 (awarded 1916)	Rolland, R.	1866–1944	France
1916	Heidenstam, C. von	1859–1940	Sweden
1917	Gjellerup, K.	1857–1919	Denmark
	Pontoppidan, H.	1857–1943	Denmark
1919 (awarded 1920)	Spitteler, C.F.	1845–1924	Switzerland
1920	Hamsun, K.P.	1859–1952	Norway
1921	France, A.	1844–1924	France
1922	Benavente, J.	1866–1954	Spain
1923	Yeats, W.B.	1865–1939	Rep. of Ireland
1924	Reymont, W.	1868–1925	Poland
1925 (awarded 1926)	Shaw, G.B.	1856–1950	UK
1926 (awarded 1927)	Deledda, G.	1871–1936	Italy
1927 (awarded 1928)	Bergson, H.	1859–1941	France
1928	Undset, S.	1882–1949	Norway
1929	Mann, T.	1875–1955	Germany
1930	Lewis, S.	1885–1951	USA
1931	Karlfeldt, E.A.	1864–1951	Sweden
1932	Galsworthy, J.	1867–1933	UK
1933	Bunin, I.A.	1870–1953	France
1934	Pirandello, L.	1867–1936	Italy
1936	O'Neill, E.G.	1888–1953	USA
1937	Martin du Gard, R.	1881–1958	France
1938	Buck, P.	1892–1973	USA
1939	Sillanpaa, F.E.	1888–1964	Finland
1944	Jensen, J.V.	1873–1950	Denmark
1945	Mistral, G.	1889–1957	Chile
1946	Hesse, H.	1877–1962	Switzerland
1947	Gide, A.P.	1869–1951	France
1948	Eliot, T.S.	1888–1965	UK
1949 (awarded 1950)	Faulkner, W.	1897–1962	USA
1950	Russell, B.	1872–1970	UK
1951	Lagerkvist, P.	1891–1974	Sweden
1952	Mauriac, F.	1885–1970	France
1953	Churchill, W.	1874–1965	UK
1954	Hemingway, E.	1898–1961	USA
1955	Laxness, H.	1902–	Iceland
1956	Jimenez, J.	1881–1958	Spain
1957	Camus, A.	1913–1960	France
1958	Pasternak, B.	1890–1960	USSR
1959	Quasimodo, S.	1901–1968	Italy
1960	Perse, S-J.	1887–1975	France
1961	Andric, I.	1892–1975	Yugoslavia
1962	Steinbeck, J.	1902–1968	USA
1963	Seferis, G.	1900–1971	Greece
1964	Sartre, J.P.	1905–1980	France
1965	Solochov, M.A.	1905–1984	USSR
1966	Agnon, S.Y.	1888–1970	Israel
	Sachs, N.	1891–1970	Germany
1967	Asturias, M.A.	1899–1974	Guatemala
1968	Kawabata, Y.	1899–1972	Japan
1969	Beckett, S.	1906–1989	Rep. of Ireland
1970	Solzhenitsyn, A.	1918–	USSR
1971	Neruda, P.	1904–1973	Chile
1972	Boll, M.	1917–1985	Fed. Rep. of Germany
1973	White, P.	1912–	Australia
1974	Johnson, E.	1900–1976	Sweden
	Martinson, H.	1904–1978	Sweden
1975	Montale, E.	1896–1981	Italy
1976	Bellow, S.	1915–	USA
1977	Aleixandre, V.	1899–1984	Spain
1978	Singer, I.B.	1904–	USA
1979	Elytis, O.	1911–	Greece
1980	Milosz, C.	1911–	USA
1981	Canetti, E.	1905–	UK
1982	Marquez, G.G.	1928–	Columbia
1983	Golding, W.	1911–	UK
1984	Seifert, J.	1901–	Czecho-slovakia
1985	Simon, C.	1913–	France
1986	Soyinke, W.	1934–	Nigeria
1987	Arias Sanchez, O.	1941–	Costa Rica
1988	Mahfouz, N.	1911–	Rep. of Egypt
1989	Cela, C.J.	1916	Spain

Peace

Year	Name	Lifespan	Country
1901	Dunant, J.H.	1828–1910	Switzerland
	Passy, F.	1822–1912	France
1902	Ducommun, E.	1833–1906	Switzerland
1903	Cremer, W.	1838–1908	UK
1904	Institute of International Law	Founded 1873	USA
1905	Suttner, B. von	1843–1914	Austria
1906	Roosevelt, T.	1858–1919	USA
1907	Moneta, E.	1833–1918	Italy
	Renault, L.	1843–1918	France
1908	Arnoldson, K.P.	1844–1916	Sweden
	Bajer, F.	1837–1922	Denmark
1909	Beernaert, A.M.	1829–1912	Belgium
	Balluet, P. Baron de Constant de Rebecque	1852–1924	France
1910	Permanent International Peace Bureau	Founded 1891	Switzerland
1911	Asser, T.M.	1838–1913	Netherlands
	Fried, A.H.	1864–1921	Austria
1912 (awarded 1913)	Root, E.	1845–1937	USA
1913	La Fontaine, H.	1845–1943	Belgium
1917	International Committee of the Red Cross	Founded 1863	Switzerland

Year	Name	Lifespan	Country
1919 (awarded 1920)	Wilson, T.W.	1856–1924	USA
1920	Bourgeois, L.V.	1851–1925	France
1921	Branting, K.H.	1860–1925	Sweden
	Lange, C.L.	1869–1938	Norway
1922	Nansen, F.	1861–1930	Norway
1925 (awarded 1926)	Chamberlain, Sir A.	1863–1937	UK
	Dawes, C.G.	1865–1951	USA
1926	Briand, A.	1862–1932	France
	Stresemann, G.	1878–1929	Germany
1927	Buisson, F.	1841–1932	France
	Quidde, L.	1858–1941	Germany
1929 (awarded 1930)	Kellogg, F.B.	1856–1937	USA
1930	Söderblom, J.	1866–1931	Sweden
1931	Addams, J.	1860–1935	USA
	Butler, N.M.	1862–1947	USA
1933 (awarded 1934)	Angell, Sir N.	1874–1967	UK
1934	Henderson, A.	1863–1935	UK
1935 (awarded 1936)	Ossietzky, C. von	1889–1938	Germany
1936	Saavedra Lamas, C.	1878–1959	Argentina
1937	Cecil of Chelwood, Viscount	1864–1958	UK
1938	Nansen International Office for Refugees	Founded 1921	Switzerland
1944 (awarded 1945)	International Committee of the Red Cross	Founded 1863	Switzerland
1945	Hull, C.	1871–1955	USA
1946	Balch, E.G.	1867–1961	USA
	Mott, J.R.	1865–1955	USA
1947	The Friends Service Council	Founded 1647	UK
	The American Friends Service Committee	Founded 1672	USA
1949	Boyd Orr of Brechin, Lord J.	1880–1971	UK
1950	Bunche, R.	1904–1971	USA
1951	Jouhaux, L.	1879–1954	France
1952 (awarded 1953)	Schweitzer, A.	1875–1965	France
1953	Marshall, G.C.	1880–1959	USA
1954 (awarded 1955)	Office of the United Nations High Commissioner for Refugees	Founded 1951	Switzerland
1957	Pearson, L.B.	1897–1972	Canada
1958	Pire, G.	1910–1969	Belgium
1959	Noel-Baker, P.J.	1889–1982	UK
1960 (awarded 1961)	Lutuli, A.J.	1898–1967	South Africa
1961	Hammarskjöld, D.H.	1905–1961	Sweden
1962 (awarded 1963)	Pauling, L.C.	1901–	USA
1963	International Committee of the Red Cross	Founded 1863	Switzerland
	League of Red Cross Societies		Switzerland
1964	King Jr., M.L.	1929–1968	USA
1965	United Nations Children's Fund (UNICEF)	Founded 1946	USA
1968	Cassin, R.	1887–1976	France
1969	International Labour Organization (ILO)	Founded 1919	Switzerland
1970	Borlaug, N.	1914–	USA
1971	Brandt, W.	1913–	Fed. Rep. of Germany
1973	Kissinger, H.A.	1923–	USA
	Le Duc Tho (declined the prize)	1910–	Vietnam
1974	Mac Bride, S.	1904–	Ireland
	Sato, E.	1901–1975	Japan
1975	Sakharov, A.	1921–	USSR
1976 (awarded 1977)	Williams, B.	1943–	UK
	Corrigan, M.	1944–	UK
1977	Amnesty International	Founded 1961	UK
1978	El Sadat, M.A.	1918–1981	Rep. of Egypt
	Begin, M.	1913–	Israel
1979	Mother Teresa	1910–	India
1980	Perez Esquivel, A.	1931–	Argentina
1981	Office of the United National Commissioner for Refugees	Founded 1951	Switzerland
1982	Myrdal, A.	1902–1986	Sweden
	Garcia Robles, A.	1911–	Mexico
1983	Walesa, L.	1943–	Poland
1984	Tutu, D.M.	1931–	South Africa
1985	International Physicians for the prevention of Nuclear War		USA
1986	Wiesel, E.	1928–	USA
1987	Arias Sanchez, O.	1941–	Costa Rica
1988	The United Nations Peace-Keeping Forces	Founded 1945	USA
1989	Dalai Lama (Tenzin Gyatso)	1935	Tibet

Economic Sciences

Year	Name	Lifespan	Country
1969	Frish, R.	1985–1973	Norway
	Tinbergen, J.	1903–	Netherlands
1970	Samuelson, P.	1915–	USA
1971	Kuznets, S.	1901–1985	USA
1972	Hicks, J.R.	1904–	UK
	Arrow, K.J.	1921–	USA
1973	Leontief, W.	1906–	USA
1974	Myrdal, G.	1898–1987	Sweden
	Hayek, F.A. von	1899–	UK
1975	Kantorovich, L.	1912–1986	USSR
	Koopmans, T.C.	1910–	USA
1976	Friedman, M.	1912–	USA
1977	Ohlin, B.	1899–1979	Sweden
	Meade, J.	1907–	UK
1978	Simon, H.A.	1916–	USA
1979	Schultz, T.W.	1902–	USA
	Lewis, A.	1915–	UK
1980	Klein, L.R.	1920–	USA
1981	Tobin, J.	1918–	USA
1982	Stigler, G.J.	1911–	USA
1983	Debreu, G.	1921–	USA
1984	Stone, R.	1913–	UK
1985	Modigliani, F.	1918–	USA
1986	Buchanan, J.M.	1919–	USA
1987	Solow, R.M.	1924–	USA
1988	Allais, M.	1911–	France
1989	Haavelmo, T.	1911	Norway

Rank	Company	Country	Sales $M	Headquarters	No. of employees	Product/Working field
1	C. Itoh	Japan	105 668	Osaka	9 700	Industrial conglomerate
2	Mitsui	Japan	101 415	Tokyo	103 000	Industrial conglomerate
3	Shell	UK/Netherlands	97 492	London	205 000	Industrial conglomerate
4	General Motors	USA	97 034	Detroit	657 000	Vehicle manufacturer
5	Marubeni	Japan	94 815	Tokyo	9 000	Industrial conglomerate
6	Sumitomo	Japan	93 485	Osaka	129 000	Industrial conglomerate
7	Mitsubishi	Japan	90 620	Tokyo	152 000	Industrial conglomerate
8	Exxon	USA	87 437	New York	156 000	Oil and gas producers
9	Ford Motor	USA	68 302	Dearborn	385 000	Vehicle manufacturer
10	British Petroleum	UK	56 939	London	130 000	Oil industry
11	Nissho Iwai	Japan	55 013	Osaka	5 900	Industrial conglomerate
12	International Business Machines	USA	51 689	Armonk	394 000	Business machine manufacturers
13	Mobil	USA	48 833	New York	178 000	Energy operations
14	Toyota Motor	Japan	47 782	Aichi	61 600	Automobile manufacturer
15	Sears, Roebuck	USA	46 180	Chicago	450 000	General merchandise retailers
16	Nippon Telegraph & Telephone	Japan	38 321	Tokyo	309 000	Telecommunications
17	General Electric	USA	37 482	Fairfield	302 000	Consumer goods and power systems
18	Philip Morris	USA	35 818	New York	225 000	Tobacco, beer and food products
19	Hitachi	Japan	35 614	Tokyo	135 000	Electric and electronic goods
20	Daimler Benz	West Germany	35 348	Stuttgart	199 000	Vehicle and engine manufacturer
21	Texaco	USA	32 768	New York	68 000	Integrated oil company
22	American Telegraph and Telephone	USA	32 031	New York	365 000	Telephone systems operators
23	Nissan Motors	Japan	30 589	Tokyo	59 700	Automobile manufacturers
24	Toyo Menka Kaisha	Japan	29 777	Osaka	3 300	Industrial conglomerate
25	Matsushita Electric Industrial	Japan	29 340	Osaka	76 000	Electric and electronic goods
26	E.I. Du Pont de Nemours	USA	29 047	Wilmington	140 000	Diversified energy company
27	Volkswagon	West Germany	28 621	Wolfsburg	115 000	Automobile manufacturers
28	Tokyo Electric Power	Japan	27 956	Tokyo	40 000	Electric utility
29	Deutsche Bundepost	West Germany	27 268	Bonn	502 000	Postal and telecommunications
30	Unilever PLC/Unilever NV	UK/Netherlands	26 977	London	254 000	Food, detergent goods, etc.
31	Siemens	West Germany	26 944	Munich	359 000	General engineering and electronics
32	Chevron	USA	26 796	San Francisco	79 000	Integrated oil company
33	Chrysler	USA	25 051	Detroit	109 000	Vehicle manufacturer
34	Nichimen	Japan	25 024	Osaka	3 700	Industrial conglomerate
35	Philips Lamps Holding	Netherlands	24 476	Eindhoven	336 000	Electric and electronic goods
36	K. Mart	USA	24 432	Troy, Michigan	320 000	Discount store operators
37	Kanematsu-Gosho	Japan	24 209	Tokyo	3 100	Industrial conglomerate
38	Toshiba Corporation	Japan	23 674	Tokyo	87 000	Electric and electronic goods
39	Nestle	Switzerland	22 263	Vevey	150 000	Chocolate, milk, food, etc.
40	Veba	West Germany	21 203	Dusseldorf	76 700	Oil, electricity, chemicals, etc.
41	Electricite de France	France	21 081	Paris	123 000	Electricity
42	BASF	West Germany	21 079	Hamburg	6 000	Chemicals and plastics
43	Fiat	Italy	20 704	Turin	140 000	Vehicle manufacturer
44	Honda Motor	Japan	20 531	Tokyo	27 800	Vehicle manufacturer
45	Elf Aquitaine	France	19 780	Paris	36 000	Oil, gas, sulphur
46	Bayer	West Germany	19 459	Leverkusen	94 800	Chemical products
47	Hoechst	West Germany	19 360	Frankfurt	177 000	Chemicals, dyes, plastics
48	Amoco	USA	19 232	Chicago	53 500	Integrated oil company
49	Renault	France	18 998	Boulogne	129 000	Vehicle manufacturer
50	Peugeot	France	18 352	Paris	99 700	Vehicle manufacturer

The world's largest corporations

Rank	Company	Country	Sales $M	Headquarters	No. of employees	Product/Working field
51	NEC	Japan	17 536	Tokyo	39 600	Electric and electronic goods
52	Kroger	USA	16 836	Cincinnati	163 000	Supermarkets and drugstores
53	United Technology	USA	16 370	Hartford	203 000	Aircraft equipment
54	Occidental	USA	16 298	Los Angeles	62 500	Energy resources
55	Proctor & Gamble	USA	16 207	Cincinnati	62 000	Personal care and detergent goods
56	Atlantic Richfield	USA	16 044	Los Angeles	39 400	Oil, gas
57	Nippon Steel	Japan	15 594	Tokyo	70 500	Steel products
58	Nippon Oil	Japan	15 355	Tokyo	7 700	Oil goods
59	Petroleos de Venezuela	Venezuela	15 237	Caracas	42 100	Oil industry
60	Wal-Mart	USA	15 214	Bentonville	188 000	Department stores
61	RJR Nabisco	USA	15 030	New Jersey	70 700	Tobacco, food, beverages
62	Kansai Electric	Japan	14 647	Osaka	24 800	Electric utility
63	Boeing	USA	14 639	Seattle	90 000	Aircraft
64	J C Penney & Co	USA	14 616	Dallas	177 000	Department stores
65	USX Corp	USA	14 144	Pittsburgh	53 500	Steel, oil, gas
66	Spar	Netherlands	14 116	Amersfoot	8 000	Wholesaler, retailer
67	Tenneco	USA	14 100	Houston	97 000	Oil, gas
68	Volvo	Sweden	14 082	Gothenburg	76 200	Vehicle manufacturer, oil, gas
69	Thyssen AG	West Germany	13 909	Duisburg	131 000	Iron and steel
70	American Stores	USA	13 607	Salt Lake City	122 000	Supermarkets and drugstores
71	Total	France	13 526	Paris	3 250	Oil and petrol
72	Bosch	West Germany	13 288	Stuttgart	61 400	Electric engineering and car equipment
73	Fuyitsu	Japan	12 809	Tokyo	46 000	Computers and communications
74	Dow Chemicals	USA	12 753	Midland	49 800	Chemicals and plastics
75	Eastman Kodak	USA	12 685	Rochester	123 000	Photographic and chemical products
76	McDonnell Douglas	USA	12 533	St. Louis	89 000	Aircraft
77	OIAG	Austria	12 461	Vienna	102 000	Oil, gas, metals, steel
78	Mazda	Japan	12 104	Hiroshima	27 600	Vehicle manufacturer
79	Saint Gobain	France	12 072	Paris	10 000	Construction and packaging material
80	Chubu Electric Power	Japan	11 866	Nagoya	20 000	Electric utility
81	Bell South	USA	11 697	Atlanta	97 600	Telephone system operators
82	Rockwell	USA	11 558	Pittsburgh	102 000	Aerospace and electronic equipment
83	NyNex	USA	11 521	New York	98 200	Telephone system operators
84	INI	Spain	11 488	Madrid	500	State holding
85	Usinor Sacilor	France	11 226	Paris	10 000	Iron and steel
86	Pepsi Co	USA	10 950	New York	133 000	Beverages, food, restaurant
87	Lockheed	USA	10 792	Burbank	71 800	Aerospace and defence
88	Daiei	Japan	10 721	Kobe	14 400	Supermarkets
89	Ruhrkohle	West Germany	10 618	Essen	136 000	Coal
90	Petrofina	Belgium	10 616	Brussels	22 500	Oil
91	Federated Department	USA	10 600	Cincinnati	127 000	Department Stores
92	Allied Signal	USA	10 598	Morristown	114 000	Aerospace and automobiles
93	Electrolux	Sweden	10 262	Stockholm	2 500	Electrical appliances
94	Phillips Petrol	USA	10 222	Bartlesville	28 400	Integrated oil
95	BMW	West Germany	10 194	Munich	43 300	Vehicle manufacturers
96	Westinghouse	USA	10 179	Pittsburgh	133 000	Electric and electronic goods
97	Dayton	USA	10 179	Minneapolis	108 000	Retail and department stores
98	Ciba Geigy	Switzerland	9 959	Basel	22 200	Chemicals, pharmaceuticals, plastics
99	Canadian Pacific	Canada	9 886	Montreal	130 000	Oil, steel, timber
100	May Department Stores	USA	9 834	St. Louis	76 100	Department stores

Source: Times 1000 World's Top Companies 1987
Dun and Bradstreet Principal International Businesses 1987

United Nations

The United Nations (UN) and its charter formally came into existence on October 24th, 1945. It was designed to maintain international peace and security and to develop international co-operation in each of the following spheres – economic, social, cultural and humanitarian. Its main body, the General Assembly, has representation from all 150 member states, with up to five delegates but one vote, and is responsible for any question within the scope of the charter.

The other organs are the Security Council, the Economic and Social Council, the Trusteeship Council, and the International Court of Justice; in addition there are fifteen independent international organizations co-ordinated by and with the UN.

The General Assembly has seven main committees on which member states are represented. They are (1) Political and Security, (2) Economic and Financial, (3) Social, Humanitarian and Cultural, (4) Decolonization, (5) Administration and Budgetary, (6) Legal, (7) Special Political, relieving the burden on the first. In addition, the General Assembly appoints ad hoc committees for special purposes and is further assisted in its work by such subsidiary bodies as the Board of Auditors, the International Law Commission and so on. Its main administrative body is the Secretariat, headed by the Secretary General, who is appointed for a 5-year term. The Security Council is a body of fifteen members, five permanent – China, France, U.S.S.R., U.S.A. and U.K. – and ten elected for a period of two years. The Council bears the primary responsibility for the maintenance of international peace and security and for securing prompt action to this end; such as the deployment of an international peace-keeping force. The Council may have situations or circumstances considered to threaten peace referred to it by the General Assembly, any member of the UN, the Secretary General and, in some circumstances, by non-members. The Economic and Social Council consists of 54 members elected for three years and is responsible for the creation of conditions of stability for peaceful and friendly relations among nations and thus for carrying out the functions of the General Assembly's second and third committees. It is responsible, too, for the co-ordination of the policies and activities of the United Nations' special agencies. The Trusteeship Council has a basic membership of representatives from the five countries who are permanent members of the Security Council. The Council has administered, under various forms of Trusteeship, twelve territories of which ten have assumed some form of independence and continued to exercise some supervisory powers, receiving reports from and making periodic inspections of the trust territories in order to further their total political, economic, social and educational advancement. The territories which have been under UN trusteeship are:

Original Trust Territory	Subsequent Status
Tanganyika (UK)	Independent, 9 December 1961; merged with Zanzibar, 26 April 1964, as Tanzania.
Rwanda-Burundi (Belgium)	Two independent states, 1 July 1962.
Somaliland (Italy)	Independent, 1 July 1960.
Cameroons (UK)	Northern part joined Nigeria, 1 June 1961; southern part joined Cameroon, 1 October 1961.
Cameroons (France)	Independent republic of Cameroon, 1 January 1960.
Togoland (UK)	United with Gold Coast to form Ghana, 6 March 1957.
Togoland (France)	Independent republic of Togo, 27 April 1960.
Western Samoa (NZ)	Independent, 1 January 1962.
Nauru (Australia, NZ and UK)	Independent, 31 January 1968.
New Guinea (Australia)	Merged with Papua to form Papua New Guinea, independent 16 September 1975.
Pacific Islands (U.S.A.), incorporating the Carolines, Marshalls and Marianas (excepting Guam)	The Northern Mariana Islands became a Commonwealth territory of the U.S.A. on 9 January 1978, but remain legally part of the Trust Territory until the termination of the trusteeship. The other islands are grouped in three territories which are to enter into 'free association' with the U.S.A.

International Court of Justice is a principal judicial organ of the UN, comprises of fifteen judges, no two of whom may be nationals of the same state, who meet at The Hague. The statute of the Court is an integral part of the UN charter, and accordingly all members of the UN are pro-facto parties to it and the judgments of the Court are both final and binding on them. Specialized agencies. There are fifteen of these, each independent with its own membership, budget and headquarters. These agencies set standards within special spheres and provide assistance in various forms.

International Labour Organization (ILO). Founded in 1919 in connection with the League of Nations and re-established in 1946. It is specially concerned with social justice, hours of work, unemployment, wages, industrial sickness, foreign workers, and so on.

International Monetary Fund (IMF) (1945) exists to promote international monetary co-operation, international trade and stability of exchange rates, including the elimination of foreign exchange restrictions which hamper world trade. All states have a membership with the Fund, each is assigned a quota which determines its voting rights and each may be given financial assistance.

Food and Agricultural Organization (FAO) (1946) to raise levels of nutrition, improve production and distribution of agricultural products.

World Health Organization (WHO) aims to secure the highest possible standards of health for all people and so provides both advisory and technical assistance to and co-operation with member states in all relevant fields – nutrition, control of disease, family planning, and child care are some examples. It promotes some 850 research centres throughout the world and many other services of which drug evaluation is an example.

United Nations Educational, Scientific, and Cultural Organization (UNESCO) is designed to contribute to peace and security by promoting education, culture and knowledge through all means of mass communication and thus to further respect for justice, for the rule of law and for human rights without the distraction of race, sex, language or religion. Its three organs are: the General Conference, with responsibility for deciding programme and budget; the Executive to supervise the carrying out of the programme; and the Secretariat responsible for the day-to-day functioning.

UN Member States	Year of Joining	UN Member States	Year of Joining	UN Member States	Year of Joining
Argentina	1945	France	1945	Philippines	1945
Australia	1945	Greece	1945	Poland	1945
Belgium	1945	Guatemala	1945	Saudi Arabia	1945
Bolivia	1945	Haiti	1945	South Africa	1945
Brazil	1945	Honduras	1945	Syria	1945
Byelorussia	1945	India	1945	Turkey	1945
Canada	1945	Iran	1945	U.K.	1945
Chile	1945	Iraq	1945	Ukrainian Soviet Socialist Republic	1945
China	1945	Lebanon	1945	Uruguay	1945
Colombia	1945	Liberia	1945	U.S.A.	1945
Costa Rica	1945	Luxembourg	1945	U.S.S.R.	1945
Cuba	1945	Mexico	1945	Venezuela	1945
Czechoslovakia	1945	Netherlands	1945	Yugoslavia	1945
Denmark	1945	New Zealand	1945	Afghanistan	1946
Dominican Republic	1945	Nicaragua	1945	Iceland	1946
Ecuador	1945	Norway	1945	Sweden	1946
Egypt	1945	Panama	1945	Thailand	1946
El Salvador	1945	Paraguay	1945	Pakistan	1947
Ethiopia	1945	Peru	1945	Southern Yemen	1947

UN Member States	Year of Joining	UN Member States	Year of Joining	UN Member States	Year of Joining
Burma	1948	Ivory Coast	1960	Swaziland	1968
Israel	1949	Madagascar	1960	Fiji	1970
Indonesia	1950	Mali	1960	Bahrain	1971
Albania	1955	Niger	1960	Bhutan	1971
Austria	1955	Nigeria	1960	Oman	1971
Bulgaria	1955	Senegal	1960	Qatar	1971
Cambodia	1955	Somalia	1960	United Arab Emirates	1971
Finland	1955	Togo	1960	Bahamas	1973
Hungary	1955	Zaire	1960	German Democratic Republic	1973
Ireland	1955	Mauritania	1961	Germany, Federal Republic of	1973
Italy	1955	Mongolia	1961	Bangladesh	1974
Jordan	1955	Sierra Leone	1961	Grenada	1974
Laos People's Democratic Republic	1955	Tanzania	1961	Guinea-Bissau	1974
Libyan Arab Jamahiriya	1955	Algeria	1962	Cape Verde	1975
Nepal	1955	Burundi	1962	Comoros	1975
Portugal	1955	Jamaica	1962	Mozambique	1975
Romania	1955	Rwanda	1962	Papua New Guinea	1975
Spain	1955	Trinidad and Tobago	1962	Sao Tomë and Principe	1975
Sri Lanka	1955	Uganda	1962	Surinam	1975
Japan	1956	Kenya	1963	Angola	1976
Morocco	1956	Kuwait	1963	Seychelles	1976
Sudan	1956	Malawi	1964	Western Samoa	1976
Tunisia	1956	Malta	1964	Dijibouti	1977
Ghana	1957	Zambia	1964	Vietnam	1977
Malaysia	1957	Gambia	1965	Dominica	1978
Guinea	1958	Maldives	1965	Solomon Islands	1978
Benin	1960	Singapore	1965	St. Lucia	1979
Burkina Faso	1960	Barbados	1966	St. Vincent and the Grenadines	1980
Cameroon	1960	Botswana	1966	Zimbabwe	1980
Central African Republic	1960	Guyana	1966	Antigua and Barbuda	1981
Chad	1960	Lesotho	1966	Belize	1981
Congo	1960	Yemen	1967	Vanuatu	1981
Cyprus	1960	Equatorial Guinea	1968	St. Kitts-Nevis	1983
Gabon	1960	Mauritius	1968	Brunei	1984

The European Community

The European Community had its origins in 1950 when it was proposed that France and Germany should pool their coal and steel industries under one authority and were joined in this by Belgium, Italy, Luxembourg and the Netherlands, forming the European Coal and Steel Community (ECSC). These six countries were the original members of the European Economic Community (EEC) which was, with Euratom (the European Atomic Energy Community), established in 1957 by two respective treaties signed in Rome.

In 1970, the Six invited Britain, the Irish Republic, Denmark and Norway to apply for membership of the Community and in 1972, with the exception of Norway, they were accepted by the Treaty of Accession. Subsequently Greece, Spain and Portugal have similarly signed Treaties of Accession.

In 1965 the original Six, already with a common Parliament and Court of Justice for their three communities, agreed to merge their other bodies into a single Commission and a single Council.

The Commission has seventeen members, who are appointed by agreement from the twelve Member States for a four-year period and who are pledged to independence of any national or other particular interest. Its principal work is to initiate Community policy and action. In addition it acts as a mediator between member governments in Community affairs, guardian of Community Treaties and maintains information offices in all parts of the world.

The Council of Ministers consists of ministerial representatives from each Member State according to the subject under discussion. It is the main decision-making body working on proposals submitted by the Commission. It also acts with executive powers by issuing Regulations and Directives to all Member States.

The European Parliament, before 1979, had its members nominated by their national parliaments, but since then there have been direct elections to a Parliament; the number of seats held varies from United Kingdom's eighty-one to Luxembourg's six.

The Parliament must be consulted on all major issues and has powers of supervision over both the Commission and Council of Ministers, and a measure of control over the Community's annual budget. Its members serve on specialized committees and sit in political groups – Socialists, Christian Democrats, and so on.

The European Court of Justice, superseded the Court of Justice of the ECSC, and now serves all three communities. It performs a wide range of functions, from safeguarding the law, in the interpretation and application of the various Community Treaties, making judgements on the legality of decisions by the Commission or Council, to hearing cases brought by Member States, Community institutions, firms and individuals. Its decisions are binding on all Member States.

Council of Europe

The Council of Europe, founded in 1949, aims to achieve greater unity between its 22 members, to safeguard their European heritage, facilitate economic, social, cultural and educational progress, and to further human rights and freedom.

The organs of the Council are:

(1) The Committee of Ministers, the chief organ, is made up of the foreign ministers of member countries meeting twice yearly. Their decisions take the form of agreements, known as Conventions.

(2) The Parliamentary Assembly is the other body. It holds 3 week-long sessions each year, debating reports on agricultural, social, political, legal, and regional planning affairs, together with reports from other organizations.

The Council's principal achievement has been the European Convention of Human Rights under which was established the European Commission and the European Court of Human Rights. Accordingly, all inhabitants of Member States are protected against their own governments in such areas as the right to live, freedom from torture, the right to liberty, respect for private and family life, freedom of thought, religion and expression, and freedom of association.

There are, in addition, some 124 other Conventions and these include the social security code, guidelines on extradition, and the legal status of migrant workers.

Member Nation	Date of Adherence	Member Nation	Date of Adherence
Belgium	1950	Sweden	1950
Denmark	1950	Turkey	1950
France	1950	United Kingdom	1950
Germany (Fed. Rep.)	1950	Austria	1957
Greece	1950	Cyprus	1961
Iceland	1950	Malta	1966
Ireland	1950	Switzerland	1972
Italy	1950	Liechtenstein	1978
Luxembourg	1950	Portugal	1978
Netherlands	1950	Spain	1978
Norway	1950		

North Atlantic Treaty Organization

The North Atlantic Treaty Organization (NATO), the brainchild of Canada's Secretary of State for External Affairs in 1948, came into being with the signing of the North Atlantic Treaty in 1949 by twelve of the sixteen members – Belgium, Canada, Denmark, France, Iceland, Norway, Italy, Luxembourg, the Netherlands, Portugal, U.K. and the U.S.A. At a later date Greece, Turkey, Spain and West Germany were admitted. The Treaty sought to provide a collective defence organization, linking a group of European nations with the U.S.A. and Canada, to maintain military preparedness with the aim of preventing war. The North Atlantic Council is the principal authority of the organization and is composed of permanent representatives heading delegations of advisers and technical experts in all aspects of defence and related subjects. The Council has meetings twice a year. The Defence Planning Committee of the Organization in the field of defence has the same authority as the Council and is similarly organized with its permanent representatives meeting regularly and holding meetings of defence ministers twice a year.

Commonwealth

The Commonwealth is a voluntary association of independent states comprising of a quarter of the world's population and covering a third of the world's total area. It is made up of 49 members with their associated states and dependencies "autonomous . . . equal in status . . . united by a common allegiance to the Crown". Queen Elizabeth II is the Head of State for 18 of the members and is personally represented as such by a Governor General, appointed by the Queen on the advice of the country concerned. Twenty-four member states are republics, while three others have monarchs – Brunei, Lesotho and Tonga – and two, Malaysia and Western Samoa, equivalent Heads of State. The Commonwealth operates principally through consultation, and through its Secretariat as its agency for communication. The first level of consultation is the meeting of Commonwealth heads of government, held every two years in the different Commonwealth capitals to discuss international developments and co-operation among its members. In the interim years, meetings of ministers and senior government officials take place to ensure continuity of contact and to deal with issues of trade, education, health, law, science, agriculture and so on. The Commonwealth is served, too, by its Secretariat as its agency for communication between Commonwealth governments on common issues, promoting consultations, distributing information, arranging meetings and providing technical assistance for economic and social development. The Secretariat operates through its different divisions such as international affairs, technical co-operation, and particularly in the development of human resources through education and training and specific programmes. The Commonwealth Institute is in London and acts as an information centre with permanent exhibitions on all Commonwealth states, an arts centre, a theatre, library, and an education department with a teachers' resource centre.

Warsaw Pact

The Warsaw Pact refers to a treaty of friendship and non-aggression of 1955 entitled 'Treaty of Friendship, Co-operation and Mutual Assistance'. It was signed by the U.S.S.R. and its satellite countries – Bulgaria, Czechoslovakia, the German Democratic Republic (East Germany), Hungary, Poland and Romania.

The aim was to set up a joint military command structure and to ensure that all countries of the pact would come to the defence of any one member who suffered aggression.

Organization of the Petroleum Exporting Countries

Organization of the Petroleum Exporting Countries (OPEC) was created in 1961 as a permanent organization to co-ordinate policies to protect the interests, individually and collectively, of their members, especially in respect of export prices and quotas and in their dealings with the major oil companies. Its 13 members are Algeria, Ecuador, Gabon, Indonesia, Iran, Iraq, Kuwait, Libya, Nigeria, Qatar, Saudi Arabia, United Arab Emirates, and Venezuela. The supreme authority is the conference of Ministers of Oil, Mines and Energy of member states meeting about twice a year to formulate policy. Management of the organization, and implementation of conference policy resolutions, is in the hands of a Board of Governors with a Secretariat under its control, based in Vienna, to carry out the executive functions.

World Council of Churches

The World Council of Churches was constituted in 1948 to promote unity between the many Christian churches. It embraces 307 member churches with adherents in more than 100 countries and serves virtually all Christian traditions except the Roman Catholic Church which sends observers to the Assembly.

The Council has its focus in the Assembly, to which the member churches send their delegates and which meets every ten years to determine broad policies. More detailed decisions are made by the Central Committee of 150 members, elected by the Assembly and which meets annually and in turn a smaller Executive Committee to serve specific working groups and programmes. The implementation of the policies laid down by the Assembly and the co-ordination of the activities of these working groups and programmes are the responsibility of the General Secretariat, based in Geneva.

International Red Cross

The International Red Cross has now three elements – the International Committee of the Red Cross, the League of Red Cross and Red Crescent Societies, and the National Red Cross Societies. Together they operate under the umbrella of 'The International Red Cross Conference' which meets every four years and is the supreme decision-making body.

The International Committee of the Red Cross was founded in 1864 with the signing of an International Treaty – The Geneva Convention – by delegates of sixteen states. Its purpose was to secure international agreement for the protection of the wounded and those captured in war, and for doctors, nurses, hospitals and ambulances providing care for them and for whom the special emblem, the Red Cross, was chosen. In 1949, delegates from 61 countries added three further conventions to cover wounded, sick and shipwrecked members of the armed forces in an entirely new area – civilian victims, civilian detainees, and inhabitants of occupied territories. The Committee has 25 members, all of whom must be Swiss. The headquarters are in Geneva where a staff of over 400 are employed.

The League or International Federation of Red Cross and Red Crescent Societies was founded in 1919 to maintain the same co-operation between National Societies in peace as in the war period (1914-1918) and to provide a permanent liaison body to encourage the development of National Societies for the alleviation of suffering as a contribution to peace. The League, based in Geneva, has a secretariat of 170 from thirty nationalities. National Red Cross or Red Crescent Societies now exist in almost every country, subscribing to the principles and activities of the International Committee and in affiliation with the League which helps National Societies in their peace work for "the improvement of health, the prevention of disease and the mitigation of suffering". National Societies' activities vary according to the nature and needs of their country communities, but in general include emergency relief at times of disaster, health services and social assistance to individuals, first aid courses and blood services. In war time, Societies serve as auxiliaries to the medical services of the armed forces and aid prisoners, refugees, and civilian internees.

International Olympic Committee

The International Olympic Committee's broader purpose is to provide for the development of those physical and moral qualities which are the basis of sport, but its specific function is to organize the 4-yearly Olympic Games, the Olympic Winter Games and the Games for Paraplegics.

Interpol

Interpol (International Criminal Police Organization) aims to promote the widest possible mutual assistance between police forces for the pursuit and arrest of criminals and to develop institutions for the prevention of crime.

Regional Organizations

These organizations are associations of states in specific regions to promote and to co-ordinate policies for their mutual benefit – economic growth, political stability, social and cultural development, and so on.

The following are illustrations of these regional bodies:

The League of Arab States;
The Organization of African Unity;
The Organization of American States (West Indies, Central and South American Countries and U.S.A.);
The Association of South East Asian Nations (ASEAN);
The Caribbean Community and Common Market (CariCom);
Central American Common Market (CACM);
Colombo Plan (Commonwealth and other countries in Pacific and Indian Oceans).

Danger labels

liable to explosion · danger of fire (inflammable liquids) · danger of fire (inflammable solids) · substance liable to spontaneous ignition · danger of emission of inflammable gases on contact with water · oxidizing substance or organic peroxide · toxic substance

harmful substance · corrosive substance · radioactive substance · radioactive substance · radioactive substance · radioactive substance

Washing labels

- White cotton and linen without special finishes
- Colourfast cotton, linen or viscose without special finishes
- White nylon, white polyester, white cotton
- Coloured nylon. polyester, acrylic/ cotton mixtures, cotton or viscose with special finishes
- Non-colourfast cotton, linen and viscose
- Acrylics, acetates blends with wool, polyester/wool blends
- Woollen blankets, wool mixtures. Shrink resistant machine-washable

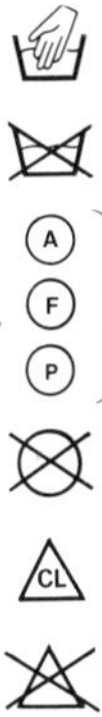

- hand wash only
- do not wash
- dry clean
- do not dry clean
- chlorine household bleach may be used
- do not bleach

- tumble drying is beneficial
- do not tumble dry
- drip dry
- hang out to dry
- dry flat

- cool iron
- warm iron
- hot iron
- do not iron

Plimsoll line

This is a series of lines painted on the outside of a cargo ship's hull showing the various safe levels to which a ship can be loaded. The Plimsoll line or load line is a circle through which is a horizontal line and the ship should not be loaded to the point that the line sinks below sea level. The loading levels are marked for loaded and unloaded draught in sea and fresh water, winter and summer and tropical or northern waters. It was Samuel Plimsoll (1824–1898), a member of parliament, who instigated the Merchant Shipping Act in 1876 which introduced the measure to the United Kingdom.

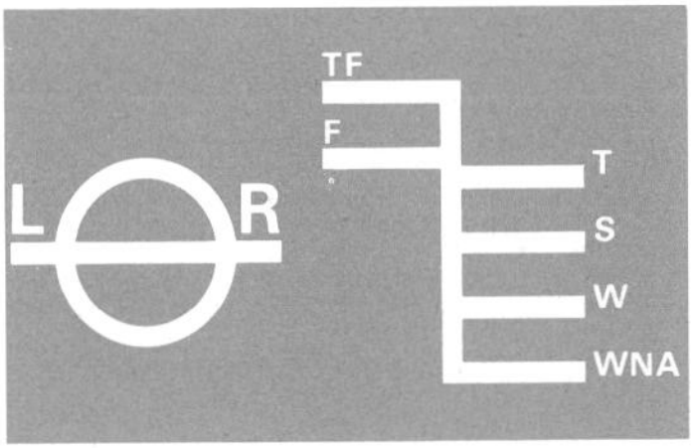

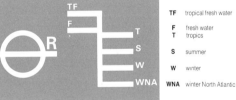

TF tropical fresh water
F fresh water
T tropics
S summer
W winter
WNA winter North Atlantic

Other symbols

no smoking · smoking · information · ladies · gentlemen · escalator · parking · camp site · caravan site · airport · irradiated food

Euro Tunnel

The Euro Tunnel is an Anglo-French project which will connect road and rail networks of Europe by creating a fixed link between Coquelles, near Calais in France, and Folkestone, in England, where work began on a 350-acre terminal site in 1988. Almost 5 000 people are employed in the construction of this project which is composed of three tunnels requiring 700 000 segments of pipe which will give 150 kilometres of tunnels when completed. One of the first suggestions of a channel tunnel was presented to Napoleon in 1802 by Albert Mathieu, a French engineer, but the idea was abandoned on the resumption of the Napoleonic wars. In 1958 the Channel Tunnel Study Group was established; many proposals followed and finally, in 1986, the Anglo-French Channel Tunnel Treaty was signed.

The Tunnel will open in 1993 and the journey will take approximately 30 minutes for those travelling by the shuttle trains at up to 100 miles per hour.

Eiffel Tower

A metal tower, 300 metres high, built in the Champ de Mars, Paris, for the centennial exposition in 1889. Composed mainly of iron lattice-work, it was the highest building in the world until 1930. The framework is composed essentially of four uprights which rise from the corners of a square measuring 100 metres on each side - the area it covers at the base is nearly 2.5 acres. There are three platforms situated at different levels and at the top is the lantern with a gallery 5 metres in diameter.

The Tower is the most famous work of French engineer Alexandre Gustave-Eiffel, who had wide experience in construction of large metal bridges before he undertook the building of the Tower. Although an outstanding engineering achievement, the Tower was originally disliked but has become one of the great Parisian landmarks.

Statue of Liberty

A statue of a woman, 46 metres high, and holding a torch in her hand, which stands on Liberty Island in New York harbour. Designed by the French architect and sculptor Frédéric Bartnoldi, famous for his patriotic monuments, it was given to the Americans in 1884 by the French to commemorate the French and American Revolutions. Every traveller entering or leaving New York harbour sees the statue which has become a US national monument.

Panama Canal

A canal connecting the Atlantic and Pacific Oceans, built across the isthmus of Panama which is a narrow strip of land linking North and South America. In 1880 the French Canal Company, under Ferdinand de Lesseps, attempted the construction of a sea-level waterway over difficult country with steep gradients and equatorial forest. The building was halted by bankruptcy in 1889.

In 1903 the USA acquired construction rights from newly-independent Panama and, using a different design involving a lock system, the Canal was opened in 1914. It is 82 kilometres long and transit time is normally between 8 and 10 hours.

The Canal is vital to strategic and commercial intercoastal shipping of the United States and to Europe as it opens up the west coast of North and South America for trading, especially in bulk commodities.

In 1980 a treaty between the USA and Panama transferred ownership of land within the canal zone to Panama and the Canal itself is to be ceded to Panama by the year 2 000.

Kremlin

The best known Kremlin, the Moscow Kremlin, (now a public museum of Russian architecture) was originally built of wood in 1156. The present structure, shaped like an isosceles triangle and built of bricks, was erected between 1485 and 1495. Many notable buildings are contained within the Kremlin Square, amongst them the Cathedral of St. Michael where the Tsars are buried. During Stalin's rule the Kremlin was closed to the public, opening again in 1955.

Sydney Opera House

The opera house in Sydney is one of the city's premier landmarks. It was opened by Queen Elizabeth II in 1973. As befits a building of such stature it had a colourful history.

In 1957 the state government estimated from the plans that the opera house would be built in 5 years and cost $7 million. In fact, it cost nearly fifteen times as much ($102 million) and took 16 years to build.

Such delay and expense caused much political embarrassment and the Danish-born architect Jørn Utzon left before the project was completed.

The design itself is meant to resemble white billowing sails, the roofs being lined with Swedish ceramic tiles, and rising to over 67 metres. It is estimated that there are over one million of these tiles in the building, weighing over 158 000 tons. In addition, there are over 350 kilometres of tensioned cables and 6 200 square metres of French tinted glass.

Famous constructions

Stockholm Globe Arena

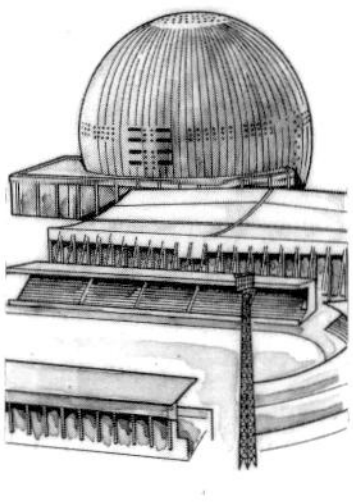

The Stockholm Globe Arena is a unique development situated in a new city district called Hovet. The centrepiece of this development is the Globe, which is the largest spherical building in the world, seating 16 000 spectators. It is a multi-arena built as an amphitheatre to house every conceivable sporting and cultural event.

The Globe is 110 metres in diameter and 85 metres high. Its height ensures good acoustics and the domed ceiling, which contains 150 skylight windows, can be converted into a gigantic screen for the presentation of picture and light shows. The seating is arranged on three levels, enabling all spectators to have a perfect view and also to be closer to the event than in any other world arena.

Sakkara

An immense mortuary area, about 1.6 kilometres wide and 8 kilometres in length, near Memphis, an ancient city of Lower Egypt. The most famous monument at Sakkara is the 'Step Pyramid', the first known pyramid and the world's first stone monument: architecturally the immediate fore-runner of the Giza Pyramids. It was designed in 2630 BC for King Zosar (Djosar) of the 3rd Dynasty by Imhotep, the first named architect in history.

The structure form of the 'Step Pyramid' was revolutionary at the time, and built in response to theological considerations and the result of a series of experiments. It forms a gigantic staircase with six broad steps ascending to the sky and is an amazing piece of ingenuity. Its complex plan involves numerous galleries 25 metres below ground, some of which were reserved for queens and royal children and where an amazing hoard of royal tableware has been found, consisting of 40 000 stone vases in addition to many other items.

Among the many other private tombs is the smallest of the pyramids - the Pyramid of Unas - where the first texts of funerary hieroglyphs, known as the 'Private Texts', were discovered.

Trans-Siberian railway

The world's longest railway, known as Siberia's lifeline, runs from Moscow to Vladivostok and covers 9 335 kilometres. Construction began in 1891 across difficult terrain. The line was hastily built and when first completed was not safe for trains going faster than 20 miles per hour. The original single track line, built between 1891 and 1905, has been replaced by a double-track and electrified line over its entire length. The completed journey, with nearly a hundred stops, takes nine days.

Taj Mahal

Mogul Emperor Shah Jahan erected a monument by the River Jumna as a tomb for his beloved wife, Mumtaz Mahal, crown of the palace, who died in childbirth in 1631. It is one of the greatest sights of the world, built mainly in white marble, beautifully carved in open traceries and designs largely in-laid with semi-precious stones. The proportions and balance of the main structure, together with four supporting minarets, show Mogul architecture at its peak. It took twenty-one years to complete and employed 20 000 labourers and craftsmen.

Shah Jahan had always intended to build a black mausoleum on the opposite river bank for his own tomb, but his son seized power and imprisoned him. He died in 1666 and was laid to rest with his wife in the Taj Mahal.

Design

Design is the capacity to change ideas and images into substance; in the manufacturing world it is universal.
The early history of design was directed at the visual image and whilst this is still so it has also advanced to embrace the quality, character and function. Design embraces both the producer and the consumer. A critical element is co-ordinating supply and demand, which involves professional designers, manufacturers and, perhaps most important, the consumer.

Pompidou Centre

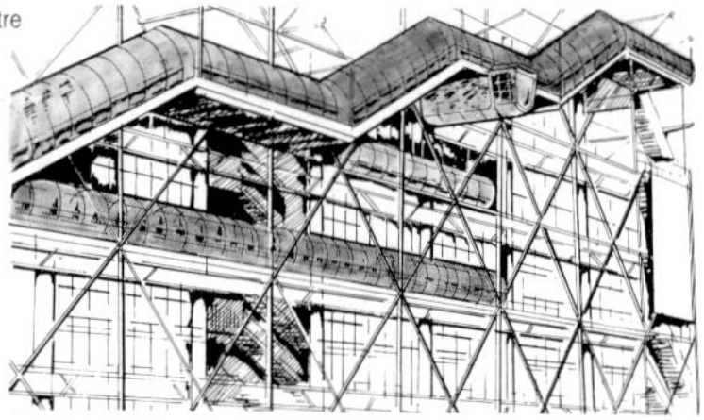

Architecture

Architects can be thought of as design subcontractors, buildings now being treated as products. Classicism is the cultural tradition in architectural design as it provides a fixed base with a flexibility of detail. Architects have moved away from this mould, producing striking buildings such as the Globe in Stockholm, Sydney Opera House and the Pompidou Centre. In the latter the architect used an integral part of the building, the brightly coloured air conditioning pipes, as a supportive structure on the exterior.
In this world of jet travel, telecommunication, system building and fast-track construction the traditional virtues are returning with a revival of classicism. An example of this is seen in Britain when Mercury Communications commissioned a new telephone kiosk but no designer succeeded in improving on the traditional British Telecom kiosks which have a place of affection in the community so that over 1 000 of them are listed buildings.

British Telecom kiosk

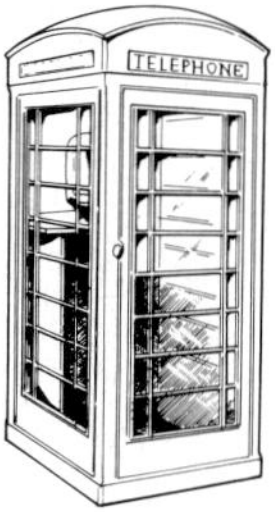

Logo

Recognition of the 'value' of ideas goes back to the 1830s when there were campaigns to copyright textile patterns after new rotary printing machines allowed mass production. Today, with the development of logo designs, the recognition of value has a new meaning. Identical clothing communicates different ideas about the product, its wearer and its value simply by the logo. Some food and drink brands have internationally known logo designs so that they are instantly recognizable – Heinz, KitKat, Coca-Cola, MacDonalds. Such logo designs are worth a great deal, they are essentially a symbol which consumers can identify with to separate a product such as Coca-Cola from its competitors. Large amounts of money are spent via advertising on adding value and imagery to products and the brand or company logo is thus a visible shorthand through which all these values and images are relayed. With many products, however, the logo is an important part of an overall packaging design.

Coca-Cola and Coke are registered trade marks which identify the same product of The Coca-Cola Company.

With Coca Cola for instance it can be argued that the bottle shape is an important part of the design. In other cases it is the pack design, whether the graphics or pack form, that are the only thing that distinguish one product from another.
Finally the advertising itself is also a major focus of design. It has the most direct ability to change and influence consumers ideas about a product. To a brand or Company it can often be a more tangible asset than the product which it advertises.

Furniture

More than any other piece of furniture it is the chair that conveys wealth, rank and taste in private as well as public places. It imposes a unique problem as no other object combines the demands of the human form with the discipline of technology, structure and manufacture. The desire to create a chair as an aesthetic object has often resulted in designs which are sculptural rather than practical. From the development of steam bending wood in mass-production to the use of tubular steel and the invention of plastics, innovation in production has stimulated novelty in form.

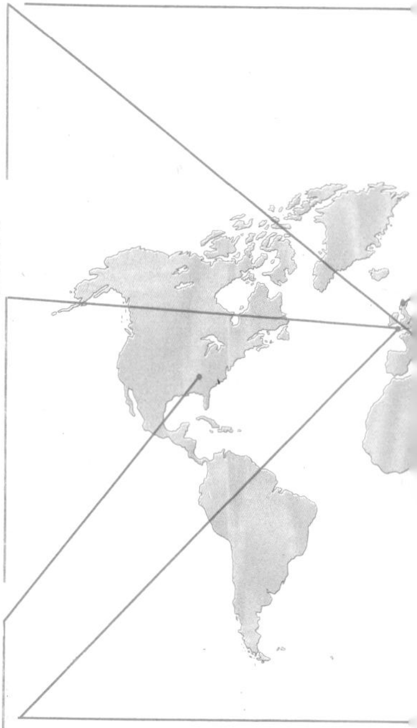

Wedgwood China

Famous design

Rennie Mackintosh –
Art Nouveau furniture

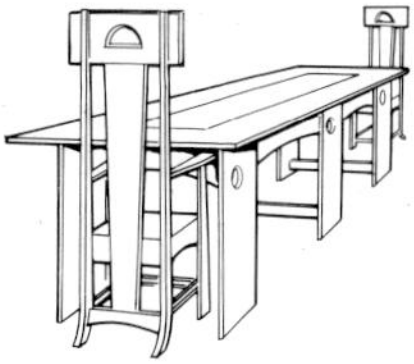

Scandinavian furniture was among the first type of merchandise to be identified by the public as 'Designed'. During World War II their designers were able to provide well-made, well-conceived furniture at a time when other countries were struggling to provide for their basic needs. The Ant chair was designed by Arne Jacobsen for the canteen of a pharmaceutical factory in Copenhagen. It was stackable, light and cheap and has been continually produced since 1952.

Ant Chair

Cars

This has now affected cars, which are being differentiated by design, but in the future they may depend solely on the prestige of the brand. The 'deux chevaux' has been in production longer than any other vehicle in history (except the safety bicycle), following its design by Pierre Boulanger in 1939. It was originally conceived as an artisan car to carry two people and 50 kg of luggage at 50 kilometres per hour. Its interlinked suspension provided a soft and even ride.

Porsche

Today, the 'deux chevaux' and the Porsche can be found side by side in the showroom. One is a practical design and the other is an aesthetic exercise. Car design, as like all other areas of design can therefore have functional or aesthetic benefits. Many modern mass market cars trade on their functional design which enables them to perform better in particular areas than their competitors. A car which has a good aesthetic design can, however, be just as important if it makes the car look good or the driver feel good.

Deux Chevaux (2CV)

Domestic Appliances

The first household appliances were crude and industrial-looking; later to stimulate sales in a stagnant market they were given a modern streamlined appearance. In the 1950s designers created distinct domestic white and brown goods. The present trend is towards greater decoration and colour co-ordination. The market for most popular appliances has reached saturation point, forcing manufacturers to promote new lines, shapes and technical innovations. The design of appliances reflects the erosion of the distinction between the kitchen and dining rooms. The design encourages utilitarian objects styled to be acceptable on the dinner table without disguising their functional purpose. Some designs, however, have a purely functional benefit such as the cordless kettle, which removes the need for a flex.

Alexis teapots

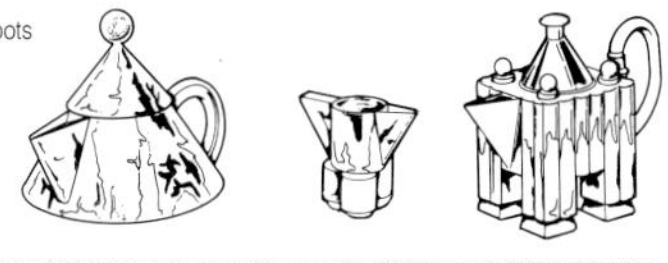

Ceramics

Josiah Wedgwood used design to advance his business. The traditional practice was to send batches of products either to the merchants or the markets. He employed designers to produce a wide range of standardized products, then in 1773 and 1788 Wedgwood catalogues appeared with texts in English, French, German and Dutch. By 1809 a Wedgwood showroom opened in London enabling customers to study catalogues and inspect samples and the producer to anticipate the orders.

Josiah Wedgwood used design to achieve a competitive advantage. Within the world of ceramics experiments in technology and innovations have continued into this century but a taste for hand decoration on quality items still persists.

The use of design was further exploited by the Parisian window designers. While promenading, window shopping was encouraged by the spacious boulevards and large windows of the department stores. The passers-by were attracted to the range and novelty of the goods displayed, encouraging them to 'come in and buy'.

This has covered only a few aspects of design, showing its development and present day advancement. Look around you and see the number of everyday items in which the function and form has been influenced by the designer.

How to use the conversion tables

The figures in the blue columns can be read as either the metric measure or the imperial equivalent:
10 inches is equal to 25.40 cm; and 10 cm is equal to 3.937 inches

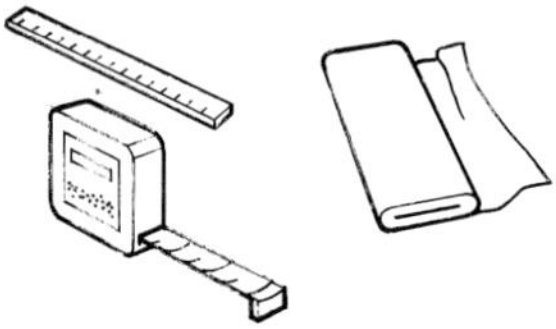

Length

To convert centimetres to inches (UK, USA): multiply by 0.394

To convert inches (UK, USA) to centimetres: multiply by 2.54

To convert metres to feet (UK, USA): multiply by 3.280

To convert feet (UK, USA) to metres: multiply by 0.305

Centimetres	Units	Inches
2.54	1	0.394
5.08	2	0.787
7.62	3	1.181
10.16	4	1.575
12.70	5	1.969
15.24	6	2.362
17.78	7	2.756
20.32	8	3.150
22.86	9	3.543
25.40	10	3.937
50.80	20	7.874
101.60	40	15.748
152.40	60	23.622
203.20	80	31.496
254.00	100	39.370

Metres	Units	Feet
0.305	1	3.280
0.610	2	6.562
0.914	3	9.842
1.220	4	13.123
1.524	5	16.404
1.829	6	19.685
2.134	7	22.966
2.439	8	26.246
2.743	9	29.527
3.048	10	32.808
6.096	20	65.616
12.192	40	131.232
18.288	60	196.848
24.384	80	262.464
30.480	100	328.080

Area

To convert hectares to acres, multiply by 0.405.

To convert acres to hectares, multiply by 2.471.

To convert km² to sq. miles multiply by 0.386

To convert sq. miles to km² multiply by 2.590

To convert cm² to sq. inches multiply by 0.155

To convert sq. inches to cm² multiply by 6.4516

Hectares	Units	Acres
0.405	1	2.471
0.809	2	4.942
1.214	3	7.413
1.619	4	9.884
2.023	5	12.355
2.428	6	14.826
2.833	7	17.297
3.237	8	19.769
3.642	9	22.240
4.047	10	24.711
8.094	20	49.421
16.187	40	98.842
24.281	60	148.263
32.375	80	197.684
40.469	100	247.105

km²	Units	sq. miles
2.590	1	0.386
5.180	2	0.772
7.769	3	1.158
10.359	4	1.545
12.949	5	1.931
15.539	6	2.317
18.129	7	2.703
20.718	8	3.089
23.308	9	3.475
25.898	10	3.861
51.796	20	7.723
103.592	40	15.445
155.388	60	23.168
207.184	80	30.890
258.980	100	38.613

cm²	Units	sq. inches
6.451	1	0.155
12.903	2	0.310
19.355	3	0.465
25.506	4	0.620
32.258	5	0.775
38.710	6	0.930
45.161	7	1.085
51.613	8	1.240
58.064	9	1.395
64.516	10	1.550
129.032	20	3.100
258.064	40	6.200
387.096	60	9.300
516.128	80	12.400
645.160	100	15.500

Areas of continents

	Sq. Km	Sq. Miles
Asia	43 608 000	16 833 000
Africa	30 335 000	11 709 000
North & Central America	25 349 000	9 785 000
South America	17 611 000	6 798 600
Antarctica	13 340 000	5 149 240
Europe	10 498 000	4 052 000
Australasia	8 923 000	3 444 278

Areas of oceans

	Sq. Km	Sq. Miles
Pacific Ocean	165 384 000	63 838 000
Atlantic Ocean	82 217 000	31 736 000
Indian Ocean	73 481 000	28 364 000
Arctic Ocean	14 056 000	5 426 000

Areas of countries

	Sq. Km	Sq. Miles
Argentina	2 777 815	1 072 598
Australia	7 682 300	2 965 370
Brazil	8 511 965	3 285 620
Canada	9 922 385	3 830 840
China	9 597 000	3 704 440
Cuba	121 046	46 739
Djibouti	23 000	8 880
Egypt	1 000 250	386 095
France	543 965	209 970
Ghana	238 305	92 016
Iceland	102 819	39 701
India	3 166 830	1 222 395
Indonesia	1 919 400	741 138
Italy	301 245	116 280
Jamaica	10 991	4 244
Japan	369 700	142 705
Kenya	582 646	224 977
Madagascar	587 041	226 674
Malaysia	329 749	127 326
Mexico	1 972 547	761 660
Nepal	145 391	56 139

	Sq. Km	Sq. Miles
New Zealand	268 704	103 754
Pakistan	887 747	342 785
Panama	78 046	30 135
Papua New Guinea	461 691	178 272
Peru	1 285 216	496 261
Philippines	300 000	115 839
Saudi Arabia	2 149 690	830 060
Singapore	618	238
South Africa	1 184 825	457 345
Spain	504 782	194 912
Sri Lanka	65 610	25 334
Sudan	2 505 813	967 570
Sweden	449 790	173 620
Thailand	514 000	198 470
United Kingdom	244 755	94 475
Uruguay	186 926	72 177
U.S.A.	9 363 130	3 614 170
U.S.S.R.	22 400 000	8 646 400
Venezuela	912 050	352 170
West Germany	248 665	95 985
Zaire	2 344 885	905 430

Distance

To convert Kilometres to miles (UK, USA), multiply by 0.621

To convert miles (UK, USA) to Kilometres, multiply by 1.609

Kilometres	Units	Miles
1.609	1	0.621
3.219	2	1.243
4.828	3	1.864
6.437	4	2.486
8.047	5	3.107
9.656	6	3.728
11.265	7	4.350
12.875	8	4.971
14.484	9	5.592
16.093	10	6.214
32.187	20	12.427
64.374	40	24.855
96.561	60	37.282
128.748	80	49.710
160.934	100	62.137

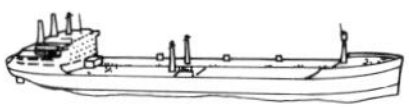

To convert kilometres to nautical miles multiply by 0.54

To convertnautical miles to kilometres multiply by 1.852

km	Units	Nautical Mile
1.852	1	0.54
3.704	2	1.08
5.556	3	1.62
7.408	4	2.16
9.260	5	2.70
11.112	6	3.24
12.964	7	3.78
14.816	8	4.32
16.668	9	4.86
18.52	10	5.40
37.04	20	10.80
74.08	40	21.60
111.12	60	32.40
148.16	80	43.20
185.20	100	54.00

Tyre pressures

lb. per sq.in.	18	20	22	24	26	28	30	32	34	36	38	40
Kg. per sq.cm.	1.26	1.40	1.54	1.68	1.83	1.96	2.10	2.24	2.38	2.52	2.66	2.80

Distances between cities

Kilometres / Miles	Cairo	Calcutta	Copenhagen	Hong Kong	Johannesburg	London	Los Angeles	Moscow	New York	Paris	Rio de Janeiro	Rome	Sydney	Tokyo
Cairo		3541	1992	5064	3894	2180	7580	1803	5605	1994	6149	1325	8959	5947
Calcutta	5699		4395	1653	5256	4946	8152	3438	7921	4883	9366	4486	5678	3195
Copenhagen	9860	7072		5388	5732	592	5594	970	3845	638	6345	951	9968	5403
Hong Kong	8150	2659	8671		6669	5980	7232	4439	8047	5984	11001	5769	4582	1786
Johannesburg	6267	8459	9225	10732		5637	10362	5692	7979	5426	4420	4811	6860	8418
London	3508	7961	952	9623	9071		5442	1552	3463	212	5778	889	10558	5942
Los Angeles	12200	13120	9003	11639	16676	8758		6070	2446	5645	6310	6331	7502	5475
Moscow	2902	5534	1561	7144	9161	2498	9769		4666	1545	7184	1477	9008	4651
New York	9020	12747	6188	12950	12841	5572	3936	7510		3626	4832	4280	9935	6741
Paris	3210	7858	1026	9630	8732	342	9085	2486	5836		5708	687	6539	6038
Rio de Janeiro	9896	15073	10211	17704	7113	9299	10155	11562	7777	9187		5725	8389	11551
Rome	2133	7219	1531	9284	7743	1431	10188	2376	6888	1105	9214		10143	6127
Sydney	14418	9138	16042	7374	11040	16992	12073	14497	15989	16922	13501	16324		4861
Tokyo	9571	5141	8696	2874	13547	9562	8811	7485	10849	9718	18589	9861	7823	

Weight

To convert kilograms (kg.) to pounds (lb.), multiply by 2.205

To convert pounds (lb.) to kilograms (kg.), multiply by 0.454

To convert grammes to ounces multiply by 0.0353

To convert ounces to grammes multiply by 28.35

To convert tonnes to UK tons multiply by 0.984

To convert UK tons to tonnes multiply by 1.016

kg.	Units	lb.	Grammes	Units	Ounces	Tonnes	Units	UK tons
0.454	1	2.205	28.4	1	0.0353	1.016	1	0.984
0.907	2	4.409	56.7	2	0.0705	2.032	2	1.968
1.361	3	6.614	85.1	3	0.106	3.048	3	2.953
1.814	4	8.819	113.4	4	0.141	4.064	4	3.937
2.268	5	11.023	141.8	5	0.176	5.080	5	4.921
2.722	6	13.228	170.1	6	0.212	6.096	6	5.905
3.175	7	15.432	198.5	7	0.247	7.112	7	6.889
3.629	8	17.637	226.8	8	0.282	8.128	8	7.874
4.082	9	19.842	255.2	9	0.317	9.144	9	8.858
4.536	10	22.046	283.5	10	0.353	10.161	10	9.842
9.072	20	44.092	567	20	0.705	20.321	20	19.684
18.144	40	88.185	1134	40	1.411	40.642	40	39.368
27.216	60	132.277	1701	60	2.116	60.963	60	59.052
36.287	80	176.370	2268	80	2.822	81.284	80	78.737
45.359	100	220.462	2835	100	3.527	101.605	100	98.421

Volume

To convert gallons (UK) to gallons (USA): multiply by 1.201

To convert gallons (USA) to gallons (UK): multiply by 0.833

To convert litres to gallons (USA): multiply by 0.264

To convert gallons (USA) to litres: multiply by 3.785

To convert litres to gallons (UK): multiply by 0.22

To convert gallons (UK) to litres: multiply by 4.55

Gallons (UK)	Units	Gallons (USA)	Litres	Units	Gallons (USA)	Litres	Units	Gallons (UK)
0.833	1	1.201	3.785	1	0.264	4.55	1	0.22
1.665	2	2.402	7.570	2	0.528	9.10	2	0.44
2.498	3	3.603	11.355	3	0.792	13.65	3	0.66
3.331	4	4.804	15.140	4	1.056	18.20	4	0.88
4.163	5	6.005	18.925	5	1.320	22.75	5	1.10
4.996	6	7.206	22.710	6	1.584	27.30	6	1.32
5.829	7	8.407	26.495	7	1.848	31.85	7	1.54
6.661	8	9.608	30.280	8	2.112	36.40	8	1.76
7.494	9	10.809	34.065	9	2.376	40.95	9	1.98
8.327	10	12.009	37.850	10	2.640	45.50	10	2.20
16.654	20	24.019	75.700	20	5.280	91.00	20	4.40
33.307	40	48.038	151.400	40	10.560	182.00	40	8.80
49.961	60	72.056	227.100	60	15.840	273.00	60	13.20
66.614	80	96.075	302.800	80	21.120	364.00	80	17.60
83.268	100	120.094	378.500	100	26.400	455.00	100	22.00

Temperature

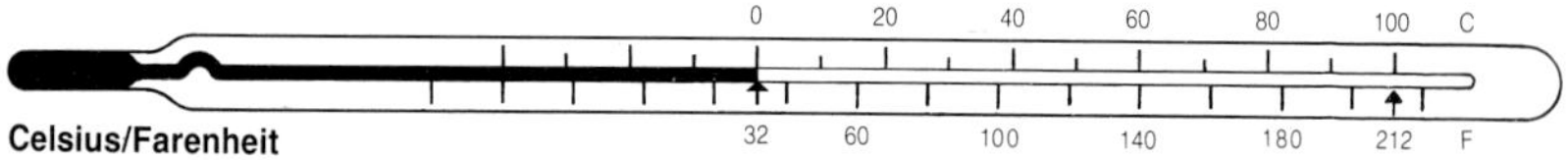

Celsius/Farenheit

To convert celsius to farenheit [(°C x 9) ÷ 5] + 32 = °F

To convert farenheit to celsius [(°F − 32) × 5] ÷ 9 = °C

Scales for different purposes

The Kelvin Scale was developed by William Thompson Kelvin (1824–1907), a Scottish physicist who was a professor at Glasgow University. In 1848 Kelvin determined that there was a point in temperature at which the motion of all particles ceased. Indeed, it is at this point that hydrogen, the lightest of all gases, ceases in motion. Kelvin called this point in temperature 'absolute zero' and he then devised a scale of temperature that took this, rather than the freezing point of water, as its base – so 1 Kelvin = 1°C.

The Beaufort Scale was devised in 1805 by Admiral Sir Francis Beaufort (1774–1857) and is one of the most widely used scales of wind speed. The wind speeds are based upon easily observable indicators and runs from 0 (calm) to 12 (hurricane). The main indicators refer to the way in which smoke rises, trees sway or bend, and the damage incurred to buildings. The scale has also been calibrated to wind speed in knots, such that a gale, Beaufort scale 8, has a wind speed of 34–40 knots and a hurricane, Beaufort scale 12, has a speed greater than 64 knots.

The Richter Scale was devised by Charles Francis Richter (1900–1985) in 1935 as a measure of earthquake intensity. Its original use was confined to earthquakes in California, based upon the local geological structure, though it is now used extensively world-wide. The scale uses a logarithmic scale from 0 to 9, with 9 being the most severe earthquake. Seismic recordings taken from a number of locations calculate the epicentre of the earthquake and its intensity. On average there are only nineteen events a year that register greater than a 7 on the scale. The highest ever recorded is 8.6.

Clothes

Approximately equivalent sizes

Men's Shirts

UK & USA	14½	15	15½	16	16½	17
Continental	37	38	39	40	41	42

Men's Suits

UK & USA	38	40	42	44	46	48
Continental	48	50	52	54	56	58

Ladies' Coats and Jackets

UK	10/30	12/34	14/36	16/38	18/40
USA	8	10	12	14	16
Continental	36–40	38–42	40–44	42–46	44–48

Men's Shoe Size

UK	7	8	9	10	11
USA	7	8	9	10	11
Continental	40	41	42	43	44

Ladies' Shoe Size

UK	3	4	5	6	7
USA	4	5	6	7	8
Continental	36	37	38	39	40

Paper sizes

International

2A = 1189 × 1682 mm (46.8″ × 66.2″)
A0 = 841 × 1189 mm (33.11″ × 46.81″)
A1 = 594 × 841 mm (23.39″ × 33.11″)
A2 = 420 × 594 mm (16.54″ × 23.39″)
A3 = 297 × 420 mm (11.69″ × 16.54″)
A4 = 210 × 297 mm (8.27″ × 11.65″)
A5 = 148 × 210 mm (5.83″ × 8.27″)
A6 = 105 × 148 mm (4.13″ × 5.83″)
A7 = 74 × 105 mm (2.9″ × 4.13″)

British

Double Crown = 508 × 762 mm (20″ × 30″)
Royal = 508 × 635 mm (20″ × 25″)
Medium = 457 × 584 mm (18″ × 23″)
Demy = 445 × 572 mm (17½″ × 22½″)
Large Post = 419 × 533 mm (16½″ × 21″)
Crown = 381 × 635 mm (15″ × 20″)
Foolscap = 343 × 432 mm (13½″ × 17″)

Wines and spirits

	Proof (Sikes)	Volume of alcohol
Table wines	14°-26°	8%-15%
Port, sherry	26°-38½°	15%-22%
Whisky, gin	70°	40%

Derivations of Words

Length	Germanic langaz meaning long. In Old English quality of being long
Centi	this element denotes 100
Metre	from Greek metron a means of measuring, a measure, a rule, a verse
Inch	from Latin uncia meaning twelfth part
Feet	a lineal measure originally based on the length of a man's foot
Hect	Greek hekaton meaning hundred
Area	Latin area – vacant piece of level ground
Acre	Old English aecer – piece of tilled or arable land, field or a unit of square measure of land
Distance	Latin distantia
Pressure	Latin pressus – weight of pain from the action of pressing
Mile	derived from a Roman measure of 1,000 paces
Weight	Old Norse weht – a weighing device
Kilo	from Greek Khilioi – thousand in weights and measures
Gram	Greek grámma small weight. Adopted as the unit of weight in the metric system
Pound	Latin pondo meaning in weight, a unit of weight or the gold unit of value
Volume	Latin volumen meaning roll of writing, size or bulk
Litre	derived from French litron an obsolete measure
Gallon	the ultimate origin is unknown perhaps of Celtic origin
Pint	perhaps Germanic origin
Quart	Latin quartus meaning fourth – one fourth of a gallon
Grade	from early modern French meaning a step or a stage

Wheel 4000 B.C.

Wheels were first used as far back as 4000 B.C. as a means of transport in Asia Minor and the Caucasus. The wheel is believed to be a development from the use of trees as rollers. Early wheels were made of three planks of wood formed to make a disk, with a central axle hole, and held together by copper and wood clamps. The earliest spoked wheels occurred in Mesopotamia in 2000 B.C.

Glassware 1500 B.C.

Early glass vessels were made as far back as 1500 B.C. in Egypt. Vases and bowls were either sculptured out of a solid block of glass or formed on a core consisting of a bag of sand. Glassware production became far greater with the invention of the blowpipe by the Syrians in the 1st century B.C. A hollow iron tube was dipped into molten glass and air blown down the tube resulted in the formation of a glass or vase.

Paper 3500 B.C.

This is one of the earliest 'inventions' by mankind. Paper was first made from papyrus reeds on the banks of the River Nile. Reeds were laid out across each other, crushed and dried to form a flat surface on which things could be drawn and written. By 1300 B.C. parchment from untanned skin was being used in Egypt, but in the 2nd century B.C. the Chinese began making paper from a pulp of cloth, wood and straw. It was the growth of newspapers in the 17th century in Europe that led to the major developments in paper production.

False Teeth 700 B.C.

The Etruscans of northern Italy were making dentures from human or carved animal teeth around 700 B.C. Such 'natural' false teeth soon became foul but they were cheap and popular well into the 19th century. The longer lasting porcelain teeth were produced in France in 1788 and the correctly fitting dentures were first made in 1851 in France when dental plates of vulcanite (hardened rubber) were formed from a cast of the gums.

Porcelain 851

The earliest examples of pottery date back as far as 7000 B.C. and originated from Iran. Porcelain, however, was not invented until 851, during the T'ang dynasty in China. Porcelain is fired at a higher temperature than stoneware (1200°C and 1400°C), the Chinese using china clay and feldspar in its production. Although quantities of porcelain have been found in excavated tombs that date back to the 9th century, it was not until 1300 that porcelain was made in large quantities.

Compass

Compass 11th century

In the 1st century A.D. the Chinese used the magnetic iron-bearing stone, magnetite, to act as a compass. The earliest true compass, however, was in the 11th century and took the form of a needle thrust through a straw, floating in a bowl of water. This was not practical at sea and, in 1250 in the Mediterranean, the needle was attached to a marked card balanced on a central pivot. By the 16th century the needle was mounted on gimbals (self-aligning bearings) and by the 20th century the gyroscope compass had been developed, the first being produced in Germany in 1908.

Spectacles

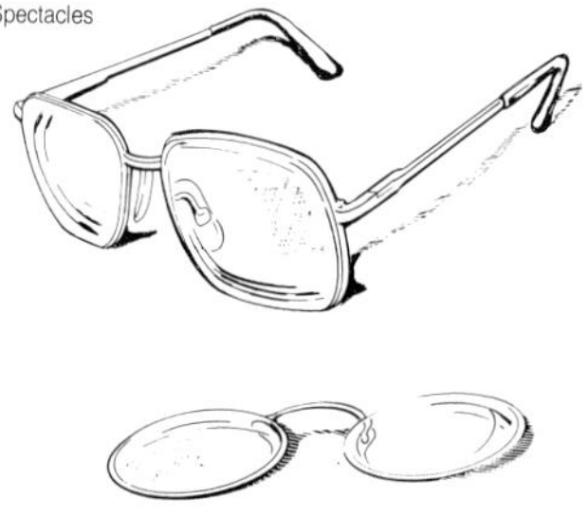

Clock – mechanical 1290

Shadow clocks (2000 B.C.) and water clocks (1360 B.C.) had been used for many years by the Egyptians, but the first European mechanical clocks were used in English and Italian monasteries around 1290 A.D. They consisted of a series of wheels and cords, a falling weight causing a wheel to make a forward jerk, thus moving a finger or pointer. The original inventor remains unknown though one of the earliest known European clocks was that of Giovanni de Dionde finished (after 16 years) in 1364. In China, however, there are records of a machine clock 9 metres high, driven by powered water, described in writings of 1088 A.D. by the Chinese imperial tutor, Su Sung.

Spectacles 13th century

The idea of wearing lenses to correct vision was first put forward by English scientist Roger Bacon in the 13th century. Although he did not develop the idea, it is believed that his work led to the appearance of spectacles in the late 13th century in Venice. Spectacles at this time were used for correcting close-up vision and were also being used in China at the same time. It was not until the mid-15th century that concave lenses for correcting short-sightedness were developed in Italy. In 1775 Benjamin Franklin, the American statesman, had the first pair of bi-focals made for him and in 1887 Dr. Eugen Frick of Zürich developed the first contact lens.

Printing Press 1450

Johann Gensfleisch, known as Gutenburg, invented the printing press around 1450. In 1438 he had invented a mould for making type of individual letters. These moulds were fitted on an adjustable frame. By 1450 Gutenberg began commercial printing, the earliest books being the Donatus Latin Grammar and the 42-line Latin Bible (so-called because of the number of lines per column).

Pendulum – clock

Clock – pendulum 1657

In 1581 Galileo is said to have observed and timed the swings of a lamp following an earth tremor and from this formed the theory of isochronism (things being performed in equal times). His son, Vicenzio, followed his father's work but without completing a clock. It was the Dutch astronomer and physicist Christiaan Huygens who, however, designed the first practical pendulum clock in 1657. Built by Salomon Coster, the first pendulum clock was correct to within five minutes a day.

Piano

Piano 1709

The first piano was built in 1709 by Bartolomeo Cristofori in Italy. It differed from the harpsicord, which already existed, in that its strings were struck by a hammer, not plucked, thus enabling the volume of sound to be varied. In 1770 Johann Stein, a German, devised the 'escapement' which ensured that the hammer moved away from the string and let it vibrate freely. In 1783 an English firm, Broadwood and Sons, devised the sustaining and soft pedals to give even more variation in sound.

Lightning Conductor 1752

The American scientist and statesman Benjamin Franklin invented the lightning conductor in 1752. After flying a kite in a thunderstorm he realised that lightning was a discharge of electricity. He therefore attached wires down the sides of buildings in Philadelphia and lightning, when it struck, took the easiest path to the ground via the wire.

Submarine 1776

The first true submarine was invented by an American engineer, David Bushnell, in 1776 even though Cornelius Drebbel, a Dutchman, had developed a submersible craft in 1620. Bushnell's submarine was made of wood and driven by a hand-turned propeller. Buoyancy tanks were flooded for submerging and the 'Turtle', as it was known, was used to fix explosives to the British ship 'Eagle' in the American War of Independence.

Steam Engine 1777

In 1698 Thomas Savery, an English engineer, patented the design for a pump that drew water by means of a sucking action. The water was then expelled by steam under pressure. In 1712 Thomas Newcomen, a Devon blacksmith, invented an atmospheric engine which created a vacuum when water was sprayed into a cylinder. The first true steam engine, however, was invented by James Watt in 1777 and was used in Cornish mines. He devised a separate condenser into which steam was passed from the cylinder. This made his engine far more efficient than that of Newcomen.

Balloon 1783

In 1782 the French papermaker brothers, Joseph and Étienne Montgolfière, made their first experiments with balloons. In Spring 1783 they repeated their experiments, but this time outside. On November 21st 1783 Jean Pilâtre de Rover and the Marquis d'Arlandes manned a Montgolfière balloon for the first time. The flight lasted 26 minutes, with the balloon landing five miles away from its start at the Bois de Boulogne. French physicist J.A.C. Charles led the rapid development of balloons and ballooning became an instant craze.

Air Balloon

Gas Lamp

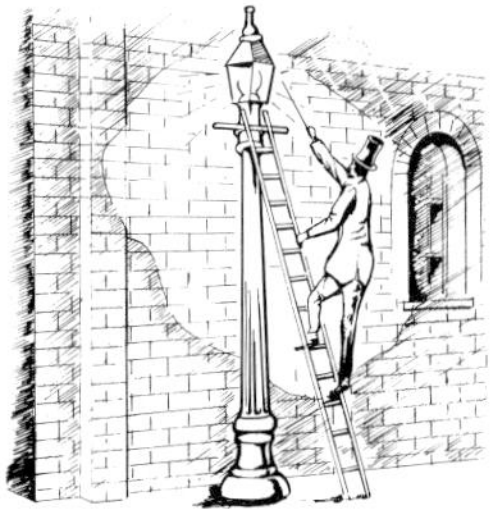

Gas Lamp 1799

Philippe Lebon, a French civil engineer, first patented his gas lamp and heater, known as the 'Thermolampe', in 1799. This early form of gas lamp was, however, largely unsuccessful. Even so, by 1823 some 346 kilometres of London's streets were lit by over 40 000 gas lamps of one kind or another. It was not until 1893 that Carl Aver, a Viennese chemist, perfected the design using rare earth metals such as zirconium. These metals were used to form a mantle which glowed when sufficiently heated by gas. Aver established the Incandescent Gas Light Company in 1885 but his lamps, and those of others, continued to prove inadequate or too expensive. By 1903 the electric light bulb had begun to take over.

Typewriter 1808

The typewriter was invented by Pellegrino Turri of Castelmovo in Italy in 1808. It was devised so that a blind friend could write without a secretary. By 1841 Scotsman Alexander Bain had developed the inked ribbon and in 1866 Sholes and Glidden designed a modern typewriter in the USA, a design taken over by the Remington Small Arms Company. In 1961 International Business Machines made a further major development by mounting the letters on a 'golf ball', rather than on individual bars as previously.

Camera 1827

The principles involved in photography date back as far as 1000 A.D. to Arabian astronomers. It was, however, not until 1827 that Joseph Niépce, a French country gentleman, produced the earliest known photograph. This first photograph was taken on a pewter plate coated with bitumen which hardened and turned white. By 1835 William Fox Talbot, an English scientist, had extended the photographic process onto paper with his first negative-positive method, from which modern photography developed. By 1888 George Eastman developed the first camera, a Kodak, which used a flexible roll film.

Camera

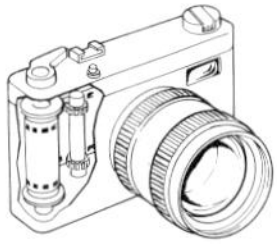

Sewing Machine

Sewing Machine 1829

This was invented in 1829 by Barthélemy Thimmonier in St. Etienne, France. The key to the machine's development was a needle which had the eye near the point, so ensuring that the whole needle did not have to pass through the fabric. In 1833 Walter Hunt, inventor of the safety pin, developed a machine which had two interlocking threads. This was the forerunner of modern developments. In 1851 Merritt Singer from New York patented a lock-stitch machine worked by a treadle.

Electric Dynamo 1832

In 1831 English scientist Michael Faraday discovered that an electric current could be generated by moving an electric conductor in the field of a magnet. Frenchman, Hippolyte Pixii, made the first practical development of this discovery with his hand-powered dynamo in 1832, operating an alternating current. In the 1850s it became clear that magnets were not necessary and in 1866 Varley and Wheatstone in England and in 1867 von Siemens in Germany both developed dynamos that did not use permanent magnets but electro-magnets instead. In 1870 Zénobe Gramme developed the first practical industrial dynamo, resulting in the production of electrical power on a large scale.

Postage Stamp 1840

Before prepaid post was introduced the recipient of a letter had to pay for its delivery. The first adhesive postage stamp was developed by the Scottish printer James Chalmers. After making some specimens he stuck one on a letter and addressed it to the General Post Office in 1839. As a result of further efforts by Sir Rowland Hill the British Post Office introduced the first official postage stamps, the Penny Black and Twopenny Blue, in May 1840. This pioneering role gives Britain the unique privilege of needing no country name on her stamps.

Safety Pin 1849

The modern safety pin was invented by Walter Hunt in New York in 1849. The simple closure concealed the point for safety and the invention enabled Hunt to settle a $15 debt.

Refrigerator

Refrigerator 1850

As far back as 2000 B.C. food was being stored in pits of ice in Mesopotamia. Modern refrigeration was invented in 1850 by James Harrison, a Scottish printer, who owned the Geelong Advertiser in Victoria, Australia. He discovered, when cleaning type, that ether made the metal cold. Ether has a low boiling point, so Harrison compressed gaseous ether into a liquid with a pump and when he released the pressure the ether turned to gas and cooled considerably. The system was first applied to a brewery in Bendigo, Victoria.

Dynamite 1867

In 1846 Italian chemist Ascario Sobrero, discovered the explosive nitroglycerine. Alfred Nobel, a Swedish inventor, set out to tame the explosive. However, during experiments in 1864 his brother, Emil, and four workmen were killed in his own factory. Three years later he absorbed nitroglycerine in charcoal and the explosive could then be set off by detonation. Nobel regarded his 'safety explosives' as a form of national defence and when he died he set aside £2 million of his fortune to the establishment of an international peace prize.

Barbed Wire 1873

In 1873 Joseph Glidden of de Kalb, Illinois, developed the first machine to mass-produce barbed wire. It was originally used to fence off areas of cattle ranges in the American West and was then used extensively amid the trenches during World War I.

Diesel engine

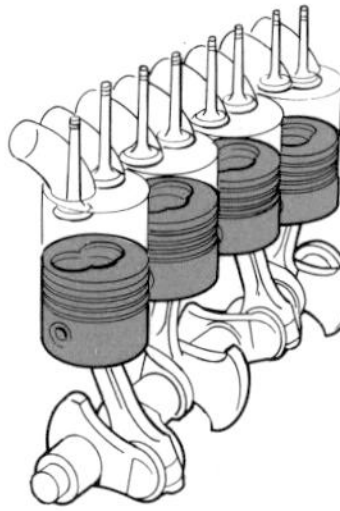

Telephone 1876

The telephone was invented by Alexander Graham Bell, a Scotsman who emigrated to the United States. For almost a year Bell worked on transmitting messages down steel strips which vibrated and gave a musical note. In this work Bell realised that the human voice could actually be transmitted by this means and on February 14th 1876 Bell successfully communicated down a length of wire to his assistant in another room. Bell was granted a patent. On the same day Elishna Gray also applied for a patent but Bell was given priority because he applied first. Bell made his first long distance call over 13 kilometres on August 10th 1876 in Canada at Paris, Ontario. By 1877 Thomas Edison had developed a telephone with a separate ear and mouth piece.

Diesel Engine 1883

The first engine was developed by two Italians, Eugenio Barsanti and Felice Matteucci, in 1856. It was, however, in 1883 that German engineer Gottlieb Daimler designed the first petrol driven engine. The use of petrol gave more power for the same weight of engine. Daimler then added a carburettor to the engine which blended petrol and air together in a fine spray. The result was an engine that by 1885 was driving the first practical motor car.

Carburettor 1884

The jet carburettor was first devised in 1884 by Edward Butler, an English inventor. He originally used it on his own petrol tricycle but this vehicle was soon abandoned. As a result the carburettor was essentially re-invented by William Maybach of the Daimler company in 1893.

Zip fastener

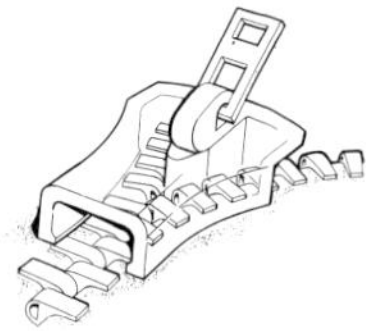

Zip Fastener 1891

Whitcomb Judson, a Chicago engineer, originally developed this invention for shoes. Nine years later, in 1900, he devised a similar fastening but this time devised for fabrics. Yet both Judson's designs were unreliable and it was not until 1906 that Gideon Sundbach, a Swedish engineer working for Judson, perfected the design through the use of interlocking metal teeth. By 1913 the invention had become commercially viable.

Vacuum flask

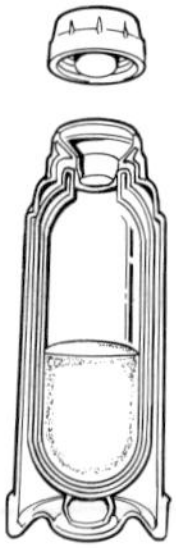

Vacuum Flask 1892

It is known that the Romans understood the heat-retaining properties of a double-walled vessel as one was found in the ruins of Pompeii. In 1892 the Scottish physicist, Sir James Dewar, had his glass-blower, Reinhold Burger, blow for him a double-walled glass vessel which he coated with mercury on the two inner walls. He then exhausted the air in the space between them to create a vacuum. The silvering hindered heat loss by radiation and the vacuum stopped heat loss by convection or conduction, so that the liquid in the glass bottle remained at a constant temperature. Reinhold Burger, as a partner in the Berlin company, marketed the vacuum flask in 1902 and patented the invention in 1904. Burger protected the glass bottle further by adding a nickel case. The name 'Thermos' was chosen in a competition.

X-ray 1895

The X-ray was first discovered by Wilhelm Röntgen in Wurzberg, Germany, in 1895. He discovered that an electric discharge in a vacuum on to a metal plate created a new form of radiation which would pass through everything except metal and bone and would darken photographic film. He named them X-rays, meaning unknown. By 1897 Walter Cannon had extended their use showing that when animals swallowed a liquid solution containing metals, the intestines showed up under an X-ray. Today the X-ray has a wide number of medical and security uses.

Paper Clip 1900

The paper clip was invented by Norwegian, Johann Vaaler, and patented by him in Germany in 1900. It is a simple loop of wire enabling sheets of paper to be held together without pins or folding.

Aeroplane 1903

All through civilization man has had a fascination with flying. Between 1891 and 1896 Otto Lilienthal, a German civil engineer, made controlled flights of up to 229 metres in gliders. In Dayton, Ohio, the Wright brothers, Wilbur and Orville, were developing a powered aeroplane. For their test site they chose Kitty Hawk in North Carolina where there were strong yet constant winds. They made over 1 000 unmanned glides before, on December 17th 1903, Orville Wright made the first powered and controlled flight in Flyer 1. Aeroplane development proceeded rapidly and by July 1909 Frenchman Louis Blériot made the first crossing of the English Channel, his plane averaging a speed of 40 miles per hour.

Aeroplane

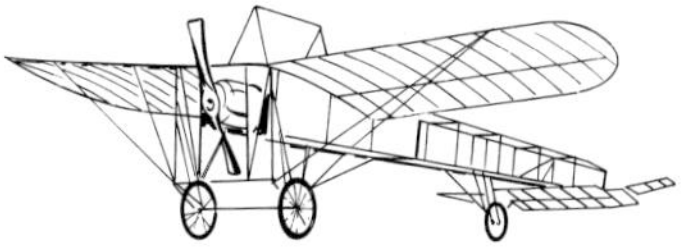

Sellotape 1928

A strip of clear cellulose film coated with a rubber-based glue was developed in 1928 by research technician Richard Drew for the Minnesota Mining and Manufacturing Corporation in America. It was sold in rolls under the name of 'Scotch Tape' as a general purpose sticky tape.

Cats eye set in road surface

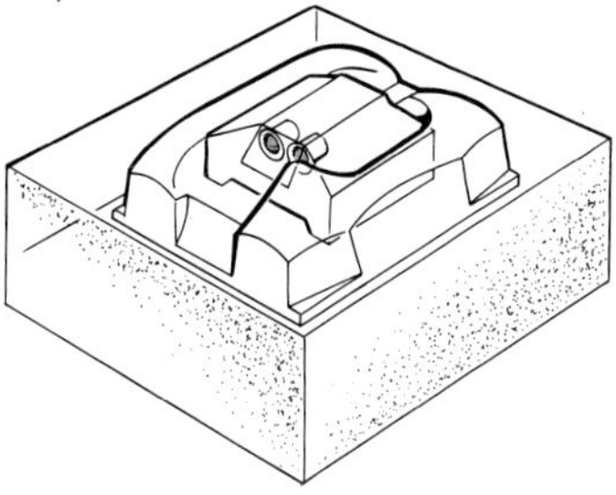

Cats Eyes 1934

This invention was patented in Britain in 1934 by a Lancastrian road contractor, Percy Shaw. He observed how cats eyes reflected the light and conceived the idea of 'cats eye' road studs which would reflect the light from a vehicle's headlights. Each eye consisted of a prism which, with an aluminium mirror behind it, reflected the light directly from where it came. Two pairs of studs, one each way, were mounted in a domed rubber pad so that the studs were just above road level. The pad was also designed so that each time it was depressed by a car wheel it wiped the prism clean. Now widely used on most modern roads, the 'cats eye' saw its first major use during the World War II, facilitating safe driving on unlit roads during the blackouts.

Helicopter 1936

The first practical helicopter was the Fa-61, designed by Heinrich Focke in Germany in 1936. It had twin counter rotating rotors set on outriggers and led the field until 1939. Igor Sikorsky, a Russian engineer who fled to the USA, developed the VS-300, the first helicopter with one single rotor plus a small tail rotor, which solved the problem of torque. The extra manoeuvrability this provided made the VS-300 a major success. In the 1950s helicopters benefitted from the development of the jet engine. The helicopter today remains unsurpassed in its role as a rescuer of victims from inaccessible places.

Jet Engine 1937

The jet engine was invented by Frank Whittle in 1937, when he built his first prototype engine. The first aircraft to fly with a jet engine, however, was the Heinkel He 178 which flew in Germany in 1939. The first flight with Whittle's engine was in 1941 in a Gloster test aircraft, but it was his engine that became the forerunner for modern aviation. The engine worked through air entering at the front, having its pressure increased and then being passed to a combustion chamber. The ignition of fuel under these pressures resulted in rapidly expanding gases which shot out of the back of the engine and created an equal thrust forwards.

Satellites 1957

In 1945 Arthur C. Clarke, the science-fiction writer, suggested that signals beamed from earth to satellites in space could be re-broadcast from there to ground stations thousands of miles from the original transmitter. His idea was a remarkable piece of foresight as on October 4 1957 the Russians launched the first artificial satellite. It was called Sputnik I and orbited at a height of up to 590 miles. Four months later the first American satellite, Explorer I, was launched. In December 1958, following experiments with communication satellites, the USA launched a satellite called SCORE (Signal Communication by Orbiting Relay Experiment). It carried a pre-recorded message from President Eisenhower of a Christmas greeting to the American people. The launch of Telstar, on July 10 1962, achieved the first live television pictures to be transmitted that brought millions of television viewers into instant contact with world events. One of the first events covered was the shooting of President John F. Kennedy. Telstar could transmit sixty telephone calls simultaneously or one television channel. Today the development of satellites enables people to receive world-wide news as it happens and to communicate with the other side of the world. The advent of the space shuttle, from which satellites can be launched, means that in the future it will be possible to refuel or recover satellites.

Sputnik I satellite

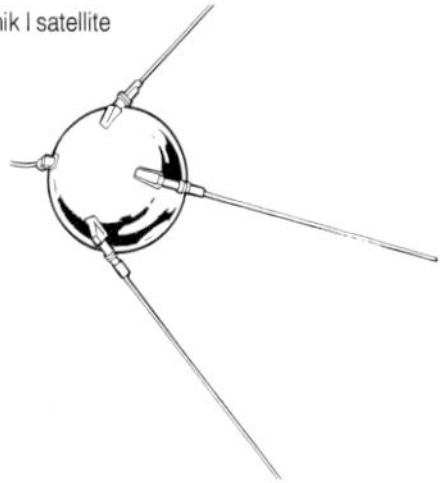

Hovercraft

Hovercraft 1959

Patented by boat designer Christopher Cockerell in England in 1955, the hovercraft was not publicly shown until 1959. Using some 'kitchen' experiments he established the principle and patented the idea. On June 11 1959 the first full-sized craft was commissioned in the Isle of Wight. By July 25th Cockerell had made the first Channel crossing by hovercraft and created a sensation with his invention. By July 1968 a regular cross-Channel hovercraft service was operating, carrying both passengers and cars.

Silicon Chip 1971

The silicon chip, or 'chip', is a tiny crystal wafer, the most versatile form of which is known as a microprocessor. These were first produced in 1971 by the American Intel Corporation and Texas Instruments. They can perform a myriad of arithmetic and storage functions and have numerous applications, not least with computers themselves. On account of their miniscule size they can fit into virtually anything that needs to be controlled, some microprocessors being able to perform 400 000 additions in a second. The pace of development has been rapid and microprocessors are becoming smaller and more powerful all the time.

Country	GNP millions of US $ 	GNP/Capita US $	Employment % 	Private House % 	Total no. of children 
USA	3,634,582	14,000	90	*	3,669,141
Japan	1,157,456	9,387	97	60	1,431,577
U.S.S.R.	721,931	2,495	59	*	*
Fed. Rep. of Germany	613,356	10,154	91	36	586,155
France	489,434	8,850	92	47	768,431
United Kingdom	425,541	7,571	87	35	656,417
Italy	348,385	6,058	90	*	623,103
Canada	336,746	12,707	88	*	367,227
China	224,263	201	*	*	*
Brazil	209,398	1,420	96	*	2,559,038
India	193,820	238	*	*	*
Australia	182,136	11,038	90	68	247,348
Mexico	175,409	2,016	*	67	2,392,849
Spain	160,926	4,074	83	*	571,018
Iran	159,215	3,190	*	70	*
Saudi Arabia	108,283	8,020	*	*	*
Sweden	94,824	11,424	98	55	98,463
Switzerland	91,100	14,234	99	28	74,684
Indonesia	83,745	469	*	*	*
South Africa	72,855	2,035	*	*	158,032
Nigeria	67,291	615	*	*	*
Argentina	64,829	2,000	96	*	663,429
Norway	54,719	13,028	97	67	51,134
Algeria	45,234	1,831	*	*	818,613
Egypt	39,421	766	95	*	1,601,265
Pakistan	34,051	310	*	8	2,435,171
Greece	33,466	3,346	96	*	116,481
New Zealand	23,368	6,872	*	*	51,798
Portugal	19,154	1,824	93	*	130,492
Cuba	15,195	1,461	*	*	182,067
Dominican Republic	8,575	1,261	*	*	*
Zaire	4,588	135	*	*	*
Brunei	3,503	1,751	98	*	6,330
Angola	2,541	261	*	*	*
Madagascar	2,379	212	*	*	*
Bahamas	1,449	658	*	*	5,439
Martinique	1,307	408	*	*	5,705
Fiji	1,124	1,124	99	*	19,464
Guadeloupe	1,114	337	*	*	6,750
Burundi	1,088	825	*	*	*
Botswana	1,073	113	*	*	*
Belize	182	226	*	*	6,150
Seychelles	147	756	*	*	1,722

* Data not available

World wealth

Passenger cars 1000 units	TV sets per 1000 inhabitants	Telephone 1000 units	Daily Newspapers per 1000 inhabitants	Population per Physician	Country
126,728	646	176,391	269	549	USA
27,144	560	63,976	575	735	Japan
9,631	307	26,667	405	267	U.S.S.R.
25,217	354	36,582	408	431	Fed. Rep. of Germany
19,300	369	33,002	191	480	France
15,884	457	29,518	421	143	United Kingdom
20,389	405	24,331	82	345	Italy
10,731	460	16,618	226	548	Canada
238	6	5,539	33	1,769	China
9,379	122	10,570	44	1,632	Brazil
339	3	3,488	20	2,545	India
6,636	428	8,329	337	524	Australia
4,854	111	6,796	*	2,136	Mexico
8,874	256	13,825	79	362	Spain
90	55	2,144	8	2,551	Iran
152	262	1,752	10	2,606	Saudi Arabia
3,081	387	7,410	524	478	Sweden
2,552	370	5,270	381	726	Switzerland
866	23	669	*	11,740	Indonesia
2,727	71	3,648	*	1,906	South Africa
213	6	623	*	9,591	Nigeria
1,388	202	3,108	*	530	Argentina
1,428	305	2,579	483	477	Norway
511	65	709	22	2,780	Algeria
890	41	800	78	760	Egypt
1,351	11	511	14	2,911	Pakistan
1,151	174	3,529	102	393	Greece
1,466	289	2,011	325	606	New Zealand
1,517	149	1,764	50	456	Portugal
165	164	493	118	722	Cuba
116	78	165	42	4,023	Dominican Republic
*	0.4	39	*	13,452	Zaire
71	145	18	*	1,869	Brunei
*	4.3	40	7	15,404	Angola
44	8	37	6	9,939	Madagascar
52	156	91	154	1,218	Bahamas
78	127	84	99	789	Martinique
31	*	51	102	2,222	Fiji
89	112	69	98	746	Guadeloupe
9	*	6	0	45,432	Burundi
12	*	18	22	7,378	Botswana
6	*	5	41	3,261	Belize
3	*	11	63	2,222	Seychelles

Source: 1988 Statesman Year Book
1987 United Nations Year Book

Tollund Man

In Northern Europe the peat bogs produce favourable conditions for the preservation of organic material. The custom of the people inhabiting this region in the Iron Age was to throw the bodies of executed captives into certain marshes to propitiate the gods. Many of the people who died violently have been preserved, among them a body discovered at Tollund in Denmark. The Tollund Man died by strangulation, probably as part of a ritual and when unearthed the rope with which he was hanged was still around his neck. The condition of the head is evidence of the peat bogs' amazing preservative qualities for the man's features have kept a serene and almost smiling expression.

Lascaux Cave

This Stone Age cave site near Montignac in the Dordogne region of France is one of Europe's greatest prehistoric finds, discovered in 1940 by four boys looking for their dog which had disappeared down a hole. An impermeable layer of chalk on the ceiling and a layer of carbonate of lime on the walls prevented water penetrating, served as a stable base for the pictures, and helped to 'fix' them. For thousands of years the cave was cut off from the outside air and therefore the pictures were preserved. The cave consists of four chambers containing many hundreds of stylised animal paintings in colours ranging from yellow to black. The way these have been painted, one on top of another, suggests many different civilizations have decorated the cave with their own paintings.

Easter Island

A volcanic island in the Pacific Ocean lying 3 600 kilometres west of Chile, first inhabited by the Polynesians from the Marquesas Islands. Famous for rongorongo hieroglyphs and huge monolithic stone statues. These statues pose one of the world's greatest unsolved archaeological mysteries – how they were moved and erected and why work ceased abruptly and their stone tools were discovered abandoned. Norwegian anthropologist, Thor Heyerdahl, believes the island was inhabited by two groups, a highly civilized pale-skinned people from South America, the 'Long Ears', and the Polynesians, the 'Short Ears'. According to this theory the stone statues were derived from ancient monolithic human statues of pre-Inca Peru, homeland of the 'Long Ears'. In about 1650 violent civil war broke out between the two groups and therefore work at the quarry ceased as all were called to battle.

Nazca Lines

The Nazca lines provide the most fascinating archaeological site in the Americas. These ground markings cross a large barren plain and cover some 500 square kilometres near Nazca on the south coast of Peru. These spectacular lines made by the Nazca culture during A.D. 100 – A.D. 800 show an understanding of sophisticated mathematics and surveying techniques. They were made by removing dark surface gravel to expose the underlying light coloured rock. Some lines are straight for long distances and may have been sacred ritual pathways; others are in the forms of drawings, depicting animals and birds, or form geometrical shapes. The lines, which can only be seen properly from the air, were probably made primarily as offerings to their gods.

Knossos

The discovery of the ruins of the palace at Knossos were made by a British archaeologist, Sir Arthur Evans, in 1900. He made the excavation and reconstruction his life's work. The palace was probably built in about 300 B.C. for the legendary King Minos and comprised 1 300 rooms and covered some 200 square metres. It consisted of buildings up to five storeys high, with thick columns supporting the various levels and majestic staircases leading to the higher levels and the roofs. It was discovered that the Cretans worshipped the bull; the horns of the sacred emblem, made of stone and painted gold, topped the roofs. The palace consisted of many royal apartments, storage rooms, workshops and shrines all built around a central courtyard.

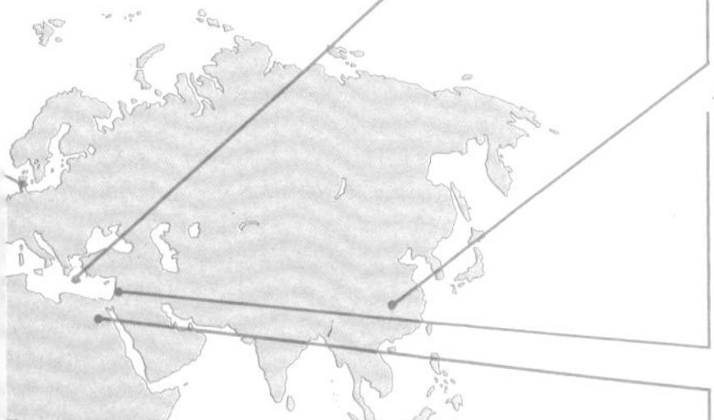

Terracotta Army

The earliest known large-scale sculpture in China occurred between 250–210 B.C. during the dynasty of the first Qin emperor when over 8 000 life-size soldiers were modelled. In 1974 a farmer digging near the emperor's tomb discovered a life-size clay soldier; on excavation archaeologists then found an estimated 6 000 infantry men in an enormous subterranean pit. Two smaller, but equally spectacular pits were found nearby. The soldiers had been hand modelled and while the clay was still malleable details of clothing and armour were shaped and fixed. The body was then baked and the missing head and hands added. The beautifully detailed heads demonstrate the genius of the Chinese potter and possibly indicate the many different races who fought in the Qin army. The Terracotta Army is now preserved and the first pit has been turned into a permanent museum.

The Dead Sea Scrolls

These Hebrew and Aramaic scrolls were accidentally discovered in 1947 in caves at Khirbat Qumran, on the shore of the Dead Sea. These scrolls, consisting of about 500 different documents, were probably produced by an ancient Jewish sect, the Essenes, and date from 250 B.C. – 70 A.D. They include texts of many Old Testament books, prayers, psalms and commentaries. They provide evidence of the accuracy of the Old Testament writings, information about the Jewish community contemporary with early Christianity and are invaluable archaeological examples of early Hebrew and Aramaic scripts. During a Jewish revolt in 66–70 A.D. the scrolls were stored in jars and hidden in the caves for safe keeping.

Tutankhamun

This was the most exciting discovery of Ancient Egypt, found in the Valley of the Kings in 1922 by British archaeologist, Howard Carter. It was the only Egyptian tomb in the valley which had not been broken open and robbed of its treasures. The preserved body of the teenage Pharaoh, Tutankhamun, of the 18th dynasty was found in a coffin of gold enclosed within a stone coffin. The face of the Pharoah was covered in a gold mask decorated with precious stones, with gold ornaments on the head. The tomb consisted of four rooms: one contained the body, the other three contained statues, chests, beds, chairs, jewellery, wines and preserved foods. The furniture was painted and covered in gold plating and inlaid with gems. This vast treasure has proved invaluable to archaeologists reconstructing Egyptian civilization at its height.

Ruins at Great Zimbabwe

Famous ancient stone ruins about 1 000 years old in Zimbabwe. Discovered by Adam Renders, a wandering hunter, in 1868 it consists of three groups: a valley, known as the 'Valley of the Ruins', covered in a maze of stone walls (about 2.5 metres high); an elliptical temple situated amongst these walls, containing a 'tower' composed of one solid mass of stonework; and on a hill to the north of this valley, the 'acropolis'. These ruins consist of many enclosures which were built end to end in order to follow the pattern of the land. Zimbabwe comes from the language of the Bantu people, who created a medieval civilization in the region and translated means 'stone houses'.

Celtic

The barbarous Celts were an Indo-European or Aryan people who originated in Switzerland and south-western Germany. They later spread and established civilizations over much of Iron Age Europe. Their movements have been traced by the remains of graves, pottery, tools and accounts left by the Romans. There were numerous Celtic tribes and chiefdoms sharing a culture that can be traced back to the Bronze Age (1200 B.C.) and the Celtic warrior aristocracy commanded considerable wealth and power. There was an upper class of warriors beneath the King, the highest rank being the Druids and the lowest the farmers. The Druids conducted sacrifices mainly in times of trouble at temples, an example of one of their temples is Stonehenge in Britain. They were also responsible for the education of young nobles.

The Celts built many round stone forts, with large family groups living close by in stockaded farmsteads. The forts were built to protect their tin mines, trading routes and for the communities' protection when under attack. The men wore tunics and cloaks, and it was the Celts who introduced trousers into western Europe.

The Celts also became outstanding craftsmen leaving behind examples of their weapons, tools, jewellery and even mirrors. Examples of their beautiful bronze work have been found in every corner of Europe and is testimony to the extent of the Celtic raids. The Celts were initially overrun by the Romans under Julius Caesar in 55 B.C. but it was not until the fifth century that the Angles and Saxons eventually drove them back into Scotland, Wales, Ireland and Cornwall. Variations of the Celtic language still remain in these areas today.

Ottoman

This was a Turkish Muslim Empire between the 14th and 20th centuries, which began through the joining of two nomadic cultures one from the Middle East with their black rectangular tents made from goats' hair and the other from Central Asia with their round white felt covered tents. It was the Ottoman branch of Turkish people led by Osman I, who ruled from 1281–1324 who set up the Empire. The fall of Constantinople in 1453 to the Ottoman rule destroyed the Byzantine empire. Constantinople was renamed Istanbul by the Ottomans and they made it the capital of their empire.

The Ottoman sultan claimed great power over Muslims both in the Mediterranean, Africa and Asia and as supreme chief of the Muslims he could claim supremacy over all the princes of Islam. The Ottoman Turks vigorously built their empire, with their power reaching its peak in the sixteenth century with the conquest of Egypt, Syria and Hungary under the rule of Suleiman I the Magnificent (1520–1566). Turkish became the main language of Anatolia with Persian as the main language of the urban and peasant society in the surrounding area, and also of literature and administration. The main religion accepted by the larger indigenous tribal groups was Islam. Most of the officials of the Empire were slaves born to Christians, and taken as children by the Turks. They were trained in the army, with the best selected for important Government posts, even as high as the Grand Vizier.

Although there were attempts to modernize the empire it started to decline in the 17th Century. Finally in World War I the Ottomans supported Germany but defeat brought the loss of their European territories and humiliation. Leading in turn to the nationalist revolution of Kemal Atatürk, who in 1922 replaced the Empire with the state of Turkey.

Aztecs

The Aztecs (or Mexica) grew out of obscurity in 100 B.C. to found an empire centred in the valley where Mexico City stands. Here they built Tenochtitlan their capital, a testimony to their elaborate and rich civilization.

They became the leaders of the neighbouring peoples and their conquests stretched up to what is now known as Guatemala. They were expert builders and craftsmen, carving huge pieces of stone with life like figures of men and animals. They also carved jade and crystal. They dyed cloth, made distinctive pottery and fine jewellery. As the Aztecs had no alphabet they kept records of the organization of their empire in picture writing in codices. They also had an accurate calendar, a carved calendar stone of 20 tons being one of their most famous relics.

Time to the Aztecs was granted by the gods in 52 year cycles. They believed that the universe had been destroyed four times. To create a new sun and moon for the fifth time, two gods had committed suicide by hurling themselves into a fire. In order to keep the sun

Old Jewish

Most of our knowledge of the Jews (or Hebrews) before Roman times comes from the Jewish Bible, known by Christians as the Old Testament. From 1550–1250 B.C. their country was divided into independent city Kingdoms, all subjects of the powerful Egyptian Pharaohs. Eventual decline of Egyptian power and the exodus from Egypt of the Jews under Moses, was the most profound moment in Jewish history. After the exodus of the Jews from Egypt to their promised land, the foundation of their code of ethics and religious practices was laid. They also began developing the worship of one supreme, invisible God and despised their neighbours for worshipping idols of several gods. From about 925 B.C. their promised land was divided into Israel in the north and Judea in the south which was centred on Jerusalem.
In 63 B.C. the Romans took the Judean capital of Jerusalem introducing four centuries of Roman rule. About 97 years later Jesus of Nazareth began his divine mission, and was arrested for his teaching, which was contrary to Jewish law and custom. He stood trial and Jewish leaders forced the Roman officer governing Judea, Pontius Pilate, to have him crucified. After the Romans finally and decisively crushed the Judeans in A.D. 70, most Jews left Judea, which then disappeared from the map.

Indus Valley

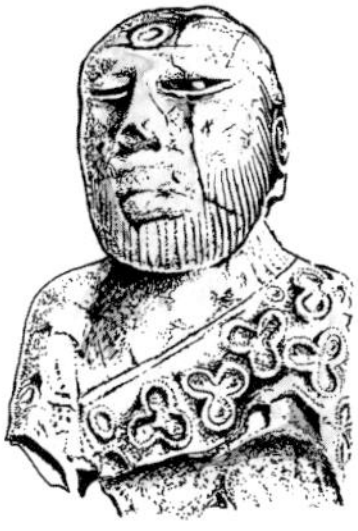

One of the three great "river civilizations" developed along the valley of the Indus River in what is now Pakistan. This civilization began about 4 500 years ago and covered a vast area of some million square kilometres.
The Indus Valley cities were planned on a grid system and laid out with precision; no previous civilization had developed such a standardized pattern of town planning. They excelled at drainage and sanitation and many of their two storey houses had private bathrooms drained by earthenware pipes, running under the streets and into huge sumps.
This civilization lasted about 1 000 years before it declined and vanished. It is thought that they were probably weakened by natural causes and that the migrating Aryan people from central Asia destroyed the civilization by war in about 1500 B.C.

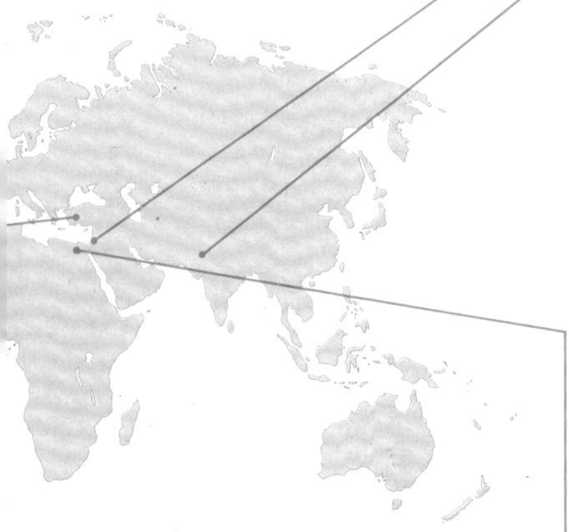

moving the Aztecs fed the remaining gods on a diet of human hearts. Consequently each year the priests sacrificed prisoners of war and as a special sacrifice one of their beautiful young men.
It was the Spaniard Hernan Cortez in 1521, who leading relatively few Spanish men conquered the country. This was achieved by the Aztecs lack of horses and guns and the neighbouring tribes turning against them. Two examples of the Aztec language that remain today are chocolate and tomato.

Egyptian

The ancient civilization of Egypt lasted for over 3 000 years, and was the first nation-state; headed by a Pharaoh. The country was divided into districts governed by local princes, with people paying a state tax of animals, crops and labour according to the size of the land that they farmed.
Much of the Egyptian achievements derived from their obsession with life after death, notably the pyramids, temples and tombs. The discovery of the Rosetta stone, showing both their demotic script and Greek, has enabled archaeologists to read about their life and religion from the many detailed paintings on tomb walls. Egypt's architects built on a grand scale, first the pyramids and then the mighty temples and tombs. The latter were made on the west bank of the Nile. This was because their equivalent of heaven was in the west with the setting sun and also it prevented the waste of valuable fertile land.
Agricultural exploitation of the narrow fertile strip each side of the Nile, necessitated technological development. This quest for agricultural efficiency made the priests and officials competent scientists and engineers. They became accurate surveyors and their study of mathematics led to an emphasis on the yearly cycle of 365 days, unknown in other early civilizations. They built the shaduf, a simple device for drawing water from the Nile, dykes, dams and irrigation canals.
The doctors were very knowledgeable and were thought to have known more about the bones, joints and muscles of the human body than the European doctors centuries later. The Egyptian doctors' knowledge was gained through cutting up and examining dead bodies while embalming them.
The Egyptians' happy and rich civilization ended when they were conquered by the Romans in 30 B.C. and when they became a province of the Roman Empire.

	Africa	Asia	Australia
4000 B.C.	Agriculture spreads southwards from Nile Delta. Egypt – white painted pottery. Early use of bronze. First harps and flutes.	Sumerians invent first wheel and wheeled vehicles. First use of potter's wheel. China – rice farming develops in Yang Shao. Thailand – copper tools.	Evidence of aboriginal inhabitants – immigrants from S.E. Asia before sea covered land between New Guinea and Australia. Aboriginal musical instruments – didgeridoos and bullroarers used. 
3100 B.C.	3100 – first Egyptian Dynasty.		
3000 B.C.	Egypt – first use of bricks for building cities.	Sumerians devise first known systems of writing on clay tablets. Thailand – bronze tools used. Indus valley – metal working reaches valley.	
2780 B.C.	2780 – first pyramid at Saqqara, world's oldest known stone building. 2772 – introduce 365-day calendar.		
2150 B.C.		2150 – Aryans first invade Indus valley. China – bronze worked in Anyang region.	
2000 B.C.	beginning of Bantu migrations from West Africa eastwards and southwards.		
1830 B.C.		1830 – founding of first Babylonian dynasty.	
1500 B.C.		1500 – Aryan people destroy Indus valley civilization by war.	
1400 B.C.	Egypt, 1292–1225, Rameses II restored Egypt's Asiatic power. Responsible for construction of temples at Abu Simbel.	1400 – Iron Age reaches Western Asia.	
1200 B.C.	1200 – horses and chariots used on Sahara trade route.		
1100 B.C.		China, 1100 – first Chinese dictionary.	
1000 B.C.		India – iron tools made in Ganges valley.	
	563 – Bantu people spreading in East Africa.	563 – birth of Siddhartha Gautama, who became Buddha. China 214 – building of Great Wall as defence against Nomadic tribes. 124 – philosophical teachings of Confucius.	
1 A.D.		4 – ? birth of Christ.	
30 A.D.		?30 – Christ crucified. China, 74–94 – opens silk trade to west. 105 – a Han official, Tsai Lun, invents paper.	
	Egypt, 127 – Ptolemy, Greek astronomer produces Ptolemy's Geography.		
200 A.D.	200 – Camels introduced from Asia come into widespread use for transport.	200 – Jewish independent state of Israel established.	
300 A.D.	300 – Berbers from Libya migrate south-westwards across the Sahara and establish Ghanian empire known as 'the Land of Gold'.		
320 A.D.		India, 320 – Revival of Hinduism. Golden age of artistic and cultural achievement.	
	439 – Vandals, a Germanic tribe, invade Africa; devastation caused gives rise to the term vandalism.		
		570 – birth of Muhammad at Mecca. Founder of Muslim religion.	
700 A.D.	700 – Bantu Africans cross River Limpopo, taking Iron Age south. Arabs conquer Tunis. Ghanians expel Berber overlords and bring empire under black rule.		
710 A.D.		710 – Muslims rule great empire from Spain in west to countries bordering China in east.	
750 A.D.		750 – Arabs learn secret of paper-making from Chinese.	

Comparing world history 4000 B.C.–750 A.D.

Europe	North and Central America	South America	
Minoan civilization settles in Crete. Skilled metal workers among first to make bronze tools and weapons. Cretan ships in Mediterranean.	Inhabited by Homo Sapiens. North America, Idaho – evidence of existence of domesticated dog.	Archaeological findings indicate inhabitation throughout South America.	**4000 B.C.**
	Central America, 3372 – first date of Mayan calendar devised by Mayan priests. Agriculture develops in Tehuacan valley, grow maize and beans.		**3100 B.C.**
Invention of weaving loom. Norway – first picture of skiing depicted by rock carving in South Norway. Crete – villages built of stone.		Peru, 3000 – villages and towns on coast.	**3000 B.C.**
			2780 B.C.
			2150 B.C.
Bronze age begins in Europe. Crete – earliest Minoan Palace built at Knossos. 2000–1450 – Minoan Crete dominate Aegean.		2000 – cotton is in cultivation.	**2000 B.C.**
			**1830 B.C.**
			1500 B.C.
1400 – Knossos destroyed by fire.			**1400 B.C.**
	1200 – Central America, Olmec civilization develops west of Yucatan. Flint tools used by Olmecs. Also invent systems of writing and arithmetic.		**1200 B.C.**
			1100 B.C.
Greeks establish colonies in Aegean. Italy – Etruscan civilization flourishes. Closely-knit family system; women great freedom and near equality with men. Greece, 776 – first Olympic Games.		Mayan civilization begins in Yucatan.	**1000 B.C.**
Roman Empire covers most of Western Europe except Britain.	Mayan civilization develops. Practise crop rotation. Inherit bar and dot system from Olmecs.	Andean peoples and Toltecs inhabit South America.	**1 A.D.**
			30 A.D.
Britain, 122 – Roman Emperor Hadrian builds wall as frontier with Scotland.			
		North Peru, 200 – Moche civilization develops in Moche valley.	**200 A.D.**
	300 – Teotihuacan culture attains importance.		**300 A.D.**
	Zapotec culture develops into classic American Indian civilizations.		
			320 A.D.
364 – Roman Empire divided into two: Western Europe and Eastern Europe.			
476 – Barbarians overthrow last Roman Emperor. End of Western Empire.			
		Peru, 700 – the Moche civilization develops; cut vast irrigation canals across desert.	**700 A.D.**
Spain, 710 – invasion of Moors from North Africa.			**710 A.D.**
	750 – Toltecs, warlike people, build empire in Mexican highlands.		**750 A.D.**

	Africa	Asia	Australia
800 A.D.	800 – growth of trans-Sahara trade between West and North Africa.		
850 A.D.	850 – Citadel built at Zimbabwe.	China, 850 – Christians forbidden to preach.	
900 A.D.	900 – the Fatimids, Egyptians, conquer North Africa.		
1000 A.D.		1096 – first Crusade to the Holy Land. 1099 – Crusaders capture Jerusalem.	
		1161 – explosives used in battle by Chinese. 1190 – Genghis Khan, warrior leader of Mongolian peoples, creates Mongol Empire.	
1210 A.D.	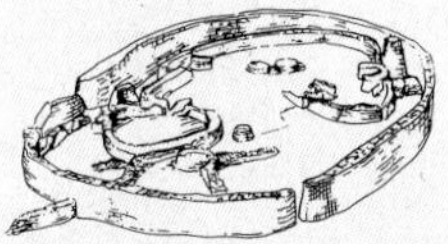	Mongolia, 1210 – Mongols invade China. Later conquer all China.	
1300 A.D.	1300 – beginning of building of the Great Enclosure of Zimbabwe. 1324 – Mansa Mensa, King of Mali, travels to Mecca. On return he builds the Great Mosque at Timbuktu.		
1400 A.D.	1400 – decline of Mali Empire.	1405 – Death of Tamerlane; Mongol Empire disintegrates. India, 1411 – Ahmed Shah builds Ahmadabed aŝ his capital.	
1450 A.D.		Turkey, 1453 – Mohammed II conquers Constantinople, end of Byzantine Empire.	
	1468 – rise of Songhai Empire. 1481 – fortified trading posts, called factories, built by Portuguese exporting gold. 1492 – Spanish begin conquest of North Africa. 1493 – Songhai Empire reaches height.		
		China, 1498 – modern toothbrush first described in Chinese encyclopaedia.	
1500 A.D.	1513 – Portuguese ascend Zambezi River and establish posts at Sena and Tete.	1513 – Vasco Nunez de Balboa discovers Pacific Ocean. 1517 – Ottomans capture Cairo – end of Mameluke Empire.	

Comparing world history 800 A.D. – 1521 A.D.

Europe

800 – Franks introduce Feudalism.

900 – Vikings develop shipbuilding and discover Greenland.

962 – Otto the Great, King of Germany, revives Holy Roman Empire.
993 – first Christian King of Sweden, Olaf Skutkonnung.

England, 1086 – Domesday Book commissioned by William the Conqueror.

1099 – Republic of Venice increasing power, based on trading and shipping.

Italy, 1184 – Leaning Tower of Pisa is built.

Italy, 1210 – Francis of Assisi founds Franciscan Order.
England, 1215 – Magna Carta, stating justice to all and no imprisonment without fair trial.
Italy, 1271 – Venetian traveller, Marco Polo, sets out for China.
England, 1300 – wool trade very prosperous.

England, 1337 – beginning of Hundred Years War between England and France.
1348 – Black Death ravages Europe.

Portugal, 1432 – Portuguese Prince, Henry the Navigator, organizes many voyages of discovery along coast of West Africa.

Germany, 1450 – invention of printing by machinery and movable letters.
England, 1453 – end of Hundred Years War between England and France.

1492 – first globe constructed by Nuremberg geographer Martin Beheim. Spanish conquer Granada; end of Muslim rule in Spain; Jews expelled.

1500 – Italian Renaissance.

1521 – Martin Luther outlawed. Beginning of Protestant Reformation.
France, manufacture of silk introduced.

North and Central America

850 – the Maya emigrate into Yucatan.
900 – Toltecs adopt Tula as their city; build terraced pyramid and colonnaded buildings.

1000 – Maya civilization begins to collapse.
North America, Viking Biarni Heriulfsson blown off course, sights North America.

Central America, Mexico, 1151 – end of Toltec Empire.

1492 – Columbus, Italian navigator, reaches America; discovery of New World.

1497 – Cabot, Italian explorer reaches Newfoundland.

South America

1000 – from this date onwards the Moche valley becomes site of Kingdom of Chimur. In South Peru vast areas being intensively cultivated.

1300 – beginning of expansion of Incas who establish an empire extending from Ecuador to Chile.

1438 – Incas establish empire in Peru.
1442 – Portuguese find gold at Rio de Oro.

1493 – first Spanish settlement in New World, Hispaniola.

1498 – Columbus discovers South America.

800 A.D.
850 A.D.
900 A.D.
1000 A.D.
1210 A.D.
1300 A.D.
1400 A.D.
1450 A.D.
1500 A.D.

Africa

1591 – Moroccans defeat forces of Songhai Empire which then disintegrates.

1650 A.D. 1650 – foundation of Cape Colony by Dutch.

1659 – French found trading station on Senegal coast of West Africa.

1700 A.D. 1700 – rise of Ashanti power, an agricultural people on Gold Coast of West Africa.

Asia

1649 – Russians reach Pacific.

Japan, 1650 – beginning of popular literary culture.

Japan, 1688 – rise of merchant class.

India, 1774 – Regulating Act. British India to be ruled by Governor General and Council.

Korea, 1778 – Christianity introduced.

Australia

1607 – De Torres, a Spaniard, probably discovers Australia.

1645 – Tasman, Dutch navigator, circumnavigates Australia and discovers New Zealand and Tasmania.

1768 – James Cook, British navigator and cartographer, discovers Australia.

1788 – British colony of Australia founded.

1620 A.D.

1650 A.D.

1650 – World population estimated at 500 million.

1700 A.D.

1760 A.D.

Comparing world history 1522 A.D.–1789 A.D.

Europe

1522 – one of Magellan's Portuguese ships completes first circumnavigation of the world.

Russia, 1547 – first Tsar of Russia rules so cruelly that he is known as Ivan the Terrible.
1559 – tobacco first introduced into Europe.
1588 – Spanish Armada defeated by the English.

Sweden, 1611–1632 – Gustavus Adolphus, King of Sweden, makes Sweden powerful.
1617 – William Harvey, English physician, announces discovery of the circulation of blood.

England, 1649 – execution of Charles I. Republic declared.

Sweden, 1659 – Swedish empire at height.
France, 1675 – Paris becomes centre of European culture.
Russia, 1682 – Peter the Great, Tsar of Russia, enlarges and modernizes his kingdom.
England, 1688 – 'Glorious Revolution'. Constitutional monarchy in England.

Russia, 1698 – tax on beards.
1700 – great age of German Baroque music, Handel and Bach.

1760 – European enlightenment; Voltaire, Diderot, Hume.

Russia, 1773 – peasant uprising.

France, 1789 – Revolution begins; abolition of feudal system.

North and Central America

1541 – Hernando de Solo, Spanish explorer, discovers Mississippi River.

1607 – Henry Hudson, English navigator, voyages to Greenland and up Hudson River.

1620 – Puritan Pilgrim Fathers from England reach Cape Cod in 'Mayflower' and found New Plymouth, later called Boston.
1625 – Dutch settlements in New Amsterdam, later called New York.
1642 – French found Montreal.

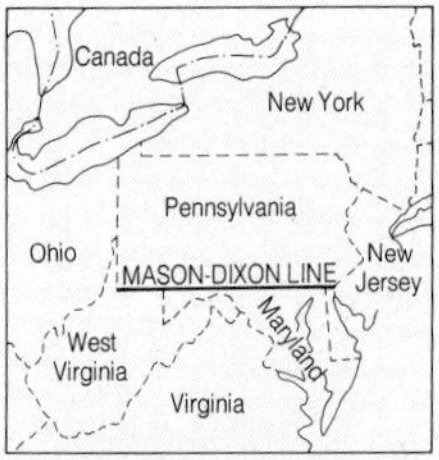

1769 – Mason Dixon line divides Free States from Slave States.
1773 – Boston Tea Party. American colonists throw shipload of tea into Boston harbour, angered by paying tax on goods.

1775–1783 – American War of Independence.
1776 – Declaration of Independence.
1778 – American states now free from Britain unite to form United States of America.

1789 – George Washington becomes first President of USA.

South America

1522–1533 – Spanish exploration of Pacific coast of South America under Francisco Pizarro.

1559 – Portuguese begin sugar cultivation in Brazil.

1620 A.D.

1650 A.D.

1693 – gold discovered in Brazil.

1700 A.D.

1713 – Slaves to Spanish America from West Africa. Height of British slave trade.

1760 A.D.

	Africa	Asia	Australia
	Egypt, 1798–1799 – Cairo taken by French led by Napoleon Bonaparte. English under Nelson defeat French fleet. 1804 – Fulanis, a Muslim people, conquer Hausa and establish Muslim Empire.	China, 1805 – Christian literature is banned.	1803 – Matthew Flinders, British navigator and hydrographer, circumnavigates Australia.
1810 A.D.	Egypt, 1811 – ruling Mamelukes massacred in Cairo by Muhammad Ali. 1818 – Shaka the Great forms Zulu kingdom in south-east Africa. 1822 – Liberia founded as colony for freed slaves.	India, 1818 – Britain defeats Marathas and becomes effective ruler of India. Far East, 1819 – Sir Stamford Raffles founds British colony at Singapore, free trade port.	1824–1825 – Hamilton Hume, Australian explorer discovers Murray River.
1830 A.D.	North Africa, 1830 – French begin conquest of Algeria. 1835–1837 – 'Great Trek' of Boer colonists from Cape, leading to foundation of Transvaal. 1838 – Boers in Natal defeat Zulus at Battle of Blood River. 1846–1847 – Bantu-British war in South Africa; defeat of Bantus.	1830 – Russians begin conquest of Kazakhstan. 1839 – British occupy Aden. China, 1842 – first opium war between Britain and China. Britain annexes Hong Kong.	
1850 A.D.	1853 – Livingstone, Scottish missionary and explorer, begins explorations in Africa. 1856 – Livingstone explores east Africa, central Africa and discovers Lake Nyasa. 	China, 1850 – T'ai-p'ing rebellion; immense loss of life. Revolt against Manchu dynasty. India, 1853 – first railway and telegraph lines in India. Japan, 1854 – Commodore Perry, US Navy, forces Japan to open trade with the west. China, 1856 – Anglo-Chinese war. India, 1857 – mutiny. Indian troops and civilians in north India revolt against British. India, 1858 – end of mutiny. India Bill transfers Government of India to British Crown.	1850 – granted responsible government to colony. 1851 – gold rush to Australia. 1854 – Eureka Stockade, Ballarat, Victoria, gold miners' rebellion.
1860 A.D.	Egypt, 1869 – Suez Canal opens, built by a Frenchman, Ferdinand de Lesseps.		1860 – Burke, Irish explorer emigrated to Australia in 1853, organizes with fellow explorer Wills to cross Australia from south to north. 1861 – both Burke and Wills die of starvation on return journey of expedition.

Comparing world history 1792 A.D. – 1869 A.D.

Europe

1792 – France declared a Republic.

1815 – Napoleon defeated at Waterloo and exiled to St. Helena.

1830 – revolutionary movements in France, Italy, Germany and Poland.
1833 – slavery abolished throughout British Empire.
Ireland, 1845 – potato famine, many emigrate to USA.

France, 1852 – fall of French Republic.

1854 – beginning of Crimean War. Russia against Britain, France and Turkey.

England, 1860 – Florence Nightingale founds Nursing School at St. Thomas' Hospital, London.

Russia, 1861 – Emancipation of serfs.

North and Central America

1803 – Louisiana Purchase. France sells Louisiana to USA.

1804–1805 – explorers Meriwether Lewis and William Clark explore north-west USA and reach Pacific.

1818 – 49th parallel fixed as boundary between USA and Canada.

1833 – first railroads to USA.

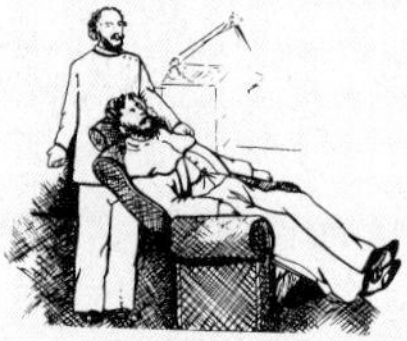

1844 – anaesthetic used by Horace Wells, US dentist.

1848 – gold discovered in California.

1852 – Henry Wells, US businessman, founds Wells Fargo, express mail service to newly developed west.

1861 – outbreak of American Civil War. Abraham Lincoln president of the USA.

1865 – end of American Civil War. Slavery abolished in USA.
President Lincoln assassinated.
North America, 1869 – Prince Rupert's Land, Manitoba and British Columbia join Canada.

South America

1810–1826 – In South America Spanish and Portuguese gain independence. Thirteen new states created.

1853 – new constitution of Argentina ends period of civil war and unrest.

1864 – War of Paraguay. Shatters economy of Paraguay.

1810 A.D.

1830 A.D.

1850 A.D.

1860 A.D.

	Africa	Asia	Australia
	1875 – Disraeli, British Prime Minister, buys Suez Canal Company shares to ensure British control of sea route to India.		
		India, 1877 – Queen Victoria proclaimed 'Empress of India'.	
1880 A.D.			1880 – Ned Kelly, notorious Australian outlaw, captured and hanged. 1885 – first time Australian soldiers fight abroad in Sudan with the British.
	1885 – King of Belgium acquires Congo.		
	South Africa – gold discovered in Transvaal. Foundation of Johannesburg. 1889 – British South Africa Company formed by Cecil Rhodes: begins colonization of Rhodesia (1890). 1899 – Boer War begins. Boers defeated and their states become part of the British Empire.		1899–1902 – Boer War; many Australian soldiers volunteer for service in South Africa.
1900 A.D.	1900 – Britains conquer Nigeria.	1900 – Boxer uprising. Fanatical Chinese nationalists attack Europeans and foreign legations.	
			1901 – first Federal Parliament held in Melbourne. The six colonies federated to form the Commonwealth of Australia.
		1904–1905 – Russo-Japanese War.	
1910 A.D.	1910 – formation of Union of South Africa. North Africa, 1911 – Italy conquers Libya.	China, 1911 – revolution, new republic formed.	
		1917 – 'Balfour Declaration' promises Jews a national home in Palestine.	1917 – Transcontinental Railway completed, 1 051 miles long.
	Egypt, 1919 – nationalist revolt.	India, 1919 – Amritsar incident. British troops fire on rioters.	
	1922 – Egypt becomes independent of British and French influence.	Turkey, 1922 – Greek army expelled. Republic proclaimed.	
			1929 – Federal Government moves to Canberra.
1930 A.D.		1932 – kingdom of Saudi Arabia formed.	
			1939 – World War II begins. Australia becomes base for war in South Pacific.
	1941 – Germany and Italy invade Egypt – World War II.	1942 – Japan over-run Southeast Asia. USA drop atom bombs on Japan, forcing surrender.	
	1943 – end of Italian and German resistance to Allied powers.		
	South Africa, 1949 – Apartheid programme inaugurated.	India, 1947 – gains independence. 1948 – state of Israel declared. China, 1949 – Communist victory after conflict between Nationalists and Communists.	

Comparing world history 1871 A.D.–1949 A.D.

Europe

Germany, 1871 – proclamation of German Empire.
1875 – growth of Labour/Socialist parties in Germany, Belgium, Holland, Britain and Russia.

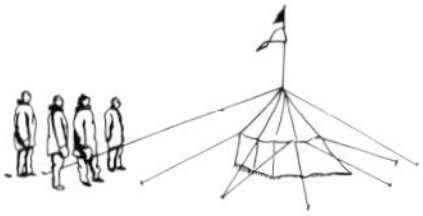

1911 – Norwegian explorer, Roald Amundson, reaches South Pole.
1914 – outbreak of World War I.
Russia, 1917 – revolution; Tsar abdicates. First Socialist state established.
1918 – end of World War I.
1919 – worldwide influenza epidemic reaches Europe.
Rutherford, English physicist, splits the atom.
1922 – Irish Free State created, Eire.

Germany, 1923 – runaway inflation.
1927 – emergence of talking pictures.

Germany, 1933 – Hitler made Chancellor. Beginning of Nazi revolution.
1939 – Germany invades Poland. Britain and France declare war on Germany.

1945 – defeat of Germany; suicide of Hitler; 50 million people killed and 34 million wounded in World War II.

North and Central America

USA, 1876 – Battle of Little Bighorn. Sioux Indians led by Sitting Bull beat General Custer and his men. One of the last battles fought to seize American West for white settlement.

1909 – Henry Ford develops conveyor belt assembly for the production of the Model-T automobile.

1927 – Charles Lindbergh, US aviator, makes first solo flight across Atlantic.

1941 – USA enters war against Germany and Japan – World War II.

1945 – United Nations established headquarters in New York.
1945 – USA drop atom bombs.

South America

1930 – military revolution in Brazil.
1932 – Chaco War between Bolivia and Paraguay over vast Gran Chaco region.

Argentina, 1946 – Juan Péron in power.

1880 A.D.

1900 A.D.

1910 A.D.

1930 A.D.

Africa

1952 – beginning of Mau Mau rebellion in Kenya, secret nationalist organization.

Egypt, 1959 – Suez crisis. Anglo-French invasion of Canal zone.

1960 A.D.

1960 – 'Africa Years'. Twenty states become independent.
South Africa, 1961 – becomes independent republic and leaves the Commonwealth.

1965 – Rhodesia illegally declares itself independent of Britain.

1970 A.D.

1976 – establishment of Transkei, first of Bantu homelands.
South Africa – Dr Christian Barnard performs first successful heart operation.

1980 A.D.

1980 – black majority rule established in Zimbabwe (Rhodesia).

South Africa, 1985 – widespread civil unrest among black community.

1986 – state of emergency declared as black unrest continues.

South Africa, 1988 – Nelson Mandela, imprisoned Black lawyer and politician, celebrates 70th birthday.

Asia

China, 1966 – cultural revolution, led by Mao Tse-Tung.

India, 1971 – Indo-Pakistan war leads to breakaway of East Pakistan to form new state, Bangladesh.
1972 – Ceylon becomes Sri Lanka.
1973 – Arab oil producing states restrict oil supply, quadruple price and spark off world economic crisis.

1979 – establishment of Islamic Republic with fall of the Shah of Iran.
Afghanistan – Soviet troops invade.

1980 – war between Iran and Iraq.

1982 – Israel invades Lebanon, expulsion of PLO, Palestine Liberation Organization, from Beirut.

India, 1984 – 2 500 killed following leak of toxic gas from Union Carbide plant at Bhopal.
1985 – Israel withdraws from Lebanon.

Afghanistan, 1988 – Soviet troops withdraw.

China, 1989 – student protest in Tiananmen Square, Beijing, leads to massacre and imposition of martial law.

Australia

1965 – forces leave for South Vietnam. First time Australians fight war without the British.

1973 – building of Sydney Opera House completed; designed by Danish architect, Jorn Utzon. The roofs are covered with more than 1 million Swedish made ceramic tiles in white and cream.
1974 – Gough Whitlam, Prime Minister, abolishes 'White Australia' policy which prevented immigration of non-whites.

1979 – new immigration scheme introduced. Parliament House Construction Authority formed to plan new Parliament House in Canberra.

1988 – Australia's bicentenary year of European Settlement. Opening of new Parliament House by Queen Elizabeth II.

1985 – World population estimated at 4 800 500 000

Comparing world history 1952 A.D.–1989 A.D.

Europe

1955 – Republic of France. De Gaulle first President.

Cyprus, 1960 – becomes independent republic.
1961 – East Germans build wall.
Yuri Gagarin of USSR, the first man in space.

Portugal, 1974 – end of dictatorship.
Cyprus – invaded by Turkey.

1976 – supersonic transatlantic passenger service begins with Concorde, Anglo-French design and construction.

Poland, 1980 – creation of independent trade union, Solidarity; martial law.
1981 – widespread demonstrations against stationing further nuclear missiles in Europe.

1983 – two Soviet cosmonauts complete 150 days flight in space.

USSR, 1985 – Gorbachev becomes leader.

1986 – nuclear accident at Chernobyl.

Poland, 1989 – democratic elections.
USSR, 1989 – Gorbachev introduces 'Perestroika'.

North and Central America

1962 – Cuba missile crisis.
1963 – President Kennedy assassinated.
1964 – US Civil Rights Bill.
1965 – USA at war with North Vietnam.

1966 – eruption of black discontent.
1968 – assassination of Martin Luther King, Black civil rights leader.
1969 – first landing on the Moon by US astronauts.

1971 – President Nixon and the Secretary of State, Kissinger, initiate policy of detente with China and USSR.

1973 – major recession.

1974 – President Nixon resigns following Watergate affair, a political scandal.

1980 – Ronald Reagan becomes President.

1983 – 'Pioneer 10' becomes first man-made vehicle to leave the solar system.

1986 – space craft 'Voyager II' at 2 000 million miles from earth sends pictures of Uranus.
1987 – Washington Summit between Reagan and Gorbachev; treaty to eliminate land-based medium-range nuclear missiles.
1988 – George Bush elected, Republican President.

1989 – 'Voyager II' sends first detailed pictures of Neptune.

South America

1959 – Cuban Revolution, led by Fidel Castro.

1970 – Salvador Allende becomes first democratically elected Marxist head of Chile.

1973 – Allende overthrown by military coup led by General Pinochet who assumes power.

1974 – large quantities of natural gas found in Nicaragua.

1979 – civil war in Nicaragua. Victorious Sandanista National Liberation Front establish provisional government.

1980 – first general election for 17 years held in Peru.
1981 – General Galtieri becomes President of Argentina following a military junta.

1982 – Argentina invade Falkland Islands. Surrender following armed conflict with UK task force.
1983 – democracy restored in Argentina. Coup in Grenada, USA invade.

1985 – democracy restored in Brazil and Uruguay.

1960 A.D.

1970 A.D.

1980 A.D.

Carthage

Carthage was founded as a commercial centre in the 9th century B.C. by the Phoenicians of Tyre on the coast of North Africa, near modern day Tunis. It grew rich, became independent of Tyre and extended its influence into regions of France and Spain.
Accordingly, Carthage came into conflict with the new growing power of Rome and the hundred years between the 3rd century and 2nd century B.C. saw the two long Punic Wars and Hannibal's triumphant crossing of the Alps. Finally Carthage was defeated and in 146 B.C. the city was completely plundered and burnt and all human habitation forbidden.
It was, however, to be a Roman outpost and administrative centre until in 765 A.D. it was captured by the Arabs and thereafter eclipsed by the growth of Tunis. Much of the remains of Roman Carthage can still be traced, including its renowned baths, temples and an amphitheatre built on the model of the Colosseum of Rome.

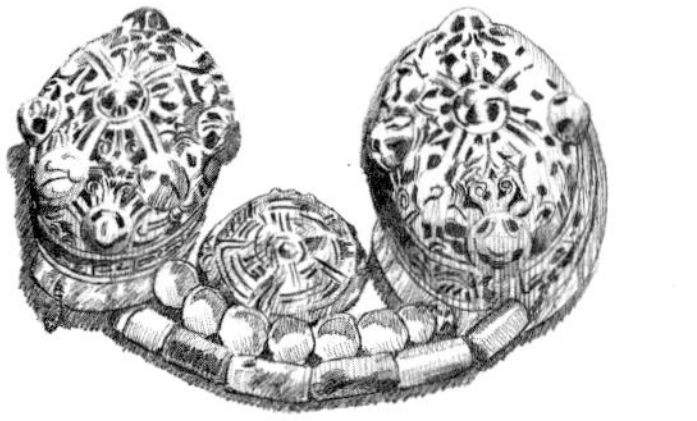

Leptis Magna

Leptis Magna was founded by the Phoenicians as a seasonal trading station on a point with a good natural harbour in North Africa, current day Libya.
It was the first to become just an ally of Rome but then, at the beginning of the 2nd century A.D., the Emperor Trojan made it a colonia (a community with full citizen rights). Later the Emperor Septimus Severus, who was born in Leptis, conferred more rights on the city and beautified it with a range of magnificent buildings.
Subject to a succession of raids, incursions to economic problems and political difficulties associated with the decline of the empire, the city too declined and by the middle of the 7th century had virtually died.
Its remains, however, are some of the best from classical antiquity and include the triumphal arch of Septimus Severus, the well-preserved Bath of Hadrian, an aqueduct, an amphitheatre, a large circus and a Forum dating from the early times of Roman occupation.

Machu Picchu

Machu Picchu, the lost Fortress of the Incas, was the refuge of those who moved deeper into the mountains to escape the plunderings of the Spanish (the Conquistadors) in the 16th century.
It remained undiscovered for 300 years and, when found in 1911, the buildings were jungle-clad ruins but later revealed to be a complex of white granite buildings, marvellously constructed to provide both for ceremony and protection. They included a magnificent royal mausoleum, shrines and temples – the Temple of the Sun, the Temple of the Moon. Worship of the Sun God, Inti, was to ensure his return in summer and for deliverance from the hated Spanish.
Today Machu Picchu is Peru's showpiece, visited by thousands to enjoy its mystery and grandeur.

Jerash

Jerash was a Roman settlement and trading centre in the north of Jordan which grew in wealth after joining the Decapolis (a federation of ten cities for trade) and continued to do so under the successive emperors. Its great prosperity was exhibited in such magnificent buildings as the Temple of Artemis, the Triumphal Arch of Hadrian and scores of churches for the large Christian community.
Jerash reached its peak in the 3rd century but then over a period of time it lost some of its trade. The power of Rome declined and in the 7th century the city was sacked by the Persians, then by the Arabs and finally in 747 it was shattered by an earthquake. This left only ruins which, exposed to the elements for centuries, became buried and forgotten.
It was not re-discovered until 1806 and excavation did not begin until 1925. Though only a small part has been excavated, such revelations as The Forum, the theatre built to seat 5 000, the Temple of Zeus, the churches and the baths are indications of the early splendour of Jerash.

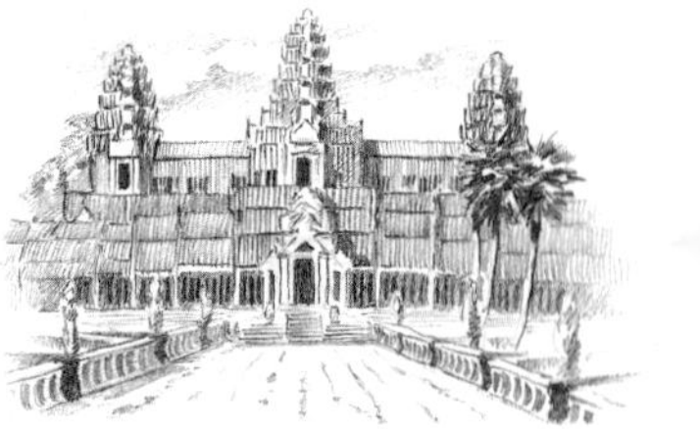

Birka

Birka, founded in the 9th century on the Swedish island of Björkö, became a thriving market and commercial centre, trading in Russian, Byzantine and Arabic goods. It was of such importance to be described as Sweden's first capital.
Archaeological evidence indicates that it began to decline about 976 and finally disappeared. Its decline was due to a combination of reasons – the better trading position of nearby Gotland, a severe drop in water level limiting navigation and the destruction and disruption caused by Danish raiders.
Birka today has only a few earthworks, but a great deal about it is known from two sources, a 'Life of St. Ansgar' – a Christian missionary who came to Birka in 829, and from some 2 000 grave mounds. These have produced domestic objects from their overseas commerce – cans, a Buddhist statuette, glass and pottery from the Rhine and objects from their trading such as scales and so on – all wonderful finds for the archaeologist.

Pompeii

Pompeii, archaeology indicates, experienced Greek and Etruscan influence before finally becoming Romanized in institutions, culture, architecture.
In A.D. 79 the city was destroyed by the violent eruption of Vesuvius, a volcano which dominates the Bay of Naples in Italy and was buried by four to six metres of volcanic ash.
The ruins were first discovered in the 16th century and excavation began in 1748 and now three-quarters of the city has been studied. The volcanic lava, engulfing people, animals and buildings served to preserve them and so unparalleled evidence of daily life has been discovered. Houses with personal possessions, workshops with their goods and equipment, shops with their stock and buildings with their inscriptions were found just as they were when abandoned by the fleeing inhabitants. These have provided not only unique information about all aspects of life in Roman times but also inspiration for European artists, architects, potters and even furniture makers.

Chernobyl

Chernobyl, a town in the Soviet Ukraine, housed a nuclear plant which was the subject of an article in the US edition of 'Soviet Life' in March 1986. In the article the Head of Nuclear Energy had said, "A serious loss of coolant accident is practically impossible".
Within one month, April 1986, Chernobyl was the scene of the world's worst nuclear accident caused by a fault deemed to be impossible – a leakage of cooling water into the graphite core leading, in turn, to a massive explosion.
The results were both devastating and disastrous – the devastation of a large area of the Ukraine caused by scattered radio-active debris, a death toll of many thousands and many more thousands seriously or fatally irradiated. There was also widespread disaster resulting from radio-active clouds being carried by winds into the atmosphere for thousands of miles and causing radio-active fall-out which affected water supplies, crops, meat and milk supplies all over Europe and beyond for a very long period.

Angkor Wat

Angkor, from the 9th to 13th century, was the royal centre of a prosperous kingdom whose kings used their vast resouces of labour and wealth both for useful projects and for temples and monuments for their own glory. These included vast systems of reservoirs and canals to provide irrigation to temples that were to become the personal mausoleum of successive kings and they identified with one of their deities. Of these, Angkor Wat, the most famous, was built by King Suryavarman II for his remains and his personal identity with Vishnu.
Angkor continued to be one of the great capitals of all Asia beyond the 13th century, but was later overrun by hostile armies and by the 19th century the city had been abandoned. In successive centuries interest was shown in the 'Lost City' by travellers and archaeologists have carried out a full programme of reconstruction through which the ancient complex of reservoirs, canals and temples has been restored to something of its original grandeur. This city can be seen when visiting Cambodia.

Pagan

In the 9th century people from the Chinese-Tibetan border established themselves in the region where, in 849 A.D., the city of Pagan was built on a site on the Irrawaddy river which controlled its navigation. It was the focus of trade routes and, as capital, it could rule a large area.
Pagan, labelled the 'City of Four Million Pagodas', reached its golden age in the 12th century, was capital of the region and supreme centre for Buddhist learning in the East. However, in 1287 Pagan was conquered by the Mongols, the empire crumbled and never regained its glory.
In 1975 a major earthquake damaged some of Pagan's priceless buildings, until then preserved by centuries of dry atmosphere, but many of its great shrines, temples and pagodas have now been restored.
The modern visitor is thus making a journey into the past as he views the marvel of this great heritage of over 2 000 buildings and this is one of the major tourist attractions of modern day Burma.

Nordic

The Vikings believed that the world was ruled by gods who lived in a heavenly place called Asgard. There were three main gods who stand out in the myths of northern Europe and whom we learn about from carved stones or scenes in metalwork. They were Odin, god of inspiration, magic and the dead; Thor, god of wind, rain, and farming; and Freyr, god of fertility and growth. Odin, the king of gods, was only seen in battles or at times of great danger. As he filled everyone with fear, only those who believed they had supernatural power dared worship him.
When Thor rode the sky in his magnificent chariot drawn by goats, there was thunder and lightning. He always carried a huge stone mallet which he used to frighten the giants and protect humans. Thor was known as a stupid, but happy god and his day, Thursday in English, became a day of meetings and great feasts.
Freyr was a god the Vikings would try to please so that their crops would grow tall and strong. After sowing their crops they pour mead or beer and scatter bread over them for Freyr's nourishment.
As well as these gods there were lesser supernatural beings who inhabited the mountains, forests and marshes. These were called elves, giants, or trolls and could help or harm men. The Vikings would worship their gods in the open air, probably beside the bountiful waterfalls, as they believed there was no danger from the evil spirits who dwelt in buildings.

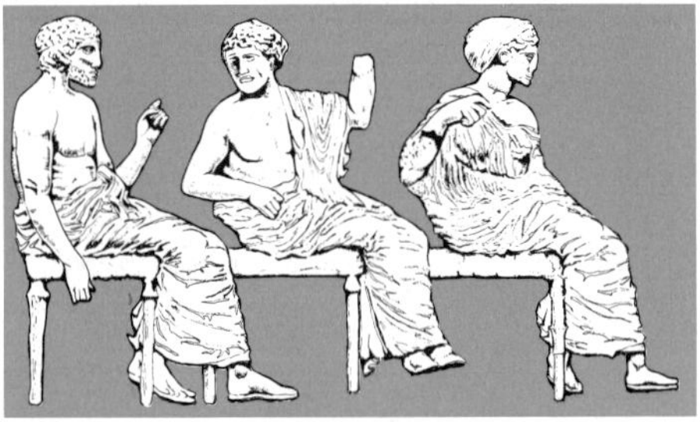

Greek

The Aryans, who invaded Greece from about 1500 to 1100 BC, brought with them nature gods conceived in human form. As Greek civilization developed, the Greeks came to regard their gods as graceful, glittering beings forming a pantheon of twelve gods – in the image of human society – living on Mount Olympus and headed by Zeus, the king and father of gods and men. This Olympian religion became the official religion of Greece and played an important part in all state functions – celebrations, games and drama alike.
Hera was the Queen of Zeus and, as such, of marriage. Poseidon ruled the sea, while Apollo was a multiple character – god of light, of music, of prophecy, and also of shepherds. Artemis was goddess of light and, in her capacity as Moon goddess, presided over childbirth. Hermes, god of travellers and athletes, was in addition the messenger of Zeus. Hephaestus, god of terrestrial but beneficent fire, and thus of the blacksmith. Other Olympians were Ares, god of war; Athena, goddess of wisdom; Demeter, goddess of the earth, of fruits and corn; Aphrodite, goddess of love and beauty; and Hestia, goddess of the home and hearth.
Beside these twelve Olympian greats, ranged others of little less importance, such as Dionysus, god of wine, and Helios, god of the sun. These lesser gods would serve as courtiers in various capacities to the Olympian greats.

Confucius

'China is Confucianism and Confucianism is China'. Although until recently the State religion, it is more a traditional philosophy and as such has powerfully moulded the outlook and the behaviour of the Chinese for many centuries. Founded in the 5th century B.C. by Confucious (551–479 B.C.) whose teachings are contained in five canonical books.
Confucius had great faith in the goodness of human nature and the central element of his teaching was the duty of the right behaviour in everyday relationships. He detailed the virtues, sincerity, kindness, respectfulness, moderation, faithfulness, as the basis of happy relationships. He also specified the five particular relationships from which a perfect community would result – ruler and subject; father and son; husband and wife; elder brother and younger; friend and friend.
Accordingly he emphasized the need for correct conduct and politeness, 'the three thousand rules of etiquette' and accepted ancestor worship as the continuation after death of earthly piety of son for parent.

Hinduism

Hinduism has over 300 million adherents in its native India, with further smaller numbers in the Far East and parts of Africa. It is a complex religion based upon a mixture of faiths, ideas and gods that have developed over time. It pervades all aspects of a Hindu's life.

The religion can be traced back to the Aryan people who invaded north-west India in 1500 BC. This is known as the 'Vedic' stage of Hinduism, named after the sacred books called the Vedas which are a collection of hymns and poems. There were many gods, mainly representing the natural forces of lightning, fire and water.

Over time a caste system developed with a social structure which had the priests or Brahman at its head. The Brahman defined a variety of cult gods and taught others that God could be found in three ways: by dedicating their work to him; by prayer and love; and by spending days on their own praying and contemplating God. The latter was the way of the priests.

In time, however, the ordinary people developed easier means of worship through the lesser gods of Vishnu and Siva. These gods form the core of modern day Hinduism, with different Hindus following Vishnu, Siva, or their wives and descendants. The religion today involves not only a multitude of pilgrimages to holy cities such as Varanasi on the Ganges river, but also an array of religious ceremonies and an intricate set of rules that govern all elements of life.

Shinto

Shinto, 'the way of the Gods', is the native religion of Japan – a nature worship especially associated with the sun. With its 3 000 gods and goddesses, mostly nature deities but some are Mikadoes and heroes of the past, Shintoism has been labelled 'the religion of the million gods'. It expressed itself all over Japan with some 20 000 shrines which were both centres for special ceremonies, such as New Moon and rice sowing, and places for everyday worship and prayers.

Through division and competition from Buddhism and Confucianism, Shintoism waned at periods. In the second half of the 19th century it was again established as the national religion, with two elements – the Sectarian section, purely religious, and the State section – intended to encourage the national spirit and patriotism. After World War II Shinto was disestablished as the State religion but it is still practised by many adherents of poor people and many marriages are still held at Shinto shrines.

Aboriginal

Dreamtime is the ever-present link between Aboriginal people, nature and the spiritual world. The rules which govern the Aborigines society and the natural world they live in were created in the period called Dreamtime – the time of creation. During this period Dreamtime ancestors broke out from their sleeping places under the ground and created the sun to warm the earth. They travelled over Australia, having many adventures and shaping landscape, people and animals into their present forms. Many of the Dreamtime ancestors were animals, plants, people or even in the form of rain, clouds and stars. During their journeys they laid down a blueprint for the Aborigines to follow. Then, worn out from their work, the ancestors sank back into the earth, creating a sacred site.

Every Aborigine is born with their own Dreaming or totem. These may be plants, like wild yams, or animals, like kangaroos, or inanimate objects, like stones. They are then considered a direct descendant of that particular ancestor and cannot harm it, so binding them closer to the natural world. Men hold the sacred knowledge about the Dreamtime and make elaborate ground paintings for totem ceremonies at sacred sites where young men are initiated by undergoing secret and painful ordeals before they learn about their heritage.

All Aboriginal art is connected with the spiritual world and although they are unable to paint their own totem, they can commemorate the myths of the Dreamtime.

Little Bighorn 1876

After the signing of the Laramie Treaty in 1868, the Sioux Indians roamed the lands north of the Platte river guaranteed to them in the treaty. In 1874, however, a strange expedition led by Colonel Custer set out for Dakota in defiance of the Government treaty, and discovered gold in the Black Hills. This area was proudly held by the Indians as the spiritual heart of their lands. While 1 000 miners illegally camped on their land the Government tried to buy it and move the Sioux Indians to a reservation. The Indians united behind their leaders Crazy Horse and Sitting Bull to defend their lands. The Government decided to teach the Sioux Indians a lesson and a force under General George Crook set out to attack them.

On June 17th Crazy Horse and his followers heavily attacked Crook's soldiers on the banks of the Rosebud forcing Crook to carry out a fighting retreat. Colonel George Custer and a column of the 7th Cavalry were sent to chase the Indians and follow their trail until their camp was found, and then to wait until a united attack could be mounted. Although Custer was a good cavalry leader he was also a glory hunter and rushed ahead. On June 25th 1876 Custer's soldiers attacked the Indian camp at Little Bighorn. The initial attack on the camp was disastrous as the Sioux had been prewarned by their scouts. Custer had no idea of how many warriors were waiting for him and he was forced northward to higher ground. With Crazy Horse at their head, the Indians engulfed Custer and his troops so that his entire command of 200 men was wiped out. At the end of "Custer's last stand" only one cavalry horse was left alive.

The Indians' great victory was short lived as by early 1877 the army forces gained control of the Indians' land.

Gettysburg 1863

The American Civil War began in April 1861 with the shelling of a fort in Charleston Harbour. The quarrel was between the northern twenty-two Federal states who opposed slavery and the eleven southern Confederate states who formed a separate union to enable them to keep slaves.

In 1863 General Robert Lee led a Confederate army into the enemy territory of Pennsylvania, the Federal troops under General George Meade moved slowly south. They converged unintentionally at Gettysburg, both armies having been uncertain of the others location. On July 1st, the first day of the battle, the Confederate army gained small successes giving them the advantage. The following day Lee ordered a charge on the Federal army drawn up on a ridge to the south of the town. Although the attacks did not go quite to plan victory still seemed possible for the Confederates. On the third and final day after a morning of heavy fighting the Federal artillery withheld their ammunition for a further attack. The Confederates thought that the lack of shell fire was indicative of short ammunition and charged the enemy's centre but when they were very close the Federal artillery opened up causing heavy casualties. The Confederates held their line despite massive fire but ultimately withdrew and began a steady retreat.

Both sides suffered heavy losses, approximately 23,000 men from the Federal army and 25,000 men from the Confederate army died. Gettysburg was the most significant battle in the US Civil War as it was the beginning of the end for General Robert Lee and his Confederate army.

Hastings 1066

Following the death of Edward the Confessor in 1066 William Duke of Normandy claimed the promised position of King of England but England's leaders had asked Harold Godwinson, already a formidable warrior against the Welsh to become King.

While Harold was defeating the invasion of his brother the King of Norway in Yorkshire, William landed his forces of thousands of men and horses on September 28th on the Sussex coast. He had crossed the English Channel to challenge Harold for the crown of England. With the news of the arrival of William, Harold rushed southwards gathering reinforcements on the way and took up a defensive position on a small hill called Senlac, a few miles north of Hastings.

In the early morning of October 14th 1066 the Normans attacked. The weary English fought on foot armed with spears and great battle axes, while the Normans assailed them with infantry, spearmen, archers and knighted cavalry. The battle was dominated by the Normans' use of archery supported by cavalry to

Trafalgar 1805

In 1805 the French Emperor Napoleon planned to invade England but realised that first he would have to overcome the sea power of Great Britain. Napoleon decided he would entice the British fleet out of the English Channel by ordering French and Spanish ships

break through the defensive ranks of the English infantry. This was successful after the Normans had feigned flight and drawn the English from their hilltop position. Both sides suffered heavy losses, but the death of Harold allowed William the crown, but not the Kingdom: this took another six years of hard campaigning.

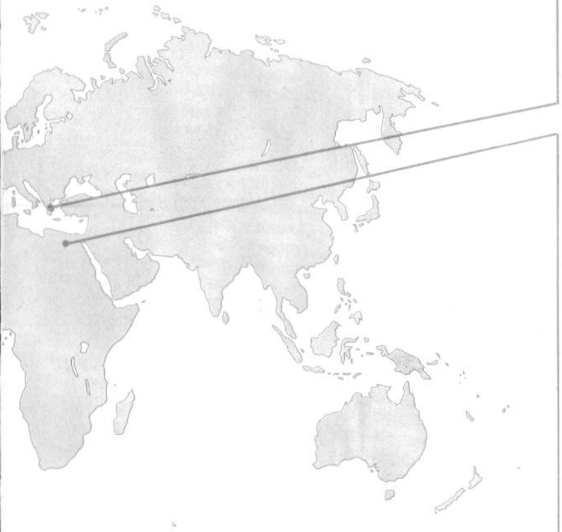

to sail for the West Indies. This would then lead the way open for his army to cross and land on the British coast. Admiral Nelson, who commanded the British fleet, fell for this plan and gave chase. The French and Spanish ships escaped and gathered at the Spanish port of Ferrol.
On September 14th 1805 Admiral Nelson took command of H.M.S. Victory and joined his fleet positioned off the coast of Spain, near Cadiz. Nelson wanted Napoleon's French and Spanish ships to put to sea to enable him to destroy the fleet, his chance came on October 19th when the French Commander was ordered to set sail for the Mediterranean.
Nelson's ships were already positioned so as to prevent their passage through the Straits of Gibraltar. The two fleets sighted one another off the Cape of Trafalgar and both Commanders then prepared for battle.
The British Fleet sailed into action and Nelson's skillful tactic of attacking enemy lines at right angles was an important element in the ultimate success of the British, this was tragically marred by Nelson's death in the hour of victory.
It was at the Battle of Trafalgar that Nelson signalled the famous message "England expects that every man will do his duty".

Thermopylae 480 B.C.

In 480 B.C. the Persian Emperor Xerxes led a mighty army into Greece, determined to conquer it. The only place at which there was any possibility of checking his progress was at Thermopylae.
The battle took place at a narrow pass, fourteen yards wide, leading from Locris to the mountains of Thessaly above Thermopylae. Here the Spartan King Leonidas made a famous stand with a force of 7 000 men which included 300 of his own Spartans who were known for their rigid military training from the tender age of seven. The Persians attempted to force the pass but with no success, owing to the fierce fighting of the Greeks and it seemed that Xerxes would have to turn back. His retreat was checked by a Greek, hoping for a rich reward, who told Xerxes of a path that led over the mountains into Thermopylae enabling the Persians to attack the Greeks from the rear. Leonidas divided his troops and sent some away keeping only 1 300 men, amongst them 300 of his Spartans; with indomitable courage he attempted for three days to hold the pass but was surrounded and outnumbered to the point of defeat. Leonidas was killed in the battle.
The dead Greeks were buried where they fell and pillars erected to their memory. Inscribed on one of these pillars were the words
"Go tell the Spartans, you who read;
we took their orders and are now dead".
The heroic stand of Leonidas and his followers inspired continued Greek resistance to the Persians and their story was told by the Greek historian Herodotus.

El Alamein 1942

El Alamein was one of the most important events of World War II. It was a war fought between the Allied Powers and the Axis Powers. The Allied Powers included United Kingdom, countries of the Commonwealth, France, Soviet Union, U.S.A. and China while the Axis Powers consisted of Germany, Italy and Japan.
The battle which was fought in the western desert of Egypt began on the night of October 23rd 1942. General Montgomery commanded the Allied troops of the Eighth Army and the German and Italian troops were led by Field Marshal Rommel. He was widely known as the Desert Fox because of his previous victories in North Africa. Allied soldiers had been fighting against the German and Italian armies for some time before the great battle. The fighting was in an area by the coast near the only road that went right through the desert. Rommel's aim was to drive the Allies out of Egypt and win the vital Suez Canal. The Allied troops managed to achieve this by stopping Rommel at Alamein.
Montgomery planned the offensive against the enemy in three stages; the first stage was bombardment to break through the enemy lines, the second stage was to advance through the gaps created by the bombardment and the third and final stage was to advance to the coast and cut off the enemy. This battle lasted twelve days until November 12th when the enemy started to retreat. This advance went on to Tunis, Italy and Western Europe until May 8th 1945 when Germany surrendered.

History of the Olympic Games

It is believed that organized athletic contests took place in Greece over 3 500 years ago and by the 6th century BC there were four major contests taking place: at Olympia, Delphi, Nemea and Corinth. The Olympic Games became the most important. They were started in 776 BC, and were held every four years until the Romans abolished them in 393 AD.

In 1887 the French Baron Pierre de Coubertin began gaining support for the resurrection of the Games. His plan was presented in 1894 and the International Olympic Committee founded. In 1896 the first games took place in Athens. The games of 1900 (Paris) and 1904 (St. Louis) were not well organized and the Interim Games of 1906 (Athens) were agreed to by Coubertin to aid the flagging movement. Their success resulted in the 1908 Games in London and subsequent Games every four years, except during World Wars I and II.

In order to keep the Games free of corruption, Coubertin felt that the competitions should be amongst amateurs. This principle remains today.

The Olympic medals

In the first Games of the modern movement in 1896, winners were awarded a silver medal and a crown of olive leaves, and runners-up a bronze medal. From 1900, however, medals have been awarded for all of the first three places – gold, silver and bronze. Victory ceremonies now take place with presentations being made to the top three athletes and the winner's national anthem being played.

The Olympic flame

At every Games since Berlin in 1936, a flame has been carried into the main stadium at the opening ceremony to light an Olympic flame that burns throughout the Games. By tradition, the flame begins its journey from Olympia in Greece, and is carried by a series of runners to the venue of the Games. This tradition was initiated by Carl Diem, the organizer of the Berlin Games, with the flame starting from the Temple of Zeus.

The Olympic flag

This was first introduced in the 1920 Games at Antwerp, though it was first presented by Coubertin in 1914. The five concentric circles of blue, yellow, black, green and red depict the five continents, and at least one of the colours appears in every nation's flag.

The Olympic Sports – Summer

The number of sports included has varied over the years. It rose to 21 sports in 1908 (including four ice-skating events) and was cut back by Courbertin in 1912 to 14 events to maintain standards. Today the Summer Games must include at least 15 of the following sports: archery, athletics (track and field), basketball, boxing, canoeing, cycling, diving, equestrian sports, fencing, hockey, football, gymnastics, handball, judo, modern pentathlon, rowing, shooting, swimming, table tennis, tennis, volleyball, water polo, weight lifting, wrestling, and yachting. Women can compete in all bar boxing, football, judo, modern pentathlon, water polo, weight lifting and wrestling. In addition the Olympic programme may also include up to two demonstration sports.

Year	Place	No. of Countries	No. of Participants	Winning Country	Memorable Events
1896	Athens	13	285	USA	A Greek shepherd, Spiridon Louis, wins marathon.
1900	Paris	19	911	France	Women compete for the first time in tennis and golf.
1904	St. Louis	10	336	USA	85 per cent of the participants are American.
1906	Athens	20	911	France	Interim Games.
1908	London	23	2064	Great Britain	For the first time entry was by nation. Marathon started at Windsor Castle and the 26 miles 385 yard race then became the standard distance.
1912	Stockholm	29	2387	USA	Electric timing introduced.
1920	Antwerp	29	2682	USA	Olympic flag and oath introduced.
1924	Paris	44	3070	USA	Paavo Nurmi (Finland) wins five athletic gold. Harold Abrahams (GB) the first European to win a sprint. Johnny Weissmuller (USA) wins three swimming golds.
1928	Amsterdam	46	2963	USA	Women compete in athletics for the first time.
1932	Los Angeles	37	1334	USA	First Olympic village constructed.
1936	Berlin	49	3972	Germany	Jesse Owens (USA) wins four athletic gold. First torch relay and television coverage.
1948	London	59	4689	USA	Francina Blankers-Koen (Holland) wins four athletics golds. Emil Zatopek (Czechoslovakia) wins 5 000 m. and 10 000 m.

Year	Place	No. of Countries	No. of Participants	Winning Country	Memorable Events
1952	Helsinki	69	4925	USA	Emil Zatopek wins 5 000 m., 10 000 m. and Marathon.
1956	Melbourne	67	3184	USSR	First time competitors entered 'en masse' in closing ceremony.
1960	Rome	87	5396	USSR	Boris Shakhlin (USSR) wins seven gymnastic golds. Abebe Bikila (Ethopia) wins Marathon – barefoot.
1964	Tokyo	93	5238	USA	Abebe Bikila wins Marathon – but with shoes.
1968	Mexico City	112	5531	USA	Bob Beamon (USA) sets long jump record – over 29ft. Dick Fosbury (USA) wins high jump with a 'flop'.
1972	Munich	122	7147	USSR	Mark Spitz (USA) wins seven swimming golds. Arab terrorists kill 11 Israeli competitors.
1976	Montreal	92	6085	USSR	African boycott of the Games.
1980	Moscow	81	5748	USSR	USA boycotted owing to USSR invasion of Afghanistan. Alexsander Ditiatin (USSR) wins eight gymnastic golds.
1984	Los Angeles	140	6854	USA	Soviet block boycott. Carl Lewis (USA) wins four athletic golds.
1988	Seoul	160	9500	USSR	Ben Johnson (Canada) breaks 100 m. world record, then disqualified for taking drugs.

The Olympic Sports – Winter

Winter sports had been included in the Summer Games between 1908 and 1920. It was therefore decided to hold independent Winter Games in 1924, despite opposition from the Scandinavian countries who felt that they would conflict with their own Nordic Games. In 1924, therefore, an 'International Winter Sports Week' was held in Chamonix, France, and in 1926 it was awarded the title of the Winter Games. Sports include skiing, ski jumping, speed skating, figure skating, ice hockey and bobsleigh.

Year	Place	No. of Countries	No. of Participants	Winning Country	Memorable Events
1924	Chamonix, France	16	265	Norway	
1928	St. Moritz, Switzerland	24	461	Norway	Sonja Henie (Norway) wins first of three gold medals for figure skating.
1932	Lake Placid, USA	17	261	USA	Figure skating held indoors.
1936	Garmisch, Germany	28	755	Norway	Alpine skiing included for first time.
1948	St. Moritz, Switzerland	28	713	Norway	
1952	Oslo, Norway	30	732	Norway	Opening day coincided with funeral of Britain's King George VI.
1956	Cortina d'Ampezzo, Italy	32	819	USSR	First Winter Games to be televised. Toni Sailer (Austria) wins all three Alpine skiing titles.
1960	Squaw Valley, USA	30	665	USSR	West and East Germany competed as one. South Africa competed for first and only time.
1964	Innsbruck, Austria	36	1312	USSR	Lydia Skoblikova (USSR) wins four speed skating golds.
1968	Grenoble, France	37	1293	Norway	Jean-Claude Killy (France) wins three Alpine skiing events.
1972	Sapporo, Japan	35	1128	USSR	Galina Kulakova (USSR) wins three ladies' skiing golds.
1976	Innsbruck, Austria	37	1261	USSR	Franz Klammer (Austria) breaks world downhill speed record – 102.8 m.p.h. John Curry (GB) wins skating with balletic performance.
1980	Lake Placid, USA	37	1067	USSR	Eric Heiden (USA) wins five speed-skating golds. USA beat USSR at ice hockey for the first time.
1984	Sarajevo, Yugoslavia	49	1278	East Germany	Marja-Liisa Hämälainen (Finland) wins three female Nordic skiing golds. Torvill and Dean star in figure skating with Ravel's Bolero. Michela Figini (Switzerland) becomes youngest female downhill skiing winner.
1988	Calgary, Canada	57	1400	USSR	Eddie "The Eagle" Edwards (GB) takes the media attention, despite losing on the ski jumps.

A true circumnavigation entails passing through two antipodal points and thereby covering a total distance of at least 40 000 kilometres, 24 860 miles.

By helicopter

The first solo round the world flight in a helicopter was completed by Dick Smith, an Australian, on July 22 1983. He flew in a Bell Model 206L Long Ranger III from and to the Bell Helicopter area at Forth Worth, Texas, USA beginning on August 5th 1982 covering a total distance of 56 742 kilometres, 35 258 miles.

By car

The fastest circumnavigation embracing more than an equator's length, driving 40 075 kilometres, 24 902 miles in 74 days 1 hour and 11 minutes. This was completed by the Canadians Garry Sowerby, the driver and Ken Langley, the navigator. It took from September 6th to November 19th 1980 driving a Volvo 245 DL westwards from Toronto, Canada, through four continents and 23 countries.

Between March 30th 1964 and April 23rd 1984 in a Citroen 2CV the entertainers Manfred Müller and Paul Ernst Luhrs drove around the world. They travelled through 83 countries covering 350 000 kilometres, 217 490 miles starting and finishing in Bremerhaven in West Germany.

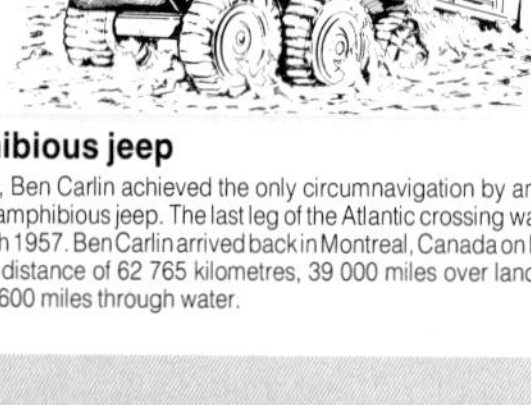

By amphibious jeep

An Australian, Ben Carlin achieved the only circumnavigation by an amphibious vehicle in his amphibious jeep. The last leg of the Atlantic crossing was completed on August 24th 1957. Ben Carlin arrived back in Montreal, Canada on May 8th 1958 completing a distance of 62 765 kilometres, 39 000 miles over land and 15 450 kilometres, 9 600 miles through water.

By flying

The fastest flight circumnavigating the world was the 37 216 kilometres, 23 125 miles flight of 36 hours 54 minutes and 15 seconds by 'Friendship One' a Boeing 747 SP flown by a Captain Clay Lacy. The flight started at Seattle, Washington, U.S.A. with 141 passengers from January 28-30th 1988. The plane refuelled only in Athens and Taipei.

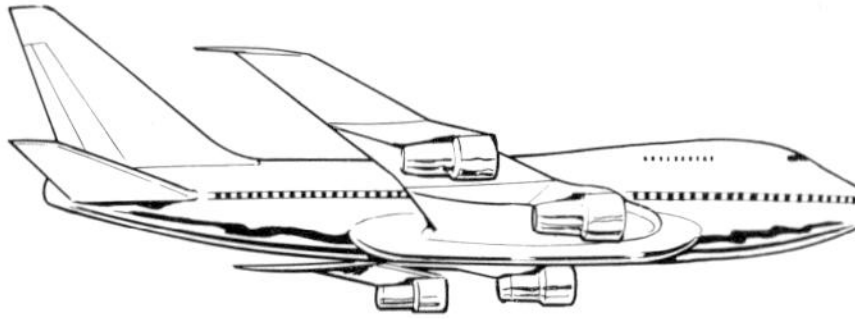

By surface transport

A British disc jockey, Simon Bates and his producer Jonathan Ruffle made an arduous record breaking round the world trip by using only surface transport such as trucks, trains, boats and buses. This attempt was organized with Oxfam, one of the best known charities working in the Third World where their aim is to provide self help. It was not solely a record making adventure as Simon Bates and Jonathan Ruffle visited Oxfam projects. They and their supporters, in Britain, raised money for these and future projects.

The journey started on June 28th 1989 from BBC Broadcasting House, London and travelling through 20 countries broadcast live each week day to British Radio One using a newly developed portable satellite dish. The journey finished on September 14th 1989 taking in total 78 days having flown one short leg from Abu Dhabi to Cairo.

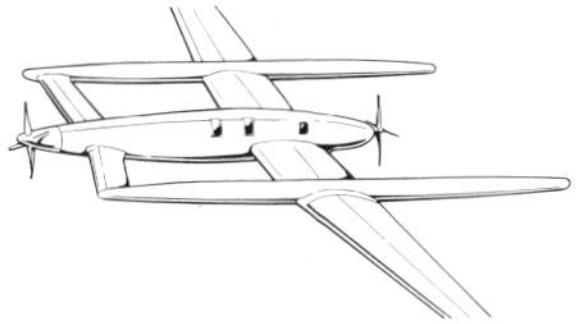

Dick Rutan and Jeana Yeager achieved the first circumnavigation without refuelling in their specially constructed aircraft 'Voyager'. The flight started from Edwards Air Force Base, California, U.S.A. starting on December 14th and finishing December 23rd 1986 taking 9 days 3 minutes and 44 seconds. They covered a distance of 40 212 kilometres, 24 987 miles averaging a speed of 186 kilometres per hour, 116 miles per hour.

The first circum-polar flight was solo. This record was made in a Piper Navajo flown by Captain Elgen Long from November 5th to December 3rd 1971. In 215 hours he covered 62 597 kilometres, 38 896 miles with the hardship of the cabin temperature sinking to −40°C, −40°F while passing over the Antarctic.

By spacecraft

Yuri Alekseevich Gagarin a Soviet cosmonaut manned the earliest space flight on April 12th 1961. He remained in orbit for 89 minutes reaching the maximum altitude during the 40 869 kilometres, 25 395 miles flight at 327 kilometres, 203 miles. This flight put his name in the record books as the first person to orbit the earth.

By foot

The first verified achievement of walking round the world was by David Kunst starting on June 10th 1970 and finishing on October 5th 1974 although the first person reported to have walked round the world was from 1897 to 1904.

By the Poles

Sir Ranulph Fiennes and Charles Burton of the British Trans-Globe Expedition achieved the first polar circumnavigation. They travelled south from Greenwich on September 2nd 1979 to the South Pole and then to the North Pole. On August 29th 1982 they arrived back at Greenwich after a 56 325 kilometre, 35 000 mile trek.

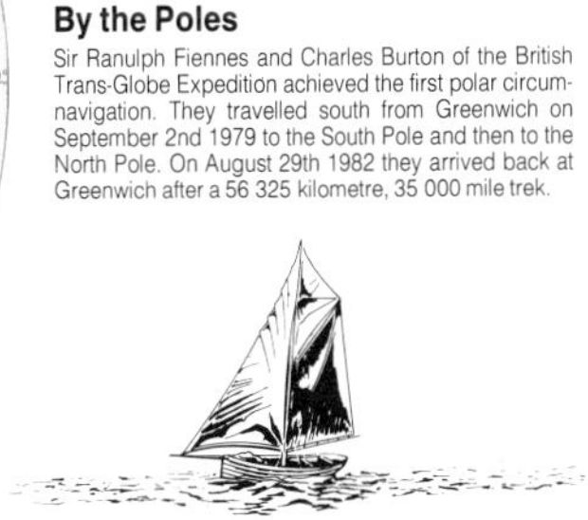

By ship

The earliest marine circumnavigation was by Juan Sebastián del Cano who took command of Magellan's expedition and remaining 17 crew after Magellan's death. The expedition started on September 20th 1519 from Seville and finished at San Lucar on September 6th 1522 covering 49 400 kilometres, 30 700 miles.

Alan Butler, a Canadian, achieved the earliest solo catamaran circumnavigation. The vessel 'Amon Re' was a 7.97 metres Bermudan sloop. The voyage started in 1980 and finished in 1986, in Barbados. This was not only the earliest solo but also the smallest catamaran to complete a circumnavigation of the world.

The earliest woman non-stop solo marine circumnavigation was by an Australian, Kay Cottee. The voyage started and finished at Sydney travelling west to east via the Horn of Africa between 1987 to 1988. The ship was a 11 metre Bermudan sloop called 'First Lady'.

New York

New York was founded by Dutch colonists and in 1625 they established New Amsterdam at the southern tip of Manhattan island. It is now the largest city in the U.S.A., and has become an important business, manufacturing and cultural centre. Manhattan was built on a grid system and the shortage of space on the island forced buildings to grow upwards not outwards. The skyline is today dominated by skyscrapers such as the twin towers of the World Trade Centre, the Empire State Building and the United Nations Headquarters.

Crowded city

The island of Manhattan, the centre of New York, is one of the most densely populated areas of the world. In its 47 square kilometres live over one and a third million people – a density of nearly 30 000 people per square kilometre. Many of the people who live here do so in extreme poverty, with most of the people who work in the area commuting in from the suburbs on a daily basis.

Agriculture

Food is mankind's raw energy resource. With the staggering increase in agricultural efficiency the world today grows enough food to support its population with plenty to spare. Unfortunately the pattern is uneven and consequently many areas of the world go short. The worlds three main crops are cereals; corn, wheat and rice.

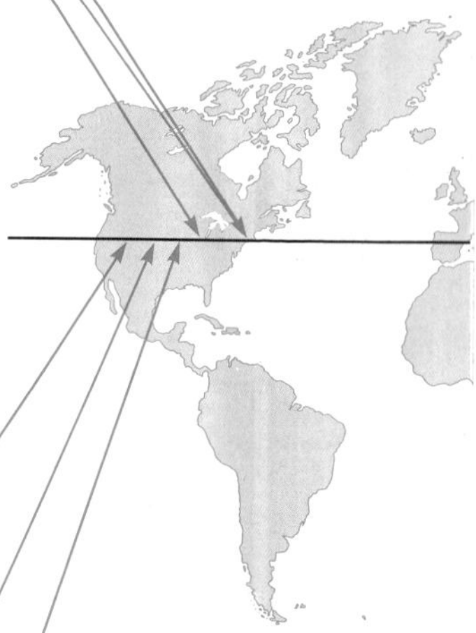

Corn

Corn is a crop of the temperate prairies and will tolerate very cold winters. The American corn belt, an area south and west of the Great Lakes, intensively produces both corn and wheat. Its combination of sunshine, rainfall and fertile soil makes it an almost perfect area for arable farming and the harvest is large enough to cater not only for American needs but also for other countries in the world.

Rockies

Along the line of 40° North lie the Rocky Mountains which rise to 4 000 metres to the west of Denver. Here too rises the Colorado River and a little further south-west it cuts through the mountains to form one of the most spectacular sights in the world – the Grand Canyon. To the east of the Rocky Mountains lie the Great Plains of central U.S.A., and the corn belt. To the west is San Francisco and the Pacific Ocean, which descends to over 10 000 metres below sea level near the Japanese coast – on the same 40° North line.

American Cowboy

The clothing worn by an American Cowboy was purely functional suiting his daily life. He was a pioneer and had to be practical. The fine wide hat served as an umbrella, shade, water bucket, drinking cup or fan and was made from expensive hardwearing felt. The shirt was made of heavy cotton or wool, with a colour or pattern that did not show the dirt. The high boots, which were ornamented and made of soft leather, had high heels to help keep the cowboy's foot in the stirrup, and the knotted handkerchief around the neck was worn so that it could quickly be pulled over the nose and mouth in the dust of a running herd. Modern day cow ranchers continue to use many of these items of clothing.

Bison

The Bison used to be widespread in number living on the plains and grasslands of North America. The turning over of land to agriculture has resulted in the majority that remain now living in parks and refuges, and this in turn can threaten the number of bison that survive. For when plentiful the bison lived in large interacting populations with breeding taking place amongst millions of individuals. Inbreeding and gene loss are, however, quite possible on reserves and this in turn weakens the species and threatens numbers. The problem lies not in the numbers of bison involved in mating, but in the need for an effective breeding system where no one bull becomes dominant. The understanding of this breeding system is thus crucial to the long-term conservation of the species.

Lonely plains

In the southern Steppes of Russia there are less than two people living in every square kilometre. These lands around the Aral Sea are almost totally deserted other than along the two main rivers that feed it from the south-east. The main centre in this direction is Tashkent some 500 kilometres away in the foothills of the Hindu Kush. To the north the population is sparse until one reaches the foothills of the Ural Mountains, again over 500 kilometres away.

Rice

Rice is the staple food of half the human race. It is an annual cereal that favours the special combination of the monsoon zones – a warm climate with a copious rainfall. Rice needs plains or terraces which can be flooded and be abundant in water in the growing season. Milling the grain removes either the outer husk (resulting in brown rice), or both the husk and bran layer which results in vitamin-B deficient white rice. While still in the husk it is known as the paddy. In China the majority of rice is grown for subsistence and therefore for their own consumption and not for export.

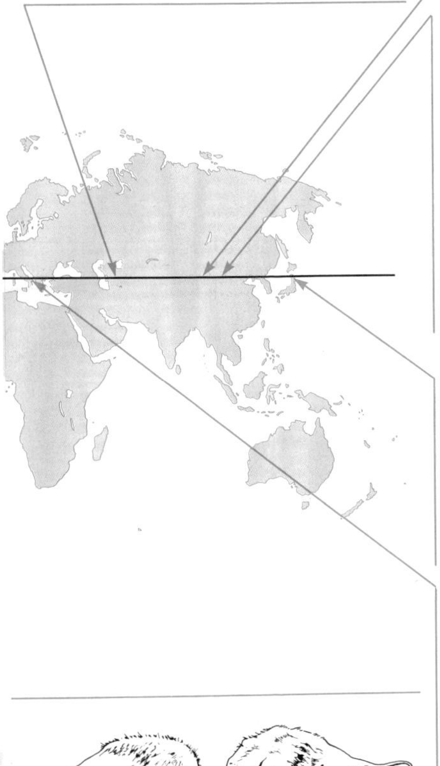

Giant Panda

The Giant Panda is one of the rarest mammals on earth, with estimates suggesting there may be as few as 500 in the wild. At this level a species is not liable to survive in the long term even though isolated populations may appear to flourish for several generations. The Giant Panda is at threat not from humans but from the dying back of bamboo, its main food. This takes place every hundred years or so but in the past the Giant Panda has always managed to supplement its bamboo diet with other foods which it finds through moving around. The increase in human settlement, however, has reduced the number of these areas and as a result small isolated populations of Giant Pandas are becoming increasingly at risk.

Japanese dress

The traditional costume of Japan, the Kimono, dates back to the 7th century and it is now worn mainly by women for formal occasions. It is an ankle length robe with wide sleeves tied at the back of the waist with a large bow or obi. There are two forms; a large butterfly bow tie for brides and girls, and a flat knot for married women. With this dress the Japanese wear flat wooden clogs which they leave in the entrance when they go indoors.

Rome

Rome is one of the greatest historical and cultural centres founded in 753 B.C. by Romulus its first King. It became the centre of the Roman Empire with many fine buildings such as the Forum, Pantheon and Colosseum reflecting the wealth of the time. Many ancient churches were constructed in early Christian times and the Vatican remains the focal point of the Roman Catholic Church. The Renaissance period of the 14th Century saw further outstanding buildings constructed such as St Peter's Basilica, with further squares, fountains and facades built during the 17th and 18th centuries.

Christians

Christianity is a religious faith based upon the teachings of Jesus Christ. Christians believe in one God and that Jesus Christ is his son who was the promised Messiah that was prophesied in the writings of the Old Testament. Christ's teachings are incorporated in the Gospel found in the New Testament section of the Bible. Holy day is Sunday when worship is held in churches and other consecrated buildings. The central act of worship is Communion, the sharing of wine and the breaking of bread, which symbolises Christ's Last Supper with his disciples before his death on the cross.

Use of drugs: Cocaine

Producer

Originally derived from a Peruvian tree but now grown worldwide; it is an alkaloid occurring in the leaves. Its cultivation provides a livelihood for otherwise deprived areas of South America (especially Columbia), Sri Lanka and Pakistan.

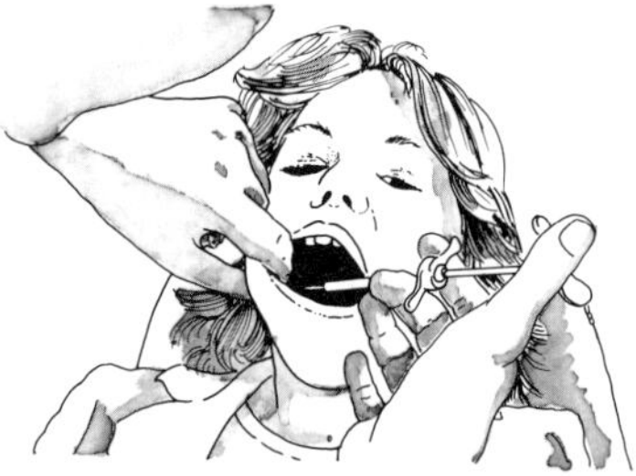

Consumer

(Anaesthesia). Cocaine and its derivative are widely used for local anaesthetic in dental surgery and other operations where local anaesthetic is preferred to general anaesthesia. It is also much used for the treatment of patients with painful terminal illnesses like cancer.

Drug Dependence

Cocaine is an addictive stimulant, taken for the euphoria which it induces by its central action on the brain. Regular use results in a condition of dependence with withdrawal causing emotional and physical distress.

Deforestation

Today about 30% of the world's land surface is covered in forest, of which one fifth is tropical. This forest is being destroyed at the rate of 2% per annum. The soil is then subject to serious erosion, with the oxygen-carbon dioxide balance of the atmosphere also being threatened. The forests are being cleared for timber, ranching and agriculture in the following countries: Brazil, Indonesia, Kenya and Nepal.

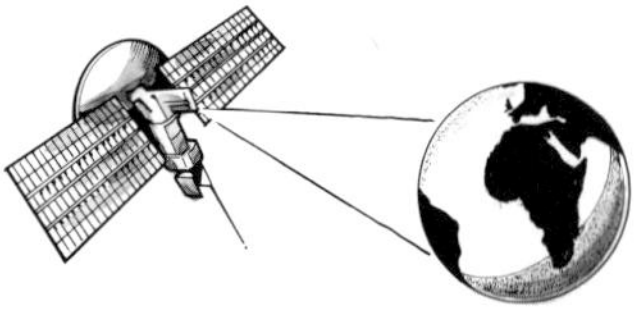

Above us

A satellite is a space craft launched into orbit around the earth, which can either re-transmit radio signals from ground to ground, or gather information and transmit it by radio signals to the ground. They are most frequently used for communication and military purposes, particularly national defence by the super powers.

Reafforestation

In Sweden and other Scandinavian countries the cultivation and management of forests is of major economic importance. Carefully managed coniferous forests especially of Pine, Spruce and Fir provide soft wood for paper pulp, furniture and house building.

Muslims

The Muslim religion is based on the belief that there is one God, Allah, and Mohammed is his prophet. The revelations received by Mohammed are recorded in the Koran. Muslims pray five times a day and must wash their face, hands and feet before prayer, and they pray facing Mecca the birthplace of Mohammed. Their holy day is Friday, and every Muslim is expected to make a pilgrimage to Mecca once during his life time.

"Tropical Snow"

Mount Kilimanjaro is a volcanic mountain in Tanzania, on the Kenyan border. It is 5895 metres high and is the highest mountain in Africa. The peak which lies within a few degrees of the equator is always covered with snow, ice and moving glaciers, whilst at the base there are native villages where bananas, coffee and maize are cultivated in a tropical climate.

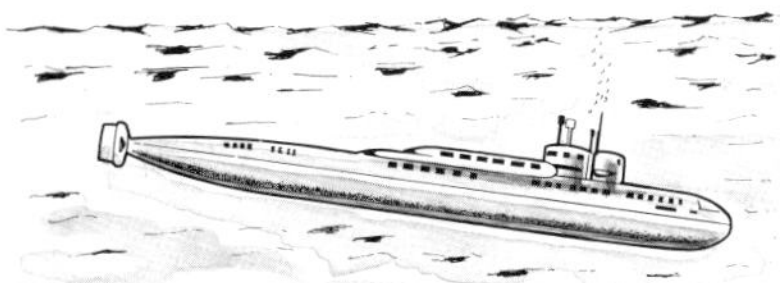

Below us

A warship designed for operation underwater. Modern submarines may be powered by nuclear reactors which do not need air to recharge batteries and can therefore remain submerged for long periods. Carrying missiles that can be launched underwater they are also undetectable and used in national defence by the super powers.

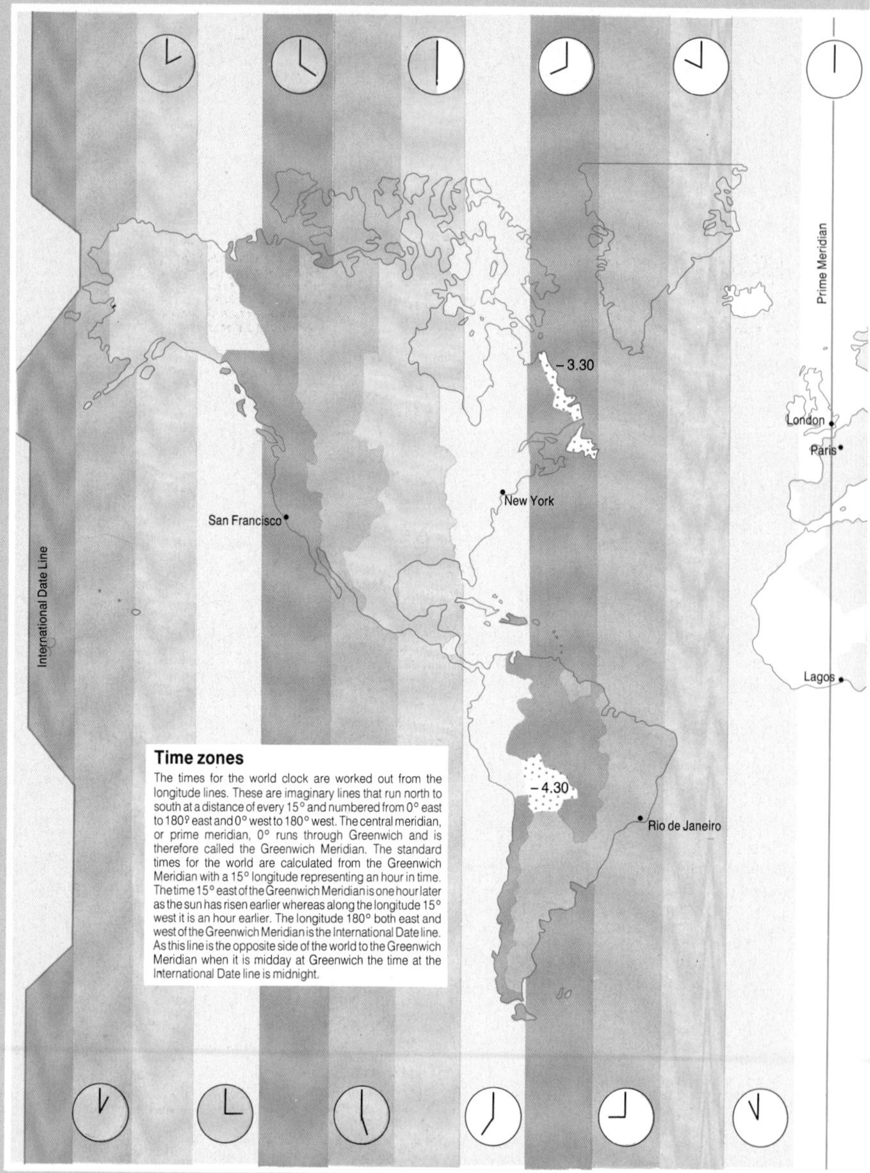

Time zones

The times for the world clock are worked out from the longitude lines. These are imaginary lines that run north to south at a distance of every 15° and numbered from 0° east to 180° east and 0° west to 180° west. The central meridian, or prime meridian, 0° runs through Greenwich and is therefore called the Greenwich Meridian. The standard times for the world are calculated from the Greenwich Meridian with a 15° longitude representing an hour in time. The time 15° east of the Greenwich Meridian is one hour later as the sun has risen earlier whereas along the longitude 15° west it is an hour earlier. The longitude 180° both east and west of the Greenwich Meridian is the International Date line. As this line is the opposite side of the world to the Greenwich Meridian when it is midday at Greenwich the time at the International Date line is midnight.

'And it was morning and night on the seventh day'